Nina Jasmin Schleicher

Chemical, Physical and Mineralogical Properties of Atmospheric Particulate Matter in the Megacity Beijing

Karlsruher Mineralogische und Geochemische Hefte

Schriftenreihe des Instituts für Mineralogie und Geochemie,

Karlsruher Institut für Technologie

Band 38

Chemical, Physical and Mineralogical Properties of Atmospheric Particulate Matter in the Megacity Beijing

by
Nina Jasmin Schleicher

KIT Scientific Publishing

Dissertation, Karlsruher Institut für Technologie,
Fakultät für Bauingenieur-, Geo- und Umweltwissenschaften, 2011
Referenten: PD Dr. S. Norra, Prof. Dr. K. Schäfer

Anschrift der Schriftleitung:
Karlsruher Institut für Technologie
Institut für Mineralogie und Geochemie
Karlsruher Mineralogische und Geochemische Hefte
D – 76128 Karlsruhe

Impressum

Karlsruher Institut für Technologie (KIT)
KIT Scientific Publishing
Straße am Forum 2
D-76131 Karlsruhe
www.ksp.kit.edu

KIT – Universität des Landes Baden-Württemberg und nationales
Forschungszentrum in der Helmholtz-Gemeinschaft

KIT Scientific Publishing 2012
Print on Demand

ISSN: 1618-2677
ISBN: 978-3-86644-827-8

Chemical, physical and mineralogical properties of atmospheric particulate matter in the megacity Beijing

Zur Erlangung des akademischen Grades eines

DOKTORS DER NATURWISSENSCHAFTEN

von der Fakultät für

Bauingenieur-, Geo- und Umweltwissenschaften
des Karlsruher Instituts für Technologie (KIT)

genehmigte

DISSERTATION

von

Dipl.-Geol. Nina Jasmin Schleicher

aus Ludwigsburg

Tag der mündlichen Prüfung: 11.05.2011
Hauptreferent: PD Dr. S. Norra
Korreferent: Prof. Dr. K. Schäfer

Karlsruhe 2011

Acknowledgments/ Danksagungen

Die vorliegende Arbeit entstand während meiner Tätigkeit als wissenschaftliche Mitarbeiterin am Institut für Mineralogie und Geochemie des Karlsruher Instituts für Technologie.

Mein besonderer Dank geht an PD Dr. Stefan Norra für die Übernahme des Hauptreferats, aber ganz besonders für die wissenschaftliche Betreuung der Arbeit. Er ermöglichte mir die Forschung in großer Freiheit und stand mir doch stets mit seinem Wissen, gutem Rat und hilfreichen Anregungen bei oftmals schwierigen Fragestellungen zur Seite. Auf seine Hilfe für den Fortgang meiner Forschung konnte und kann ich bis heute immer zählen. Stefan, vielen Dank für die gute fachliche und persönliche Unterstützung während der letzten Jahre!

Herrn Professor Dr. Klaus Schäfer danke ich herzlich für das große Interesse an meiner Arbeit und die Übernahme des Korreferats.

Frau Professor Dr. Doris Stüben danke ich für die Einstellung als wissenschaftliche Mitarbeiterin am Institut. Herrn Professor Dr. Thomas Neumann danke ich für das gute Arbeitsumfeld am Institut nach der Übernahme der Institutsleitung durch ihn.

Dr. Utz Kramar danke ich für seine fachliche Unterstützung, insbesondere bezüglich aller Röntgenanalytik am Institut und an der ANKA. Ich danke auch allen anderen Mitarbeitern am IMG für ihre Unterstützung und das gute Arbeitsklima am Institut. Für die Hilfe bei der Laborarbeit danke ich insbesondere Frau Cornelia Haug, Frau Claudia Mössner, Frau Pirimze Khelashvili und Frau Gesine Preuß. Für Wartung und Reparatur der Staubsammelgeräte und sonstigen technischen Rat danke ich Frau Gabriele Conrad. Auch den Hiwis, Inna, Jessica, Franziska, und Xu Chao, die geduldig Filter gewogen haben, danke ich für ihre Arbeit.

Beim Deutschen Wetterdienst in Freiburg bedanke ich mich für die intensive Zusammenarbeit und die Nutzung ihrer Labore. Insbesondere Herrn Volker Dietze und Herrn Mathieu Fricker danke ich für ihre Hilfe bei der Analytik in Freiburg und die vielen interessanten Diskussionen.

Furthermore, I want to thank all colleagues in Beijing, China, at the Chinese University of Geosciences Beijing (CUGB, namely Professor Cen Kuang), and the Chinese Research Academy of Environmental Sciences (CRAES, namely Professor Chai Fahe and Professor Chen Yizhen) for the good cooperation. I appreciate their support during my sampling campaigns in Beijing as well as the reliable sampling, also by the numerous Chinese students, for such a long time.

Frau Dr. Elisabeth Eiche danke ich sehr herzlich für die kritische Durchsicht des Manuskripts. Meinen Bürokolleginnen Eli, Steffi B.-S. und Xiaohui danke ich für ihre Unterstützung, Aufmunterung, und die tolle Atmosphäre im Büro, die mich jeden Tag besonders motiviert ins Institut kommen ließ.

Steffi B.-S., Eli, Harald, Dirk, Farooq, Alexander, Xiaohui, Maren, Antje, Isabel, Jan, Peter, Andreas und Steffi S. danke ich für all die Gespräche und sonstigen Aktivitäten in und außerhalb der Uni. Steffi S. danke ich zudem für die schöne gemeinsame Zeit in China.

Contents

Abstract

Worldwide, the urban population is growing fast and so is the number of megacities. By now, more people are living in urban than in rural areas. Air quality is one of the major problems affecting the inhabitants of cities and megacities. Beijing, the capital of China, is one of the most polluted cities in the world. Currently, air pollution due to particulate matter in Beijing often exceeds by far the pollution in Western European or Northern American cities as well as national and international threshold values.

The scope of this work is to develop a comprehensive understanding of the anthropogenic and geogenic aerosol pollution in Beijing. Dust storms seasonally occur in Beijing and may lead to a mixture of geogenic particles and aerosols released from numerous anthropogenic sources. The focus of this study lies on the interaction of anthropogenic and geogenic aerosol particles, their spatio-temporal variation, the identification of pollution sources as well as their impact on human health and the environment. For these purposes, spatio-temporal pattern of atmospheric particles have been investigated for a continuous sampling period of two years (September 2005 to August 2007). During this time, fine particulate matter, so-called $PM_{2.5}$, has been sampled weekly along a transect of six sampling sites through Beijing. This approach permits to trace changes of aerosol geochemistry in relation to changing land-use types and specific emission sources. Total suspended particulates (TSP) and coarse single atmospheric particles are also investigated within this study.

It is shown that air quality in the areas north-west of Beijing is highest and that the aerosol concentrations increase towards the south-eastern areas. Such a distribution suggests that anthropogenic emission sources in the southern areas significantly contribute to the pollution of the whole city. NOAA Hysplit models point to an additional impact of supra-regional pollution sources, which affect the air quality in Beijing.

Furthermore, day- and night-time aerosol concentrations are distinguished in this study. Corresponding results show that air pollution does not decrease during night time in Beijing in contrast to many European cities. Many emission sources operate day and night, such as coal combustion or domestic heating. Moreover, heavy traffic is allowed to enter Beijing only during night to prevent traffic congestion at daytime. However, an important impact on the night-time concentration of aerosols in Beijing can be attributed to the lowering of the atmospheric mixing layer, which increases the aerosol concentration due to a decrease of urban air body.

Annual courses show the high impact of dust storms on TSP concentrations but also in the fraction $PM_{2.5}$. Furthermore, especially during wintertime, heating processes and stagnant meteorological conditions lead to extremely high air pollution events. These events are comparable to dust storms with respect to mass concentration. They have, however, higher concentrations of inorganic pollutants such as Cd, As, or Pb. The lowest aerosol concentration occurs during summer, mainly due to favourable meteorological conditions, such as precipitation.

Not only the fine particle concentration of $PM_{2.5}$ is high, Beijing also suffers from a high concentration of coarse particles. Images made by Scanning Electron Microscopy and element mappings of single particles by synchrotron radiation based X-ray fluorescence analysis show that coarse particles act as scavengers for fine particles. Air pollution control measures normally start with reducing the coarse particle concentration. In this context, an open question still is how the fine particle mass concentration will respond when the coarse particle concentration is lowered.

Metal concentrations in aerosols are expressed in mass fractions (μg/g) in addition to the commonly used quantity mass per volume ($\mu g/m^3$), distinguishing the geochemical approach applied in this work from most other aerosol studies. Applying the unit μg/g, it is demonstrated that the atmospheric particles in Beijing are extremely high polluted by toxic elements, such as Zn, As, Cd, or Pb. The technique of sequential extractions is used to determine the bioavailable fraction of toxic elements in atmospheric particles, an important parameter in assessing negative health impacts. This approach shows that more than 50% of toxic metals in aerosols are mobile or even highly mobile, e.g. potentially bioavailable in human lungs.

Mitigation measures, such as closing industries and reducing traffic, had an eminent impact on the aerosol pollution in Beijing during the Olympic Summer Games in 2008. Measurements of particle mass concentrations during this abatement period reveal that the coarse particle fraction in the ambient air has been reduced more effectively than that of fine particles. This might be due to the fact, that the long-range transport plays a more important role for the fine particle size fraction. The residence time of finer particles in the atmosphere is longer and, thus, sources outside the abatement region around Beijing contribute relatively more. Black carbon, an indicator for anorganic pollution from anthropogenic combustion sources, has been strongly reduced during Olympic Games.

Simultaneously determining mass, element, black carbon and water-soluble ion concentrations from long-term measurements, in combination with the investigation of individual dust particles, as it has been done here, resulted in a truly unique and complete study. Furthermore, the investigation of bioavailable metal concentrations has contributed to the understanding of the most health-relevant atmospheric particles and their major sources. Consequently, this study gives new insights into the complex air pollution situation in a densely populated urban area. This knowledge helps to plan future mitigation measures more effectively and provides highly important data for further air quality and health assessment studies worldwide.

Zusammenfassung

Weltweit leben inzwischen mehr Menschen in Städten als in ländlichen Gebieten. Die städtische Bevölkerung und auch die Anzahl von Megastädten wachsen stetig. Luftverschmutzung stellt eines der zentralen Probleme für die Menschen in städtischen Gebieten und insbesondere auch für Megastädte dar. Peking, die Hauptstadt Chinas, gehört weltweit zu den Städten mit der schlechtesten Luftqualität. Gegenwärtig überschreitet die partikuläre Luftverschmutzung in Peking regelmäßig diejenige von westeuropäischen oder nordamerikanischen Städten sowie nationale und internationale Grenzwerte.

Das Ziel dieser Studie ist es, ein umfassendes Verständnis der anthropogenen aber auch der geogenen Luftverschmutzung in Peking zu entwickeln. Die Partikelbelastung in Peking ist stark von saisonalen Staubstürmen beeinflusst so dass sich diese geogenen Partikel mit solchen der zahlreichen anthropogenen Quellen mischen. Besonders die Interaktion von geogenen und anthropogenen Partikeln, die räumliche und saisonale Variabilität dieser Partikel, sowie die Identifikation der Quellen und die Auswirkungen auf Gesundheit und Umwelt sind von Interesse in dieser Studie. Daher wurde die räumliche und saisonale Struktur der atmosphärischen Partikelbelastung während eines kontinuierlichen Probennahmezeitraumes von zwei Jahren untersucht (September 2005 bis August 2007). Während dieser Zeit wurde Feinstaub ($PM_{2.5}$) entlang eines Transektes von sechs Probennahmenstandorten in einem wöchentlichen Rhythmus gesammelt, um die Veränderungen in der geochemischen Zusammensetzung in Abhängigkeit von wechselnden Landnutzungsstrukturen und der typischen Emissionsquellen zu ermitteln. Darüber hinaus wurden in dieser Studie auch Gesamtstaub (TSP) und Grobstaub in Form von Einzelpartikeln untersucht.

Es wurde gezeigt, dass die Luftqualität in den Gebieten nordwestlich von Peking am höchsten war und dass die Aerosolkonzentrationen in Richtung südwestlicher Stadtgebiete zunimmt. Diese Verteilung verdeutlicht die Wichtigkeit der Emissionsquellen in den südlichen Gebieten für die Luftverschmutzung der gesamten Stadt. NOAA Hysplit Modelle weisen auf einen zusätzlichen Einfluss von überregionalen Verschmutzungsquellen hin, welche auf die Luftqualität in Peking einwirken.

Darüber hinaus wurde in dieser Studie zwischen Tages- und Nachtkonzentrationen unterschieden. Die Ergebnisse zeigen, dass in Peking nachts mit keiner temporären Verringerung der Luftverschmutzung gerechnet werden kann. Viele Emissionsquellen, wie Kohleverbrennung, sind auch über Nacht aktiv. Darüber hinaus ist Schwerlastverkehr in der Innenstadt nur während der Nacht erlaubt um tagsüber Staus zu vermeiden. Einen wichtigen Einfluss auf die nächtlichen Aerosolkonzentrationen in Peking kann der niedrigeren Mischungsschicht zugeschrieben werden, welche die Aerosole im urbanen Luftkörper anreichern.

Jahreszeitliche Verläufe zeigen den großen Einfluss von Staubstürmen auf Gesamtstaubkonzentrationen (TSP), aber auch auf die Feinstaubfraktion ($PM_{2.5}$). Darüber hinaus führen, vor allem im Winter, Heizen und stagnierende meteorologische Bedingungen zu Phasen mit extremer Luftverschmutzung. Diese sind bezüglich der Massenkonzentration vergleichbar mit den Staubsturmperioden, weisen aber höhere Konzentrationen von anorganischen Schadstoffen, wie Cd, As, oder Pb, auf. Im Sommer sind die Aerosolkonzentrationen am niedrigsten, hauptsächlich wegen der günstigen meteorologischen Bedingungen, z.B. häufige und intensive Niederschläge.

Nicht nur feine Partikel ($PM_{2.5}$) treten in Peking mit extrem hohen Konzentrationen auf, sondern auch die groben Partikel. Aufnahmen mit dem Rasterelektronenmikroskop sowie die Analyse von Elementverteilungen in Einzelpartikeln mittels μS-XRF zeigen, dass grobe Partikel als "Scavenger" für feinere Partikel dienen. Luftreinhaltungsmaßnahmen beginnen normalerweise mit der Reduktion der Grobstaubkonzentrationen. In diesem Zusammenhang bleibt die Frage offen, wie sich die Konzentration der feinen Partikel durch die Abnahme der groben Partikel verändern wird.

Im Rahmen des geochemischen Ansatzes, der in dieser Arbeit verfolgt wurde, sind die Metallkonzentrationen der Aerosole nicht nur in $\mu g/m^3$ angegeben worden, wie es in Aerosolstudien üblich ist, sondern auch in $\mu g/g$. Die Angabe der Einheit $\mu g/g$ verdeutlicht, dass die Luftstaubpartikel in Peking sehr hoch an Elementen wie Zn, As, Cd, oder Pb angereichert sind, was negative Gesundheitsauswirkungen nach sich ziehen kann. In diesem Zusammenhang wurde auch die bioverfügbare Fraktion von Schwermetallen der atmosphärischen Partikel mittels sequentieller Extraktionen untersucht. Mit diesem Ansatz wurde gezeigt, dass mehr als 50% der toxischen Metalle in den Aerosolen mobil oder sogar sehr mobil und somit bioverfügbar in der menschlichen Lunge sind.

Luftreinhaltungsmaßnahmen, wie die Schließung von Industrieanlagen und Verminderung des Verkehrsaufkommens, hatten einen bedeutenden Einfluss auf die Luftverschmutzung in Peking während der Olympischen Sommerspiele 2008. Die Masse der Grobstaubfraktion in der Atmosphäre Pekings wurde stärker reduziert als der Anteil der feinen Partikel. Dies könnte darin begründet liegen, dass die implementierten Maßnahmen nur auf den Großraum Peking beschränkt waren. Durch Ferntransport aus Regionen ohne Reduktionsmaßnahmen werden aufgrund ihrer längeren Verweilzeit in der Atmosphäre vor allem feine Partikel nach Peking transportiert. Auch Ruß (BC), ein typischer Indikator für anorganische Verschmutzung von anthropogenen Verbrennungsquellen, wurde während der Olympischen Spiele stark reduziert.

Der kombinierte Ansatz, gleichzeitig die Massen-, Element-, Ruß-, und wasserlösliche Ionenkonzentrationen zu bestimmen, sowie Einzelpartikel zu untersuchen und vor allem die lange kontinuierliche Messreihe sind einzigartig. Darüber hinaus trägt die Untersuchung der bioverfügbaren Metallkonzentrationen zum Verständnis der gesundheitsrelevantesten atmosphärischen Partikel sowie ihrer wichtigsten Quellen bei. Infolgedessen hilft diese Studie dabei, neue Erkenntnisse bezüglich der komplexen Luftverschmutzungssituation in einem dicht besiedelten urbanen Raum zu erhalten. Dieses Wissen hilft dabei, zukünftige Reinhaltemaßnahmen effizienter planen zu können und liefert eine wichtige Datengrundlage für weitere Luftqualitäts- und Gesundheitsstudien weltweit.

Chapter 1

Airborne particulate matter

1.1 Basic principles and definitions

Airborne particulate matter (APM) comprises, in contrast to gaseous aerosols, particulate aerosols of different size and composition. Total suspended particulate matter (**TSP**) and **PM$_X$** (particulate matter with an aerodynamic diameter smaller than X μm) are common terms to describe the aerosol mass concentration (Seinfeld & Pandis, 2006). Within this study, TSP, PM$_{10}$, PM$_{2.5}$, and PM$_1$ samples from Beijing were investigated. Unless otherwise noted, the particle diameter refers to the aerodynamic diameter. For samples collected passively with the Sigma-2 sampler (see chapter 3.2.6) also the geometrical diameter is used as reference value and referred to as d$_g$. Passively collected APM from 2.5 to 80 μm d$_g$ (APM$_{2.5-80}$) was additionally investigated within this study.

The classification according to the **particle sizes** is not always used uniformly in literature. Within this work the terms are used with the following definitions:

Ultra-fine particles (UFP):	UFP $\leq$ 0.1 μm
Fine particles (FP):	0.1 < FP $\leq$ 2.5 μm
Coarse particles (CP):	2.5 < CP $\leq$ 100 μm
Ultra-large particles (ULP):	ULP > 100 μm

The particle size is an important parameter, because it gives information about the formation processes, determines the life-time of the particles in the atmosphere, and provides an indication where in the human respiratory tract the particles are preferentially deposited and, thus, gives a first hint about their potential health effects.

As a result of particle emission, in situ formation, and the variety of subsequent processes, the atmospheric aerosol distribution is characterized by a number of modes. The volume or mass distribution (see solid line in Figure 1.1) is usually dominated by two modes: the **accumulation** (from $\sim$ 0.1 to 2.5 μm) and the **coarse mode** (from $\sim$ 2.5 to 50 μm). The **nucleation mode** (few nanometres to 100 nm) is only of importance when the number distribution (see dashed line in Figure 1.1) is considered. Within this study, the ultra-fine particles were not investigated.

Additionally to the characteristic number and mass distribution functions, also the areas in the human respiratory system where inhaled particles are preferentially deposited are shown in Figure 1.1 according to Gieré & Querol (2010). The smaller the particle size, the deeper the particles can penetrate into the **human respiratory system** (Figure 1.1) and the more threatening they might be for human health. However, beside the particles size, also other factors, which will be discussed in section 1.3.1, govern the negative health-effects of APM.

Ultra-fine particles are removed through growth by **coagulation**, fine particles by **wet deposition** and coarse particles by **sedimentation** processes (e.g. Seinfeld & Pandis, 2006; Gieré & Querol, 2010). These main removal processes are shown in Figure 1.1. The **residence time** of APM in the atmosphere can range from minutes to days for coarse particles and from days to weeks for fine particles (Seinfeld & Pandis, 2006). The **travel distance** also varies strongly, with less than tens of km for coarse and up to hundreds or thousands of km for fine particles (Seinfeld & Pandis, 2006).

Accumulation-mode particles are the result of (i) **primary** emissions, i.e. the particles are emitted directly to the atmosphere, (ii) condensation of **secondary** sulfates, nitrates, and organics from the gas phase, which are formed within the atmosphere through gas-to-particle conversion, or (iii) coagulation of smaller particles (Seinfeld & Pandis, 2006). Particles in the coarse mode are usually produced by mechanical processes such as wind (e.g. soil minerals, sea-salt, pollen), but also by anthropogenic abrasion processes (e.g. tire and break wear). Most of the material in the coarse mode is primary but there are also some secondary sulphates and nitrates (Seinfeld & Pandis, 2006).

Urban aerosols are mixtures of primary particulate emissions from industries, transportation, power generation, construction and demolition activities, and particles from natural sources, as well as secondary material formed by gas-to-particle conversion mechanisms (Seinfeld & Pandis, 2006).

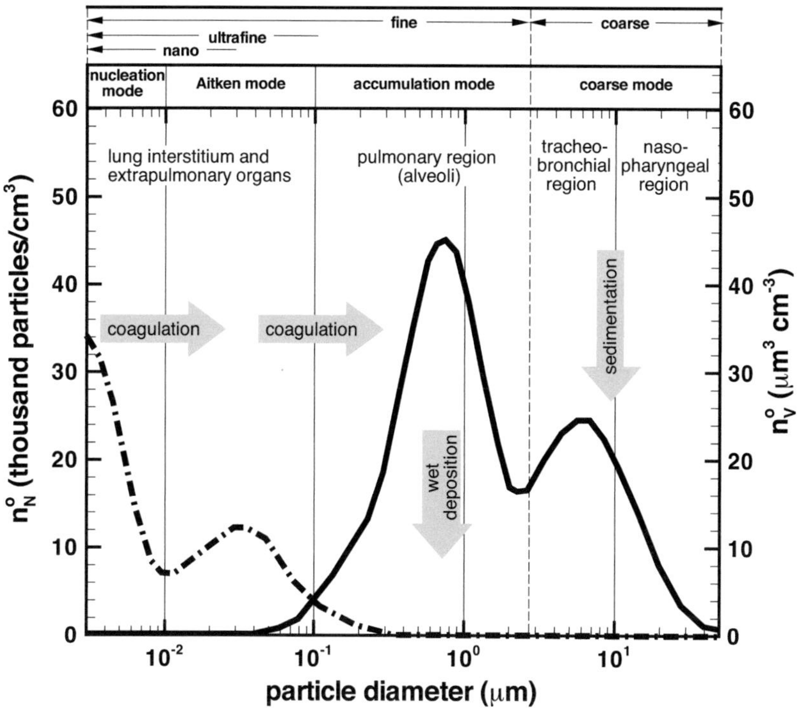

Figure 1.1: Figure according to Gieré & Querol (2010). Typical number ($n^o{}_N$, dashed line) and mass ($n^o{}_V$, solid line) size distribution of APM per cm^3 air. Particle removal processes (arrows) for the particles within the respective size range. Additionally, the areas in the human respiratory system where inhaled particles are preferentially deposited, are marked.

1.2 Air pollution in a historical perspective

Mortality is the most important health effect of APM and it is also the aspect that has been studied the longest (Anderson, 2009). A landmark in air pollution epidemiology was an air pollution episode in 1930 in a small industrial town (Meuse Valley) in Belgium, during which a large part of the population showed acute respiratory symptoms and the number of deaths was 10-times above average (Anderson, 2009). The London smog of December 1952 marks another of the early incidents, in which air pollution was considered to be hazardous to a large population and to have caused an estimated number of about 4000 excess deaths (Harrison & Yin, 2000). As a result of this episode, the Clean Air Act was introduced in the United Kingdom in 1956 leading to declining levels of APM (Highwood & Kinnersley, 2006). A comprehensive overview about air pollution in the last 50 years if given by Fenger (2009). The first global guidelines for air pollution (particulates, SO_2, NO_2 and O_3) were implemented by the WHO in 1987. In China, the Environmental Protection Law (trial implementation) was adopted in September 1979, and, in September 1987, the Law on the Prevention and Control of Atmospheric Pollution was passed by the People's Congress (Fang *et al.*, 2009). In the late 1980s, China joined the Global Environment Monitoring System (GEMS) and introduced regular systematic monitoring of air pollutants (Chen *et al.*, 2004).

The historical development of atmospheric pollution in connection with rising industrialization and per-capita income was described by Fenger (1999). Hereafter the air pollution in a developing urban area initially increases, goes through a maximum and stabilization phase and is then again reduced, when pollution abatement becomes effective. Whereas European cities are at the end of this cycle, many cities in China are probably in the stabilization phase for many air pollutants (Fenger, 1999).

1.3 Impacts of atmospheric particles on human health and the environment

1.3.1 Negative impacts of particulate aerosols on human health

Aerosol particles are known to have negative health effects on humans especially with regard to respiratory and cardiovascular diseases (e.g. Dockery *et al.*, 1996; Atkinson *et al.*, 1999; WHO, 2001). Several properties govern the toxicity of

APM: (I) overall mass concentration, (II) particle size, (III) particle surface properties and shape, (IV) chemical composition and concentration of trace metals, (V) speciation and mobility of potential toxic elements.

Overall particle **mass concentration** provides a first estimation of the magnitude of atmospheric pollution and was used already in many epidemiological studies beginning in the 1970s and 1980s (e.g. summarised by Pope & Dockery, 2006) as well as in policy making for the implementation of threshold values on national and international levels (e.g. WHO, EU, USEPA). Apart from mass concentration, **particle size** plays a decisive role for the hazards on human health, because smaller particles can penetrate deeper into the human respiratory system and are, consequently, considered as more health relevant (Dockery *et al.*, 1993; Oberdörster *et al.*, 1995; Schwartz *et al.*, 1996; Wichmann & Peters, 2000). Beside the size, also the **shape** of the particles is important, because especially sharp and long thin columnar particles penetrating the lung can cause pulmonary fibrosis and cancerous diseases (Donaldson *et al.*, 1989). The **chemical composition** and especially the concentration of potentially **toxic metals** is considered to be fundamental for the assessment of atmospheric pollution and its hazards to human health (Dreher *et al.*, 1994; Kodavanti *et al.*, 1997; Zereini *et al.*, 2001; Chillrud *et al.*, 2004). Harmful metals from ultra-fine particles can even reach the blood circulation (Nemmar *et al.*, 2002). In addition to the total amount of metals, also the **chemical speciation** of the elements is crucial due to varying toxicities of elements in different oxidation states or chemical bonds (Hill, 1997). However, it is important to keep in mind, that the specific relevance of the different particle properties and also the portions of the toxicological relevant elements and substances on health impairment are largely not known yet (Maynard, 2000; McClellan, 2002; Diabaté *et al.*, 2004) and that much research has still to be done with regard to environmental health risk assessment (Aitio, 2008; Spurgeon *et al.*, 2010).

The potential impacts on human health of some selected toxic elements and substances, which are of special relevance for this study, are described in more detail in the following section. For the investigation of the chemical composition of APM from Beijing within this study, one focus was put on these potentially toxic elements. Their concentrations, spatio-temporal variations and most important sources are discussed in chapter 4. The mobility and consequently the bioavailability and eco-toxicology of elements, depends on characteristic surface properties of atmospheric particles, the strength of the chemical bonds and on the properties of solutions in contact with APM (Smichowski *et al.*, 2005). The bioavailability of potentially toxic elements in APM from Beijing was investigated by means of sequential extractions and is discussed in chapter 5.

Toxicity of distinct elements and substances in atmospheric particles

The focus of many studies on the toxicity of APM components has often been on transition metals such as **Fe, V, Ni, Cr, Cu, and Zn** due to their ability to generate reactive oxygen species in biological tissues (Chen & Lippmann, 2009). Lippmann *et al.* (2006) found **Ni** and **V** to be significantly associated with daily mortality rates in 60 cities in the United States included in their study. Shi *et al.* (1994) observed that Cr(IV) compounds reacted with ascorbate and hydrogen peroxide and generated hydroxyl radicals, which caused DNA damage. Witholt *et al.* (2000) investigated the neurobehavioral effects of subchronic **Mn** exposure and concluded that it may increase the risk of neurobehavioral impairment in subpopulations that are in a pre-Parkinsonism state. Gwiazda *et al.* (2002) focussed on Mn exposure and concluded that an increased burden of airborne Mn might have deleterious effects on populations with sub-threshold neurodegeneration in the basal ganglia, e.g. pre-Parkinsonism. Manganese is considered as especially toxic if taken up through inhalation causing various psychiatric and movement disorders, respiratory effects such as pneumonitis and pneumonia, as well as reproductive dysfunction (WHO, 2000). Although **Zn** is essential for all organisms, toxicological studies proved that elevated Zn concentrations have negative health effects (Adamson *et al.*, 2000). Nawrot *et al.* (2006) found a significant association between environmental exposure to **Cd** and the risk of cancer. The International Agency for Research on Cancer (IARC) classifies different compounds (IARC, 2008), elements and element species according to their carcinogenic effect. **Arsenic**, **Cd** and their compounds are considered as highly carcinogenic for humans, while **Pb** and anorganic Pb compounds are classified as carcinogenic. The following guideline values for metals in atmospheric particles were established (WHO, 2000): 5 ng Cd/m^3, 150 μg Mn/m^3, and 500 ng Pb/m^3.

Raizenne *et al.* (1996) examined the health effects of exposure to **acidic air pollution** among children in 24 communities in the United States and Canada. They concluded that long-term exposure to ambient acidic particles may have a deleterious effect on lung growth, development and function. Amdur *et al.* (1978) studied the irritant potency of **sulphate salts** and found out that all considered sulphates caused a slight increase in pulmonary flow resistance and a slight decrease in pulmonary compliance. The order of irritant potency was ammonium sulphate $>$ ammonium bisulphate $>$ cupric sulphate.

Black Carbon (BC), which is mainly soot, has been associated with respiratory and cardiovascular diseases (Jansen *et al.*, 2005; Borm *et al.*, 2004; Gilmour *et al.*, 2004; Li *et al.*, 1999). Due to its small grain size, BC particles can penetrate deep into the respiratory tract of humans and, additionally, BC possesses a large surface that is able to adsorb many different harmful substances (e.g. toxic metals) which can induce diseases after inhalation.

Health studies in China

Several studies investigated the adverse health effects of air pollution in **China**. Over 400,000 premature deaths a year are blamed on high air pollution levels (Watts, 2005). Kasamatsu *et al.* (2006) studied the effects of winter air pollution on the pulmonary function of school children in Shenyang. The authors concluded that APM, associated with coal heating in winter, might have subacute health effects on pulmonary function in children. Lu *et al.* (2008) evaluated the potential toxicity of $PM_{2.5}$ in Shanghai by plasmid DNA assay. Their study showed that urban $PM_{2.5}$ caused more plasmid DNA damage compared to suburban $PM_{2.5}$ at the same dose. The authors suppose that trace elements, including Cu, Zn, Pb, Cd, Cr, Mn and Ni in $PM_{2.5}$ have synergistic interactions on plasmid DNA damage. A recent publication of Liu & Zhang (2009) reviews the correlation between ambient air pollutants (TSP, SO_2, NO_x) and children's lung function in China. The authors stated that the decline of children's lung function is related to the air pollution in China. Beijing was not included in this review, which comprised seven different Chinese cities (Chongqing, Guangzhou, Hengyang, Shenyang, Tayuan, Weihai, and Zaozhuang).

A review on air pollution and its health effects in **Beijing** was conducted by Xu *et al.* (1998). The authors summarised epidemiological studies assessing the health effects associated with both short-term and long-term exposures to air pollution in Beijing. They concluded that the studies provided coherent evidence that short-term exposures are significantly associated with adverse reproductive outcomes such as pre-term delivery and low infant weight, and excess daily morbidity and mortality, whereas long-term exposures are associated with increased respiratory symptoms or bronchitis in adults. Xu *et al.* (1994) examined the relationship between air pollution (SO_2 and TSP) and daily mortality in two residential areas in Beijing in 1989. They concluded that the association of ln(TSP) with total daily mortality was positive but not significant. Moreover, the authors analysed mortality by cause and found a doubling in TSP to be associated with total, chronic obstructive pulmonary disease, and pulmonary heart disease mortality, but only statistically significant for chronic obstruction pulmonary disease. The relation between ambient sulphate concentration and chronic disease mortality in Beijing was investigated by Zhang *et al.* (2000). The study comprised the years 1980–1992 and the authors found significant correlations between sulphate concentration and total mortality, as well as mortality due to cardiovascular disease, malignant tumour, and lung cancer. Lu *et al.* (2006) studied PM_{10} reactivity with the plasmid DNA assay method for Beijing samples. They concluded that the oxidative damage of plasmid DNA caused by Pb, Zn, and As was relatively strong. Moreover, the authors stated that Zn might be the main element responsible for the DNA damage. Wang *et al.* (2008) used Granger causality to determine whether there is a causal relationship between main air pollutants (NO_x, SO_2, CO, TSP, PM_{10}) and the mortality of respiratory diseases of the residents in Beijing. The authors stated that although TSP and PM_{10} did not have a causal relationship with the death rate of respiratory diseases, they still showed a high correlation. In a recent study, Guo *et al.* (2009) investigated the association between $PM_{2.5}$ and hospital emergency room visits in Beijing. They suggested that elevated levels of ambient air pollutants are associated with the increase in hospital emergency room visits for cardiovascular diseases in Beijing. Air quality and outpatient visits for asthma in adults during the 2008 Summer Olympic Games in Beijing were investigated recently by Li *et al.* (2010). They found out that at this time asthma visits were reduced significantly compared to the period before the pollution controls during the period of the Olympic Games.

1.3.2 Impacts of atmospheric particles on the environment

Aerosols influence the **climate** directly by the scattering and absorption of solar radiation and indirectly through their role as cloud condensation nuclei and by affecting cloud properties (Seinfeld & Pandis, 2006). The magnitude of the direct forcing of aerosols ("forcing" is the change of net flux at the tropopause, expressed in W m^{-2}) at a particular time and location depends on the amount of radiation scattered back to space, which itself depends on the size, abundance, and optical properties of the particles and the solar zenith angle (Seinfeld & Pandis, 2006). However, the impacts of aerosols on the climate are not completely understood, yet. These uncertainties in aerosol radiative forcing were discussed in detail by Schwartz (2004). Because of their size and composition, **mineral dust** particles can scatter and absorb both incoming and outgoing radiation. In the visible part of the spectrum, the light-scattering effect dominates, and mineral dust exerts an overall cooling effect. In the infra-red region, mineral dust is an absorber and acts like a greenhouse gas (Seinfeld & Pandis, 2006).

Due to its effect in scattering light, **BC** reduces visibility, an effect which is often very strong in Beijing (Sun *et al.*, 2006). With regard to influences on the climate, BC has recently be seen as a positive component of radiative forcing and, thus, to contribute to the global warming (Jacobson, 2001). Menon *et al.* (2002) modelled the climate effects of BC aerosols in China and India. They concluded that absorbing aerosols can affect the regional climate and suggested that precipitation trends in China over the past several decades, with increased rainfall in the south and drought in the north, may be related to increased BC concentrations.

1.4 Urban air quality studies

Studies on characterisation of urban aerosols have been carried out since decades in numerous cities all over the world, for example in: Chicago (Lin *et al.*, 1993), Sao Paulo (Andrade *et al.*, 1994), Helsinki (Buzorius *et al.*, 1999), Atlanta (Woo *et al.*, 2001), New York City (Qin *et al.*, 2006), Los Angeles (Lim *et al.*, 2006) as well as some Spanish cities (Moreno *et al.*, 2006). Meanwhile, such measurements belong to the standard repertoire of many environmental agencies. However, the required instrumentation is often lacking in the developing countries. Even environmental agencies of the highly polluted cities are mostly not equipped with devices for a detailed investigation of aerosol composition. Often only pilot studies on TSP or PM_{10} are available. The sampling networks are mostly restricted to gaseous air pollutants such as NO_x, CO and SO_2.

Air pollution in Chinese cities is very serious and many studies showed the adverse health effects for the affected inhabitants (see section 1.3.1). Zhang *et al.* (2008) investigated 111 Chinese cities and estimated the total economic cost caused by PM_{10} pollution to approximately US$ 29 billion in 2004, accounting for approximately 1.5% of the gross domestic product (GDP) of China. For Beijing, Zhang *et al.* (2007) calculated an economic cost between US$ 1670 and US$ 3655 million annually for the years 2000 to 2004, accounting for about 6.55% of Beijing's GDP each year. Case studies on aerosols in Chinese cities were carried out e.g. in Shengyang, Taiyuan (Ning *et al.*, 1996), Wuhan (Waldmann *et al.*, 1991), Hong Kong (Tanner & Wong, 2000), Shanghai (Shu *et al.*, 2001), Nanjing (Wang *et al.*, 2002), as well as in Beijing (He *et al.*, 2001; Okuda *et al.*, 2004; Sun *et al.*, 2004; Duan *et al.*, 2006).

More detailed information about previous studies can be found in the introduction section of each chapter, respectively.

1.5 Scope of the present work

The scope of this study is to develop a comprehensive understanding of the anthropogenic and geogenic aerosol pollution in Beijing. The characterization of the dominant geogenic and anthropogenic sources of particulate air pollution in the megacity Beijing, the complex mixing of particles from both source groups and their spatio-temporal variability, as well as the influence of mitigations measures on particulate air pollution are the main aspects considered in this study. The central questions of this work are:

- **Spatio-temporal variations of urban aerosols**
  - What is the concentration and composition of particulate matter in Beijing?
  - Which are the main parameters influencing seasonal variations?
  - Which seasons are most critical with regard to adverse health effects?
  - Which are the main parameters influencing spatial variations?
  - What are the most important geogenic and anthropogenic sources?
- **Bioavailability of toxic element concentrations**
  - Which are the elements with highest mobility and biovailability?
  - Which of those bioavailable elements are abundant and, thus, the most critical for possible negative effects on human health?
  - Which seasons have highest concentrations of bioavailable metal concentrations?
  - Which are the main sources for the bioavailable element concentrations?
- **Influence of mitigation measures on aerosol concentration and composition**
  - How fast and to what extend was the particulate air pollution reduced?
  - Which particle size class was reduced most efficiently?

  - To what extend did the weather conditions influence the particle decrease?
  - What was the chemical composition of atmospheric particulate matter during the Olympic Games and how did the composition change?
  - Which sources can be identified to contribute to the aerosol concentration during the Olympic Games?
  - How fast and to what extend did the aerosol pollution increase again after the Olympic Games?

- **Interaction between geogenic and anthropogenic particles on a single particle level**
  - Which are the typical geogenic and anthropogenic particles in Beijing?
  - Which kind of particle interactions can be found?
  - To what extend act geogenic particles as scavengers for small anthropogenic particles?

These questions will be discussed in detail in chapters 4, 5, 6, and 7, respectively.

Chapter 2

Study area: Beijing

Atmospheric pollution constitutes a big challenge for densely populated urban areas and megacities (generally defined as cities with more than 10 million inhabitants), in particular. The rapid economic growth in China (GDP: annual growth rate of about 10%, National Bureau of Statistics of China, 2009) comes along with a rapid expansion of the urban population and a huge increase in energy consumption. Urban agglomerations developed extremely fast in the last decades. While in 1980 only 19% of the Chinese population lived in cities and towns, the number increased to 26% in 1990, 36% in 2000 and 45% in 2007 (National Bureau of Statistics of China, 2009). Worldwide more people are living in urban compared to rural areas by now (UN, 2008).

2.1 Beijing and its surroundings

Beijing, the capital of the People's Republic of China, lies on the northwest border of the Great North China Plain, at 39°56' N and 116°20' E and occupies a total area of 16,410 km^3. While the city is connected to the North China Plain in the south, it is surrounded by the Yanshan Mountains in the west, north and northeast. It neighbours the Tianjin Municipality in the east, and borders Hebei Province on three sides (north, west and south). In 2005, approximately 11 million inhabitants were living permanently in Beijing (Beijing Municipal Bureau of Statistics, 2009). Additionally, migratory worker increased the population to approximately 15 million. In 2007, 67% of the total area was used for farming, 20% were building sites, while 13% were unutilized (Beijing Municipal Bureau of Statistics, 2009). The Beijing-Tianjin-Hebei economic band is located in the center of the vibrant economic area of Northeast Asia, which is one of three major rapidly developing economic zones in China (Xin *et al.*, 2010).

Anthropogenic as well as geogenic sources for air pollution in Bejing are abundant. The total energy consumption in China rose from 6×10^8 t of SCE (Standard Coal Equivalent) in 1980 to 13×10^8 in 2000 and 26.5×10^8 in 2007. Coal still is the major energy source, constituting about 70 % of the total energy consumption (2.6 bn tons SCE) in 2007 (National Bureau of Statistics of China, 2009). In Beijing, coal is used both industrially and domestically. Domestic heating in Beijing usually starts in mid-November and ends the following March (He *et al.*, 2001). Emissions from various kinds of industry contribute to overall air pollution. Traffic is considered another major anthropogenic source including direct emission products (e.g. soot) as well as particulate matter from abrasion (tire and brake wear) and resuspension processes. In 2005, there were 2.1 million vehicles in Beijing, with 1.5 million private ones, among which 0.92 million were cars. The number of vehicles increased to 3.2 million in 2008, with 2.5 million private vehicles including 1.7 million cars (Beijing Municipal Bureau of Statistics, 2009). In the last decades, many new high-rise buildings were constructed in Beijing, hindering the dispersion of air pollutants (Hao *et al.*, 2000; Liu *et al.*, 2009). Construction activities also constitute an important dust source in Beijing. In preparation for the Olympic Summer Games, which took place in Beijing in August 2008, the amount of construction sites strongly increased. The anthropogenic air pollution is superimposed by geogenic dust from bare soils transported over a short-range and from semi-arid/arid areas northwest of Beijing due to long-range transport. Dust storm episodes regularly occur in spring due to the meteorological conditions and especially if the wind direction is from NW (see section 2.2).

2.2 Meteorology

China shows a great complexity and diversity with regard to climate classification and division due to its vast territory and complex landforms. Domrös & Peng (1988) list several classification schemes compiled specifically for China. Moreover, the authors describe a climate division scheme of China by Huang Bingwei (1986) in detail, which is also published in Tietze & Domrös (1987). According to this classification, Beijing belongs to the sub-humid warm temperate climate zone (IIIB2). This zone shows a distinct seasonal contrast between a warm, partly hot summer (mean max. T: 30.8°C) and a cold winter (mean min. T.: -9.9°C). Summer and winter represent a distinct wet and dry season. In Beijing, about

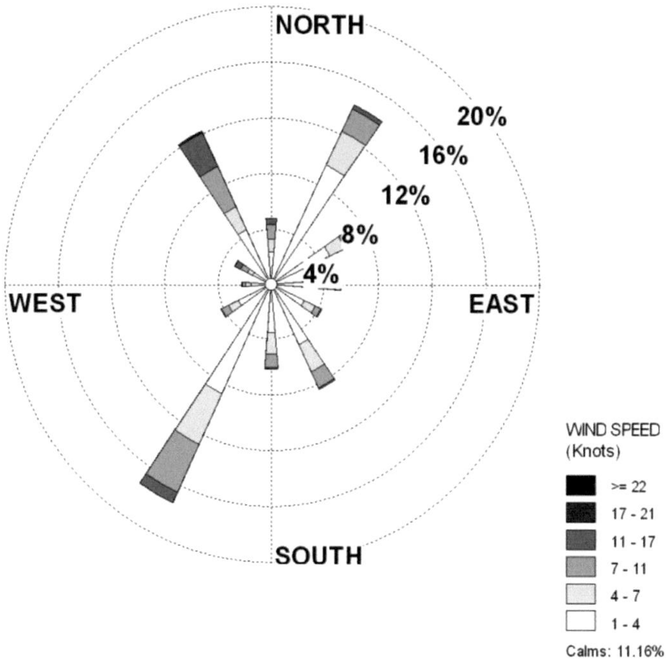

Figure 2.1: Wind rose (wind direction and wind speed) for Beijing (weather station 54511, data obtained from DWD, Offenbach) for three-hourly data from 1995 to 2004.

60% of the annual precipitation (570 mm) occurs in July and August. Table 2.2 summarises the climate data for Beijing.

The prevailing wind directions, their frequency of occurrence, and the average wind speeds are shown as wind roses (WRPLOT View Freeware, Lakes Environmental) in Figure 2.1 for a 10-year period from 1995 to 2004. Furthermore, wind roses for each season of the two-year sampling period from from autumn 2005 to summer 2007 are shown in Figure 2.2.

Table 2.1: Climate table of Beijing, China (39° 48'N 116° 28'E, 31.2 m.a.s.l.). Values are taken from Domrös & Peng (1988) and are based on the 30-year observation period from 1951-1980. The abbreviations are as follows: T.m. = mean temperature (°C), T.m.mx. = mean maximum temperature (°C), T.m.mn. = mean minimum temperature (°C), Rel.hum. = relative humidity (%), Cloud. = cloudiness, Precip. = precipitation amount (mm), SD hrs. = sunshine duration (h), W.vel. = Wind velocity (m/s), W.mx.dir. = prevailing wind direction, W.mx.pc. = percentage of prevailing wind direction (%), Calms = percentage of calms (%).

	JAN	FEB	MAR	APR	MAI	JUN	
T.m.	-4.6	-2.2	4.5	13.1	19.8	24.0	
T.m.mx.	1.4	3.9	10.7	19.6	26.4	30.2	
T.m.mn.	-9.9	-7.4	-1.0	6.6	12.7		
Rel.hum.	45	49	52	48	52	62	
Cloud.	3.0	3.8	4.8	5.3	5.5	6.0	
Precip.	3	7	9	19	33	78	
SD hrs.	6.6	7.1	7.7	8.4	9.4	9.3	
W.vel.	2.9	2.9	3.1	3.4	2.9	2.4	
W.mx.dir.	NNW	N	NNW	SSW	SSW	S	
W.mx.pc.	14	12	11	12	15	9	
Calms	18	17	14			17	

	JUL	AUG	SEP	OCT	NOV	DEC	YEAR
T.m.	25.8	24.4	19.4	12.4	4.1	-2.7	11.5
T.m.mx.	30.8	29.4	25.7	18.9	9.9	2.9	17.5
T.m.mn.	17.9	21.5	20.2	13.8	6.9	-0.6	-7.3
Rel.hum.	78	80	71	66	60	51	60
Cloud.	7.0	6.3	4.8	4.0	3.7	3.0	4.8
Precip.	113	212	57	24	7	8	570
SD hrs.	7.4	7.4	8.2	7.4	6.4	6.2	7.6
W.vel.	1.8	1.5	1.8	2.1	2.5	2.7	2.5
W.mx.dir.	S	N	N	N	N	N	N
W.mx.pc.	9	10	11	11	13	14	10
Calms	25	30	26	26	22	23	20

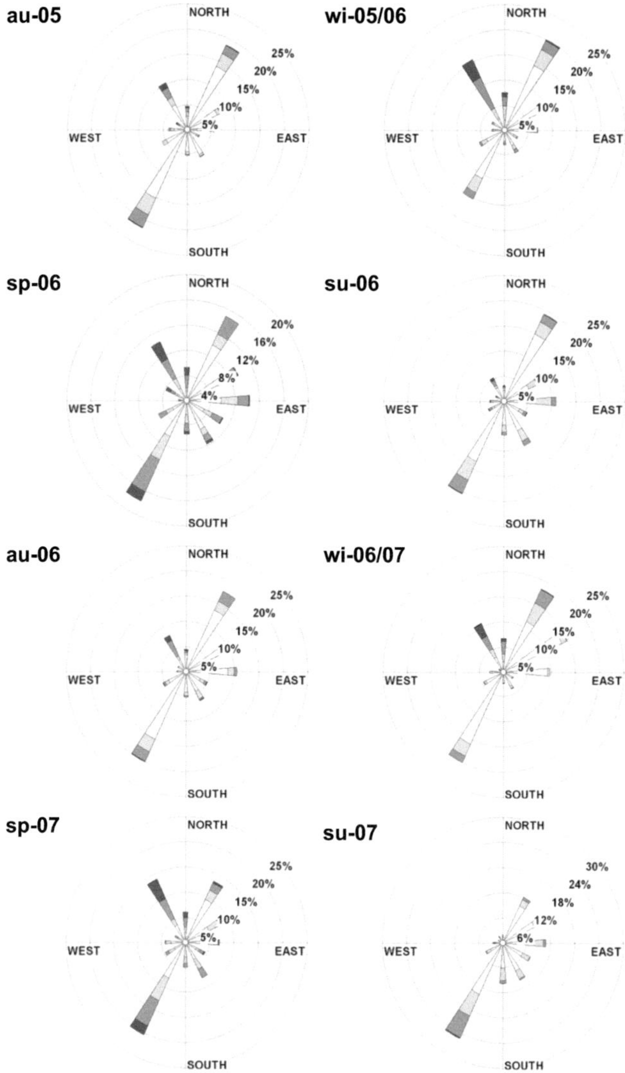

Figure 2.2: Wind rose (wind direction and wind speed) for Beijing (weather station 54511, data obtained from DWD, Offenbach) for each season from autumn 2005 to summer 2007. Legend for the wind speed see Figure 2.1). Seasons were defined meteorologically; au: autumn, wi: winter, sp: spring, su: summer.

2.3 Regional Geology

Roughly, the regional geology of China can be divided into five domains (realms), which are (Cheng, 2000):

- Tianshan–Hinggan domain
- Tarim–North China domain
- Kunlun–Qiling domain
- Sichuan–Yunnan–Qinghai–Xizang domain
- South China domain

The area around Beijing belongs to the North China domain. In the following, the most important features are briefly described according to Cheng (2000). Precambrian rocks in the Yanshan area north of Beijing are mainly composed of gneisses and granulites with banded iron formation. The Middle Proterozoic rocks occurring in the Yanshan Mountains are clastic and carbonate deposits intercalated with intermediate-basic to acid volcanic rocks. Carbonates and argilaceous-arenaceous sediments were deposited in the Cambrian and are widespread in distribution. The Carboniferous System in the North China region is widespread in distribution and is composed mainly of clastic rock formation with a few beds of carbonate rocks. The clastic rocks are coal-bearing. Permian formations are made up mainly of continental deposits and triassic rocks are made up of inland basin deposits. Intermontane basin clastic-volcanic rocks are found in the Jurassic strata. Widespread continental deposits are found in the Tertiary formations. In the Quaterny, alluvial-proluvial deposits were sedimentated. Furthermore, aeolian deposits are common. Consequently, a widespread variety of different rock types can be considered as sources for geogenic particles around Beijing.

Chapter 3

Methodology

3.1 Sampling sites and sampling procedure

3.1.1 Location and description of the sampling sites

Five sampling sites (site 1–5) were located along a transect from NW to SE through Beijing. These sites were operated during the main sampling period (period A), which comprised two years, from Sep-05 to Aug-07. Additionally, two other sites, labelled as site CUG and site CRAES were established. Site CUG was operated during the whole period A and additionally during an intensive sampling campaign in spring 2007 (period B, 27th of March to 16th of April 2007) and the Olympic period (period C, Oct-07–Dec-09). Site CRAES was established for the Olympic sampling period C. The location of the sampling sites is shown in Figure 3.1. A short description of the sampling sites and their main characteristics is summarized in Table 3.1.

3.1.2 Active sampling

During the main sampling period (period A, Sep-05–Aug-07), weekly $PM_{2.5}$ samples were collected with Mini-Volume samplers (Leckel, Berlin) at flow rates of 200 L/h at six sampling sites (sites 1–5, site CUG, Figure 3.1). At site 1 to site 5, weekly sampling was carried out separately for day- and night-time (from 7 a.m. to 7 p.m. and 7 p.m. to 7 a.m., respectively). At site CUG, sampling was performed continuously for a whole week. Additionally, weekly TSP samples were collected at site 4 with a TSP sampler at flow rates of 1 m^3/h. All samples were collected on quartz fibre filters (QF-filters). These QF-filters had a diameter of 50 mm (Macherey & Nagel) or 47 mm (Whatman), respectively.

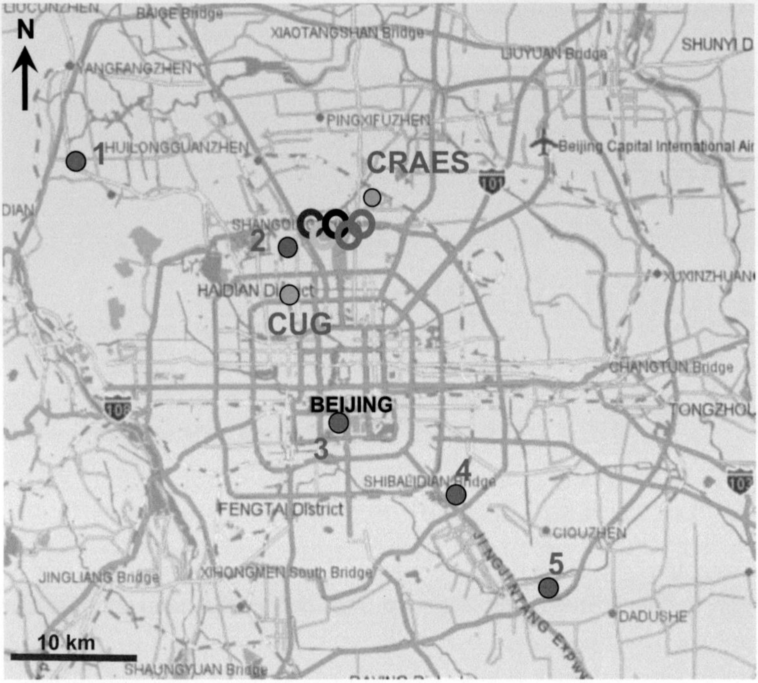

Figure 3.1: City map of Beijing with the location of the sampling sites (source of the map: http://beijing2008.go2map.com). Sites 1 – 5 operated from Sep-05 to Aug-07 (sampling period A), site CRAES from Oct-08 to Feb-09 (sampling period C), and site CUG from Apr-05 to Oct-09 (sampling periods A, B, and C). The Olympic rings mark the Olympic park ("Olympic Green"), where the Beijing National Stadium and other Olympic sites are located.

In spring 2007, during a short-term intensive measurement period (period B from the 27th of March to the 16th of April 2007), 12-hourly $PM_{2.5}$ and PM_{10} samples were collected at site CUG. Samples were collected on small QF-filters (Whatman, d = 25 mm) and the 12-hourly sampling intervals were from 7 a.m. to 7 p.m. and from 7 p.m. to 7 a.m., respectively.

From Oct-07 to Feb-09 (period C) sampling was continued at site CUG and additional samples were collected at a new site labelled CRAES. Site CRAES, at the roof of the Chinese Research Academy of Environmental Sciences (CRAES) is located 4 km north of the 5th ring road, 15 km from the city center, and 5.8 km from the National Olympic Stadium, the so-called "Bird's Nest", (Wang *et al.*,

Table 3.1: Description of the sampling sites (AGL: above ground level; CUG: China University of Geosciences, Beijing; CRAES: Chinese Research Academy of Environmental Sciences).

Site	Location	Height AGL	Period	Samples	Remarks
1	N40°04.102'	1.5 m	A	$PM_{2.5}$	agricultural
	E116°06.586'				inside patio
2	N40°00.436'	40 m	A	$PM_{2.5}$	residential
	E116°20.280'				tall buildings
3	N39°53.586'	6 m	A	$PM_{2.5}$	old Hutong quarter
	E116°23.059'				few traffic
4	N39°48.394'	1.5 m	A	$PM_{2.5}$	much traffic
	E116°28.131'			TSP	(4th ring road)
5	N39°47.197'	1.5 m	A	$PM_{2.5}$	Hotel area
	E116°38.552'				few traffic
CUG	Xueyuan Road 29	1.5 m	A	$PM_{2.5}$	University entrance
				$APM_{(2.5-80)}$	much traffic
			B	$PM_{2.5}$	(4th ring road)
				PM_{10}	sheltered by vegetation
			C	$PM_{2.5}$	
CRAES	8 Yangfang Anwai	20 m	C	TSP	Roof of the Academy
				$PM_{2.5}$	5.8 km distance to
				PM_1	Olympic Stadium

2010*b*). This sampling period was further divided into the time before (period C1, Oct-07–Jul-08), during (period C2, Jul-08–Sep-08), and after (period C3, Oct-08–Feb-09) the Olympic Games. In periods C1 and C3, samples were collected on a weekly basis at site CUG and site CRAES. During period C2, the time-resolution was increased to 24-hourly intervals for TSP and PM_1 samples at site CRAES and $PM_{2.5}$ samples at site CUG, and to 12-hourly intervals for $PM_{2.5}$ samples at site CRAES.

During sampling period A, 23% of TSP samplings were interrupted by external errors such as interruption of power supply, but only three weeks were without any record. These occasions are listed in Table 3.2 together with the share of the weeks during which the sampling operated. The weeks with sampling interruptions for $PM_{2.5}$ samples are listed in Table 3.3.

Table 3.2: List of all TSP sampling weeks of period A and the percentage of sampling duration for the corresponding weeks. (N: number; w/o: without; PCT: percentage).

	N weeks	PCT of total	Calendar weeks (ww/yy)
Weeks w/o sampling	3	2.9	06/06, 11/07, 29/07
Weeks < 25% operation	9	8.7	21/06, 41/06, 48/06, 15/07, 18/07, 20-23/07
Weeks > 25% operation	12	11.5	36/05, 44/05, 13/06, 16-18/06, 22/06, 07-09/07, 19/07, 30/07
Weeks w/o errors	80	76.9	all others
Total weeks	104	100	35/05–34/07

Table 3.3: List of all $PM_{2.5}$ sampling weeks of period A during which operational errors occurred.

	N of weeks w/o samples	Calendar weeks (ww/yy)
Site 1 – day	8	28–33/06, 21–22/07
Site 1 – night	9	20/06, 28–33/06, 21–22/07
Site 2 – day	19	39/05, 17–30/06, 05–08/07
Site 2 – night	15	16–30/06
Site 3 – day	8	37/05, 03–08/07, 24/07
Site 3 – night	7	37/05, 03–08/07
Site 4 – day	0	–
Site 4 – night	2	17–18/06
Site 5 – day	7	21/06, 29/06, 44–47/06, 22/07
Site 5 – night	8	21/06, 29/06, 44–47/06, 50/06, 22/07

3.1.3 Passive sampling

Passive sampling with a Sigma-2 device according to VDI-Guideline 2119 (VDI, 1997) was performed during all three sampling periods (A, B, and C) at site CUG (Figure 3.1). Normally, two samples were collected each week (three and four days of exposure time for each sample, respectively). During sampling period B and during the Olympic Games (period C2), the time resolution was higher with 12- and 24-hour intervals. The particles were deposited as single particles in the stilling section at the bottom of the sampler on a transparent adhesive collection plate. Further details about the method can be found in Dietze *et al.* (2006).

3.2 Chemical-physical analyses

3.2.1 Gravimetric analysis

Mass concentrations were determined gravimetrically. Gravimetric analysis with a micro-balance (Sartorius MP2, Göttingen, Germany) was performed three to five times after at least 48 h equilibration at room conditions of 40 – 45% relative humidity. Accuracy was checked against blank filters, which were also sent to Beijing but not used.

3.2.2 Acid digestion

For the analysis of trace metals one quarter of each of the quartz fibre filters was digested in teflon vessels with concentrated HNO_3 (Merck, suprapur), concentrated HF (Merck, suprapur) and concentrated $HClO_4$ (Merck, suprapur). Quality control of digestion was done by additional analysis of reference material SRM 1648 (urban PM) and GXR 2 (soil) acquired from NIST (National Institute of Standards, USA). Results of standard material were within ± 10% of the certified values for element concentrations. Methodological blank values were determined and subtracted from analytical results of loaded filter samples. Determination of elements was performed by a high-resolution inductively coupled plasma mass spectrometer (HR-ICP-MS, Axiom, VG Elemental). Element concentrations were analysed for all actively collected samples (TSP, PM_{10}, $PM_{2.5}$, PM_1) from all sampling periods.

3.2.3 Water-soluble ions

Water-soluble ions were analysed by Ion Chromatography (IC). For this analysis, one quarter of each of the collected filters was shaken for 45 minutes in 15 mL ultrapure deionized water (Milli-Q) and additionally extracted ultrasonically for 30 minutes in order to release the water-soluble parts. For anions, a Dionex ICS-1000 with ASRS-ULTRA suppressor, Ion Pac AS4A-SC column and a Na_2CO_3/$NaHCO_3$-eluent with a flow rate of 2 mL/min was used. Cations were analyzed with a Dionex DX-120 with CSRS-ULTRA suppressor, Ion Pac CS12, using a 0.1 M H_2SO_4-eluent. Standard solutions of both anions (Multi ion anion IC standard solution Specpure, Alfa-Aesar) and cations (Multi-component Cation Mix 2, Acculon) were used in different concentrations for quality control. Water-soluble anions were determined for the whole sampling period A only for site 4 ($PM_{2.5}$ and TSP) and site CUG ($PM_{2.5}$), while for all other sites only the samples from 2006 were considered. Water-soluble cations were analysed for the year 2006 for all sampling sites except site 2. Only TSP samples (site 4) were analysed for the complete period A. Concentrations in the analysed blank filters were below limit of detection (LOD) and had, therefore, not to be subtracted from the sample concentrations.

The anion (A) and cation (C) micro-equivalents were calculated with rounded atomic weights according to the following equations:

$$A = F^-/19 + Cl^-/35.5 + NO_3^-/62 + SO_4^{2-}/48 \quad (3.1)$$

$$C = Na^+/23 + NH_4^+/18 + K^+/39 + Mg^{2+}/12 + Ca^{2+}/20 \quad (3.2)$$

3.2.4 Sequential extraction procedure

For each month of a selected sampling interval from Jul-05 to May-08, one TSP filter sample, representing one complete sampling week, was selected for chemical sequential extraction (N=35). Starting in Sep-07, the TSP sampling site was moved from site 4 to the northern part of Beijing, labelled as CRAES (see Figure 3.1). The same approach was applied for $PM_{2.5}$ samples from site CUG (Feb-05–Sep-07, N=32). Of all $PM_{2.5}$ samples, only samples from site CUG were chosen for sequential extractions due to higher particle mass on the filters because of continuous sampling for the whole week without distinction between day- and night-time samples (site 1–site 5). A leaching procedure based on the scheme of Tessier *et al.* (1979) and modified by Fernández Espinosa *et al.* (2002) was applied. This methodology is summarised in Table 3.4. From each TSP filter sample a sixth part and a fourth part of each $PM_{2.5}$ filter sample was used for the sequential extraction procedure. Four fractions were distinguished (see Table 3.4), which can be classified as: (f1) water-extractable, (f2) bound to carbonates, oxides and reducible metals, (f3) bound to organic matter, oxidisable and sulfidic metals, and (f4) residual fraction. The sequential extraction procedure was carried out in polypropylene tubes. After each fractionation step the tubes were centrifuged and the respective solution was pipeted off. The remaining dust loaded filter material was dried directly in the tubes at 80°C in a water bath before applying the next step of the extraction scheme.

Element concentrations of each extraction step were measured by a high-resolution inductively coupled plasma mass spectrometer (HR-ICP-MS, Axiom, VG Elemental). Quality control was performed by including the reference material GXR 2 (soil) acquired from NIST (National Institute of Standards, USA) in

Table 3.4: Chemical fractions, assumed mobility of each fraction, reagents used and operational conditions of the sequential extraction procedure after Fernández Espinosa *et al.* (2002).

Fraction	**Mobility**	**Reagents**	**Operation**
water-soluble	highly mobile	15 mL Milli-Q	3 h shaker agitation at room temperature
carbonates, oxides, and reducible metals	mobile	10 mL $NH_2OH \times HCl$ 0.25 M	5 h shaker agitation at room temperature
bound to organic matter, oxidisable, and sulphidic metals	less mobile	7.5 mL H_2O_2 (30%) + 7.5 mL H_2O_2 + 15 mL NH_4AcO 2.5 M	water bath at 95°C until near dryness + repetition + shaker agitation at room temperature for 90 min
residual	not mobile	10 mL (HNO_3:HCl:$HClO_4$) (6:2:5)	5 h shaker agitation

the extraction procedure. The results of the standard material calculated as sum of the four extraction steps were within ±10% of the certified values for element concentrations of Ca, V, Fe, Ni, As, and within ±20% for Sr and ±30% for Mg, Co, Cd, Pb. Average blank filter values for each fraction were subtracted from the obtained analytical results of the samples.

3.2.5 Optical oil-immersion technique for BC

For analysis of BC, an optical transmission technique according to Ballach *et al.* (2001), which was modified by Fricker & Schultz (2002), was applied. A standard thermal analysis method (VDI, 1999) is used to link light attenuation by the filter samples to elemental carbon (EC) concentration. The scattering effects are minimized by immersion of the filters in a fluid similar refractive index (oil-immersion technique). The immersion fluid used for quartz fibre filters in this study is Paramount TM (ProTags). Light absorption at a constant wavelength of 650 nm was measured by a spectrometer (SPECORD 50, Analytic Jena). However, only $PM_{2.5}$ samples could be analysed with this method. TSP filters were too densely loaded to be used in the spectrometer (no detectable light could go through the sample).

3.2.6 Quantitative Optical Microscopy

The automated optical microscopic single particle analysis was performed with a computer-aided image analysis system at the German Meteorological Service (DWD) in Freiburg, Germany. This method was used to analyse single particles collected on adhesive plates with the Sigma-2 sampling device and is described in detail in Dietze *et al.* (2006). An area of 18x18 mm of the acceptor plate was embedded in an immersion fluid for transmission bright-field microscopic analysis. A motorized microscope (ZEISS Axioplan 2) with an adapted automatic scanning stage (Prior Scientific) and a CCD-digital-camera (SVS-VISTEK) was used for imaging with a 20x objective magnification. The collected particles were automatically focused by an integrated software focus of the image processing system Digitrace V.3.4 (IMATEC). Projected area and mean optical density (mean grey values) of individual particles were automatically measured in the size range between 2.5 and 160 μm. Different particles groups can be distinguished: "transparent" particles, "opaque" particles, and pollen. The mass concentrations were subsequently calculated for different size groups as described in Dietze *et al.* (2006). Their procedure can be summarized as follows. Based on the measured size-differentiated particle number N, and a known analysed area A and exposition time t, the number deposition rate D_N can be calculated: $D_N = N/(A \cdot t)$ The mass deposition rate D_M is then calculated assuming a spherical shape of the particle and standard density: $D_M = D_N \cdot (d_P/2)^3 \cdot 4\pi/3\rho_P$. Additionally, a volume correction factor $k_{vol} = 0.75$ is introduced to account for the overestimation due to a larger projected diameter. The overestimation occurs since the particles are deposited preferably with their flat side. Dividing the number deposition by a deposition velocity v_{dep} yields the ambient concentration: $C_M = D_M/v_{dep}$. Within the sampler head of the Sigma-2 device, v_{dep} is approximated by the terminal settling velocity of each particle in calm air v_{TS} (Stokes law). The size-differentiated particle mass concentration C_P can then be calculated according to:

$$C_P = D_M/v_{TS}$$

with

v_{TS}: $\rho_P \cdot d_P^2 \cdot g/(18 \cdot \eta \cdot \chi_P)$ in $cm \cdot s^{-1}$
ρ_P: particle density in $g \cdot cm^{-3}$
d_P: particle diameter in μm
g: gravitational acceleration (981 $cm \cdot s^{-1}$)
η: viscosity of air (1.81$\cdot$ 10^{-5} $Pa \cdot s$
χ_P: dynamical shape factor

The different size classes considered within this study are the following:

SC1: Particles from 2.5 to 5 μm geometrical diameter (d_g)
SC2: Particles from 5 to 10 μm d_g
SC3: Particles from 10 to 20 μm d_g
SC4: Particles from 20 to 40 μm d_g
SC5: Particles from 40 to 80 μm d_g

3.2.7 Scanning electron microscopy

Scanning Electron Microscopy (SEM) was carried out at the Laboratory for Electron Microscopy (LEM) at the Karlsruhe Institute of Technology (KIT) with a LEO 1530 Gemini coupled to electron dispersive X-ray fluorescence spectrometry (EDX) using an excitation energy of 15 kV. Minerals and other solid phases were identified by their image and the relative concentrations of elements, which were compared with spectrograms of minerals as provided by Reed (2006).

3.2.8 Synchrotron radiation based μ-XRF analyses

The collected single airborne particles were analysed for their elemental distribution by means of energy-dispersive synchrotron radiation based μ-X-ray fluorescence analysis (μS-XRF) at ANKA (Angström-Quelle Karlsruhe, Karlsruhe Institute of Technology (KIT), Germany, FLUO-beamline). The transparent adhesive collection plates could be used directly without any further preparation. The experiment was carried out under atmospheric conditions. The radiation at ANKA is delivered by a 2.5 GeV storage ring. The source at the FLUO-beamline is a 1.5 T bending magnet with a critical energy of about 6 keV. A double multilayer monochromator (W-Si multilayers in 2.7 nm period) is used at the FLUO-beamline. As focusing optics x-ray lenses (CRL) and polycapillaries can be applied. The collimated X-ray beam for the experiment had the size of approximately 2×5 μm at the sample and an energy of 22 kV was applied. The measurement of elements with an atomic number $Z < 15$ was not possible due to the measurement in air. While for all analysed elements the K-lines were used, Pb was determined with its L-lines. The measured elemental concentrations for the samples were calibrated with the reference sample StHs6 (andesitic ash from the St. Helens (USA) eruption, Stoll & Jochum, 1999). Additionally, two other reference materials, KL2 (basalt from Hawaii) and ATHO (rhyolite from Iceland), were measured for quality control. Elemental concentrations were calculated by fundamental parameters using the program PyMCA (Solé *et al.*, 2007).

3.3 Statistical methods

For all statistical analyses, the software package STATISTICA (StatSoft, USA, Version 6) was applied.

3.3.1 Descriptive statistics

Unless noted otherwise the weighted mean (ac_n) of mass or elemental concentrations was calculated according to the equation:

$$ac_n = (c_{n-1} + 2c_n + c_{n+1})/4 \tag{3.3}$$

In the appendix, the following statistical values are given for the different analysed concentrations: Number (N), average (avg), standard deviation (stdev), lower quartile (lower QT), upper quartile (upper QT), median (median), 10th percentile (10th PCT), and 90th percentile (90th PCT).

3.3.2 z-transformation

Data were normalized according to the following equation (z-transformation):

$$conc_{z-trans} = (conc_{avg} - conc_{value})/stdev \tag{3.4}$$

3.3.3 Factor analysis

Factor analysis is a statistical tool in order to group parameters based on mutual correlations among them. The main goal is to reduce the number of initial variables and to reveal patterns within the original data matrix. The algorithm allows the allocation of each variable to independent groups, called factors. With the STATISTICA software package (StatSoft, USA, Version 6) a correlation matrix is generated based on which the factors are successively extracted. An eigenvalue of 1 is used as stop criterion. Varimax standardized rotation was applied to maximize the variance of the factor loadings for each factor transforming the loadings as close as possible to +1 and -1, respectively. The communality expresses to what extent the variance of a variable can be explained by all factors. For each case (e.g. sampling week) a so called factor score was calculated which shows the estimated value of each factor for the considered case.

3.4 Enrichment factors

The calculation of enrichment factors (EFs) helps to distinguish between elements originating from anthropogenic activities and those from natural sources, and consequently, to estimate the degree of anthropogenic contamination. Since its introduction by Zoller *et al.* (1974), this method was applied for countless source apportionment studies. For this calculation it is assumed that the anthropogenic contribution of the normalizing element (e.g. Fe, or Ti) is insignificant. Within this work, EFs were calculated related to Ti with two different background values: (1) related to average crustal material (concentrations take from Reimann & Caritat (1998)), (2) related to average Chinese topsoil (concentrations taken from Chen *et al.* (2008)), according to the following equations:

$$EF_{Ti} = \frac{(Element/Ti)_{APM}}{(Element/Ti)_{crust}} \tag{3.5}$$

$$EF_{Ti} = \frac{(Element/Ti)_{APM}}{(Element/Ti)_{soil}} \tag{3.6}$$

Median concentrations of all APM samples were used for the calculation of the EFs. By convention, an average EF value < 10 is taken as an indication that a trace metal in an aerosol has a significant crustal source, while an EF value > 10 is considered to indicate that a significant proportion of an element has a non-crustal and, consequently, anthropogenic source (Chester *et al.*, 1999).

Chapter 4

Towards a better understanding of spatio-temporal variations of aerosol concentration and composition

4.1 Introduction

For the assessment of urban atmospheric pollution, it is crucial to gain detailed information about the chemical composition of atmospheric particulate matter (APM). On the one hand, potential toxic metals play a decisive role for the estimation of health effects (see section 1.3.1) and the respective burden for the inhabitants of cities and megacities can vary strongly due to an inhomogeneous spatio-temporal distribution. On the other hand, the composition of APM indicates relevant sources, which are important to know for the implementation of effective mitigation measures.

Beside total element concentrations, also water-soluble ions constitute a large part of APM mass. Studies from China showed that water-soluble ions accounted for one-third or more of the APM mass in Chinese urban regions (He *et al.*, 2001; Hu *et al.*, 2002; Wang *et al.*, 2002, 2006*b*; Shen *et al.*, 2009). The importance of water-soluble inorganic species is not only due to their high percentage in APM mass, but also due to negative impacts on human health, visibility reduction and acidic rain of some species (see section 1.3.1).

Furthermore, carbonaceous particles play an important role for the overall atmospheric pollution. Carbon in atmospheric particles occurs as carbonate, organic carbon (OC), soot, char and elemental carbon (EC) or black carbon (BC), respectively. By convention, carbon determined by thermal methods is called EC, while the carbon, determined by optical methods, is called BC (Hitzenberger *et al.*, 2006). Black carbon originates mainly from incomplete combustion processes, and in China, especially from the usage of coal and biofuels (Streets *et al.*, 2001). A major part of BC in $PM_{2.5}$ can be classified as soot. The effects of BC on the environment, climate and human health are described in section 1.3.

Within this chapter, a comprehensive data set of element and water-soluble ion concentrations as well as BC concentrations of $PM_{2.5}$, PM_{10}, and TSP samples from Beijing is presented. The focus of this chapter lies on data from sampling period A (Sep-05 – Aug-07) and B (spring 2007), whereas results from period C (Olympic Games 2008) are discussed separately in chapter 6 with particular attention on the effects of the applied mitigation measures. The results about aerosol concentration and composition in Beijing are subsequently discussed with regard to their spatio-temporal variations, possible toxicity, predominant sources and especially the distinction between geogenic and anthropogenic sources, as well as in connection to findings from other studies at different urban locations worldwide. The distinctive features of this study are the long-term period with continuous sampling over two years, the discrimination between day- and night-time samples, and the simultaneous measurements at different sites in order to gain a comprehensive knowledge about the seasonal and spatial pattern of atmospheric particles in Beijing. Furthermore, the origin of the particles with special focus on the distinction between geogenic and anthropogenic sources is another principal point of this study.

4.2 Results from long-term weekly sampling

4.2.1 Enrichment factors

The calculation of enrichment factors (EFs) is helpful in order to be able to distinguish between elements originating from anthropogenic activities and those from natural sources and to estimate the degree of anthropogenic contamination. Since its introduction by Zoller *et al.* (1974), this method was applied for countless source apportionment studies. Enrichment factors were calculated according to equation 3.5 (see chapter 3.4), and are listed in Table 4.1 for $PM_{2.5}$ samples of each of the five sampling sites (site 1–5) as well as for TSP samples from site 4. By convention, an average EF value < 10 is taken as an indication that a trace

Table 4.1: Calculated enrichment factors (EFs) referring to Ti for all analysed elements. EFs were calculated using the median concentrations of all samples from each site (see equation 3.5). Crustal concentrations were taken from Reimann & Caritat (1998).

	$PM_{2.5}$ site 1	$PM_{2.5}$ site 2	$PM_{2.5}$ site 3	$PM_{2.5}$ site 4	$PM_{2.5}$ site 5	TSP site 4
Na	2	1	2	2	3	1
Mg	1	1	2	2	1	2
Al	1	1	1	1	1	1
K	7	4	7	7	11	2
Ca	3	3	4	4	4	4
Sc	1	1	1	1	1	1
Ti	1	1	1	1	1	1
V	3	2	2	2	2	1
Cr	3	2	3	4	3	2
Mn	8	5	7	8	9	2
Fe	2	2	2	2	2	1
Co	3	2	2	3	3	1
Ni	10	5	6	16	10	2
Cu	117	118	137	228	274	30
Zn	348	275	421	499	509	85
Ga	19	13	16	23	23	4
As	739	506	850	1359	1237	175
Rb	10	7	10	10	13	2
Sr	2	2	3	4	3	2
Cd	2635	1778	2897	3911	4327	522
Sn	376	313	490	684	673	79
Sb	2085	2773	2622	4682	3617	564
Cs	32	18	26	29	36	6
Ba	3	3	4	5	5	3
Pb	865	657	968	1233	1302	148

metal in an aerosol has a significant crustal source, while an EF value > 10 is considered to indicate that a significant proportion of an element has a non-crustal and, consequently, anthropogenic source (Chester *et al.*, 1999). However, it is important to bear in mind, that nowadays, there is no element left which results solely from geogenic sources, since humans almost use any chemical element for some technical material (Hashimoto *et al.*, 1992; Nath *et al.*, 2007). Nevertheless, in the following, elements with high EFs will be referred to as elements from predominantly anthropogenic sources or short as "anthropogenic elements", whereas elements with low EFs will be classified as elements from predominantly geogenic sources or "geogenic elements".

In Beijing, elements whose occurrence was highly influenced by anthropogenic activities were Cu, Zn, As, Cd, Sn, Sb, and Pb (EFs > 100, Table 4.1). Those elements had the highest EFs in the south-eastern parts of Beijing at site 4 (As, Sn, and Sb) or site 5 (Cu, Zn, Cd, and Pb), respectively. Generally, EFs of the finer $PM_{2.5}$ samples were higher than those of total particles (Table 4.1).

4.2.2 Mass and element concentrations

Total suspended particulate matter

For the two-year period from Sep-05 to Aug-07 (period A), the average TSP concentration was 372 $\mu g/m^3$, with a maximum of 1026 $\mu g/m^3$ in November 2005 (CW 44/05) and a minimum of 140 $\mu g/m^3$ in July 2006 (CW 28/06). The TSP mass concentration in Beijing considerably varied over the course of both years (Figure 4.1). With respect to seasons, highest average concentrations were found in autumn 2005 with 524 $\mu g/m^3$. In the following year, autumn concentrations were lower with an average value of 395 $\mu g/m^3$. Winter average concentrations were also quite high (winter 05/06: 351 $\mu g/m^3$ and winter 06/07: 484 $\mu g/m^3$). Summer had lowest particle concentrations of all seasons with average values of 235 $\mu g/m^3$ in 2006, and 261 $\mu g/m^3$ in 2007. Spring TSP concentrations differed strongly between 2006 and 2007. In 2006, average spring concentrations were considerably higher with an average of 404 $\mu g/m^3$ compared to 296 $\mu g/m^3$ in 2007.

Beside the TSP mass concentrations, also the concentrations of the different elements varied strongly during the annual course. This is shown using the example of seven selected elements (Mg, Al, K, Ca, Ti, Fe and Sr) from predominantly geogenic sources in Figure 4.2 and of seven elements (Cu, Zn, As, Cd, Sn, Sb, and Pb) from anthropogenic sources in Figure 4.3. For better comparison the element concentrations were standardized using z-transformation (Equation 3.4, see chapter 3.3.2).

The maximum TSP concentration during sampling period A in November 2005 was reflected in high concentrations of all geogenic elements (Figure 4.2). Magnesium and Ca also had their maxima during this week (CW 44/05). Another concentration peak of geogenic elements occurred three weeks later at the end of November (CW 47/05, Figure 4.2). Concentrations during this week were even higher for Al, Ti, and Fe. Spring concentrations were especially high for those three elements, as well as for K. It is worth emphasizing the differences between spring 2006 and 2007 with considerably higher concentrations of geogenic el-

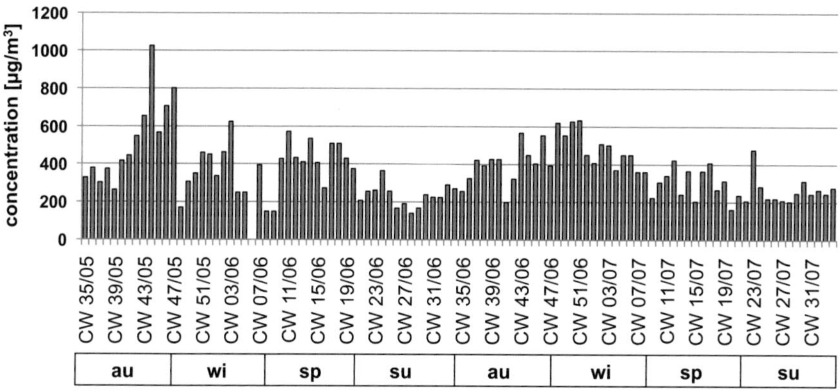

Figure 4.1: Weekly TSP mass concentrations (in $\mu g/m^3$) during sampling period A (Sep-05–Aug-07) at site 4. The seasons were defined meteorologically; au: autumn (Sep, Oct, Nov), wi: winter (Dec, Jan, Feb), sp: spring (Mar, Apr, Jun), su: summer (Jul, Aug, Sep).

ements in 2006 compared to the following year (Figure 4.2). Summer was the seasons with lowest concentrations in both years, 2006 and 2007. Moreover, also the variations from week to week were lower during this time of the year (Figure 4.2).

Anthropogenic element concentrations were also very variable during the annual course (Figure 4.3). However, those elements had a different course during the distinct seasons compared to geogenic elements. One huge difference between those two particle groups were the considerably lower spring concentrations for anthropogenic element concentrations (Figure 4.3). Generally, anthropogenic element concentrations were highest in autumn and winter (Figure 4.3). However, it is noticeable that the average concentrations in winter 06/07 were considerably higher compared to those of the previous winter.

Descriptive statistics of TSP mass and element concentrations for sampling period A are summarised in Table A.1 as volume related concentrations in ng/m^3 and in Table A.2 as mass related concentrations in $\mu g/g$.

Particulate matter smaller 2.5 μm

During the two-year sampling period from Sep-05 to Aug-07 (Period A), five $PM_{2.5}$ sampling sites were operated along a transect through Beijing from NW to SE with separated day- and night-time sampling on a weekly basis (Figure 3.1).

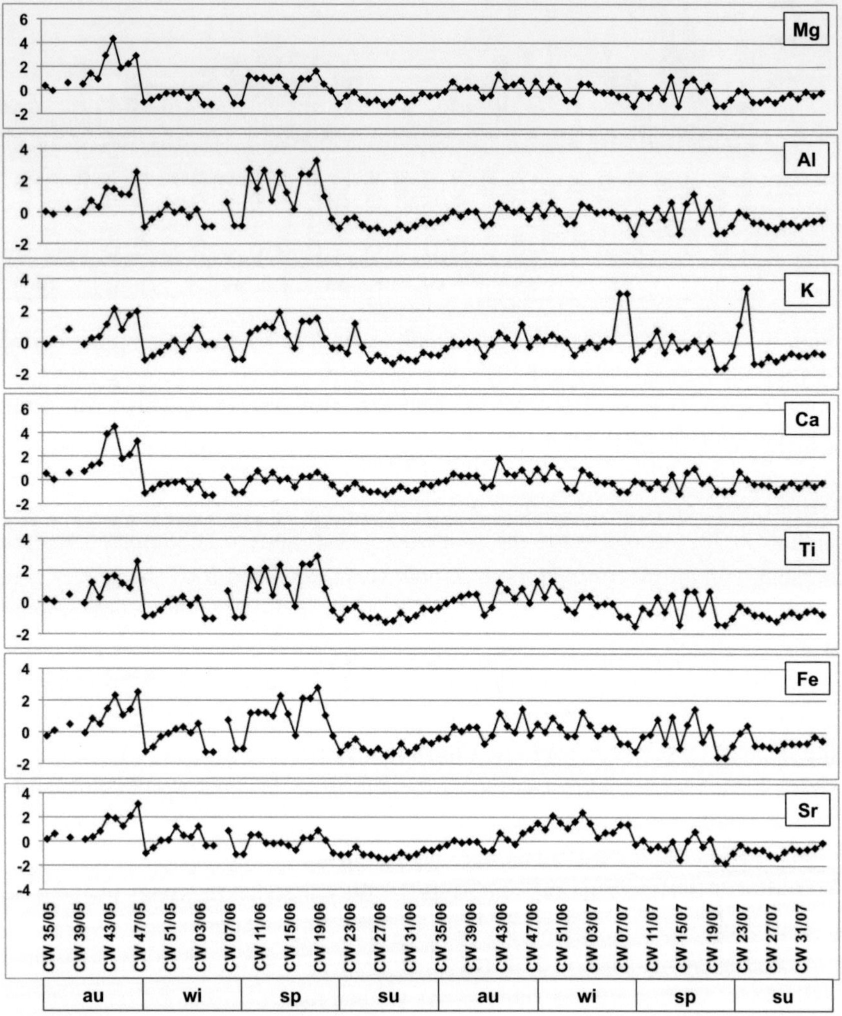

Figure 4.2: Standardized (z-transformation, Equation 3.4) element concentrations of seven selected elements from predominantly geogenic sources (Mg, Al, K, Ca, Ti, Fe, and Sr) of weekly TSP samples during period A at site 4.

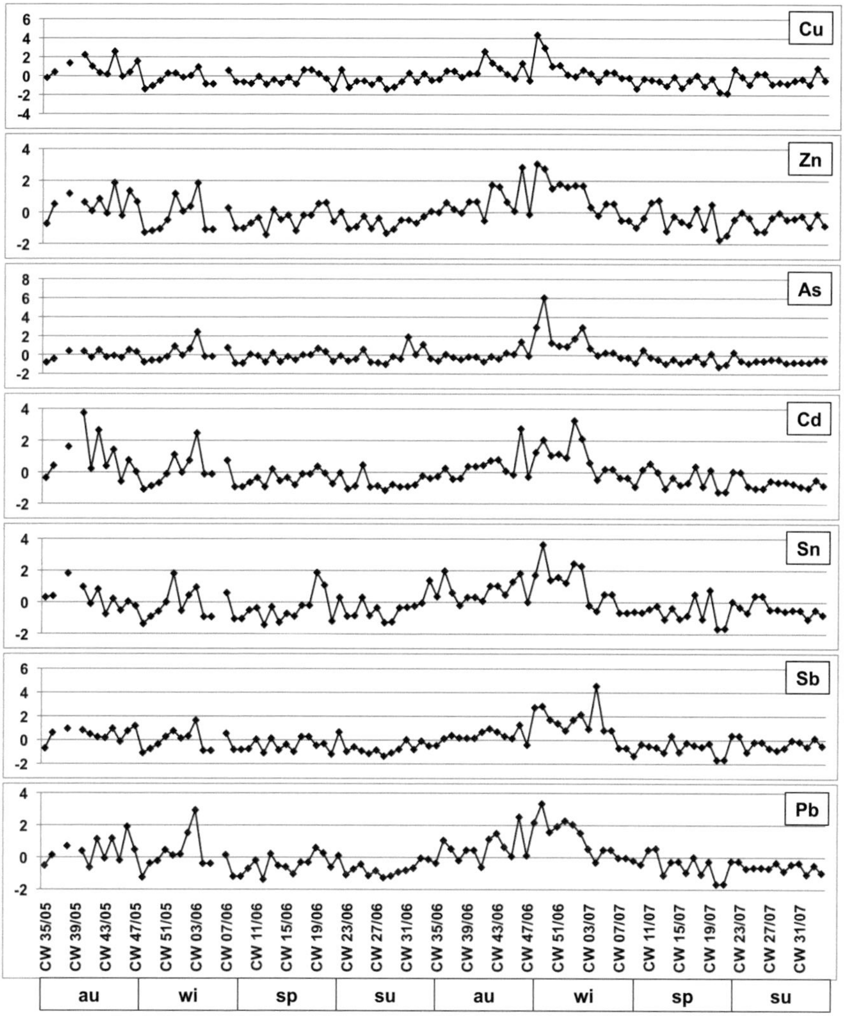

Figure 4.3: Standardized (z-transformation, Equation 3.4) element concentrations of seven selected elements from predominantly anthropogenic sources (Cu, Zn, As, Cd, Sn, Sb, Pb) of weekly TSP samples during period A at site 4.

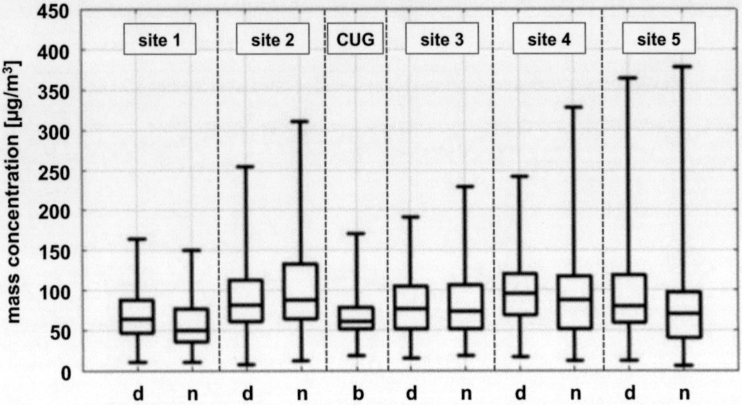

Figure 4.4: Boxplots of $PM_{2.5}$ mass concentrations from site 1 – site 5 for day (d) and night (n) samples, respectively. At site CUG, no distinction between day- and night-time samples was made (b). The lower end of the box is represented by the lower quartile, the upper end by the upper quartile, and the median value is illustrated by the line inside the box. Whiskers represent the complete data range (minimum to maximum).

Additionally, weekly $PM_{2.5}$ samples without distinction between day and night were collected at site CUG. Descriptive statistics for mass and element concentrations of the $PM_{2.5}$ samples from the different sites are listed in Tables A.3 – A.12 for volume-related data in ng/m^3 air. The same tables are also provided for mass-related data in μg/g (Tables A.13 – A.22).

Figure 4.4 shows the mass concentrations at the five sites (site 1 – site 5) separately for day- and night time as well as for weekly samples at site CUG. With the exception of site 2, day-time concentrations were usually higher than corresponding night-time values.

Site 4 was chosen as an example for a more detailed illustration of the weekly $PM_{2.5}$ element concentrations during sampling period A because at this site TSP samples were collected simultaneously and were, therefore, available for a comparison of both particle size fractions. $PM_{2.5}$ concentrations at site 4 varied from 17.2 to 242 μg/m^3 with an average value of 102 μg/m^3 during daytime and from 12.7 to 227 μg/m^3 with an average value of 96.7 μg/m^3 for night-time samples of Period A. For better comparison the element concentrations were standardized using z-transformation (Equation 3.4, see chapter 3.3.2). The same seven geogenic elements (Mg, Al, K, Ca, Ti, Fe and Sr), which were already presented for TSP samples in the previous section, are illustrated for $PM_{2.5}$ samples in Figure 4.5 separately for day- and night-time samples. Furthermore, the same annual course is shown in Figure 4.6 for seven anthropogenic elements (Cu, Zn, As, Cd, Sn, Sb, and Pb).

Element concentrations of the fine fraction smaller than 2.5 μm had a similar annual trend compared with those of total particulates (TSP results see section 4.2.2). Geogenic elements had highest concentrations in spring 2006 and lowest during both summers (Figure 4.5). Anthropogenic element concentrations were also lowest during summer and had highest concentrations in winter 2006/2007. Day- compared to night-time concentrations of the respective elements were relatively similar for both, geogenic and anthropogenic, elements.

4.2.3 Water-soluble ions

Within this study, four water-soluble anions (sulphate, nitrate, chloride and fluoride) and five water-soluble cations (sodium, ammonium, potassium, magnesium and calcium) were measured. However, only for TSP samples the concentrations of all ions were above detection limit (LOD, see chapter 3.2.3) for most of the samples. Additionally to TSP samples, at site 4 and site CUG, water-soluble anions were also analysed for $PM_{2.5}$ samples for the whole sampling period A. At all other sites, water-soluble ions were analysed for the calendar year 2006.

Sulphate (SO_4^{2-}) was the prevalent anion for both $PM_{2.5}$ and TSP samples with an average concentration of 19.4 for day- and 16.7 $\mu g/m^3$ for night-time $PM_{2.5}$ samples, and 33.9 $\mu g/m^3$ for TSP samples at site 4. The sulphate concentrations at site 1 to site 5 for the year 2006 are illustrated in Figure 4.7 for day- and night-time $PM_{2.5}$ samples, respectively, as well as for weekly $PM_{2.5}$ samples from site CUG (without discrimination between day and night). For all sites with the exception of site 2, the median day-time sulphate concentrations were higher than those during night (Figure 4.7). Site 1 in the NW of Beijing was the site with lowest sulphate concentrations, whereas site 4 in the SE of the city had the highest sulphate concentrations.

Sulphate concentrations had strong seasonal variations. The average sulphate concentrations for the different seasons are displayed in Figure 4.8 for TSP and $PM_{2.5}$ samples from site 4 as well as for $PM_{2.5}$ from site CUG for the whole sampling period A. Highest TSP sulphate concentrations occurred in winter with an average concentration 41.6 $\mu g/m^3$ for the whole season and a maximum value of 70.4 $\mu g/m^3$ for a single week (CW 03/06). Summer TSP sulphate values are also quite high with 35.6 $\mu g/m^3$ on average.

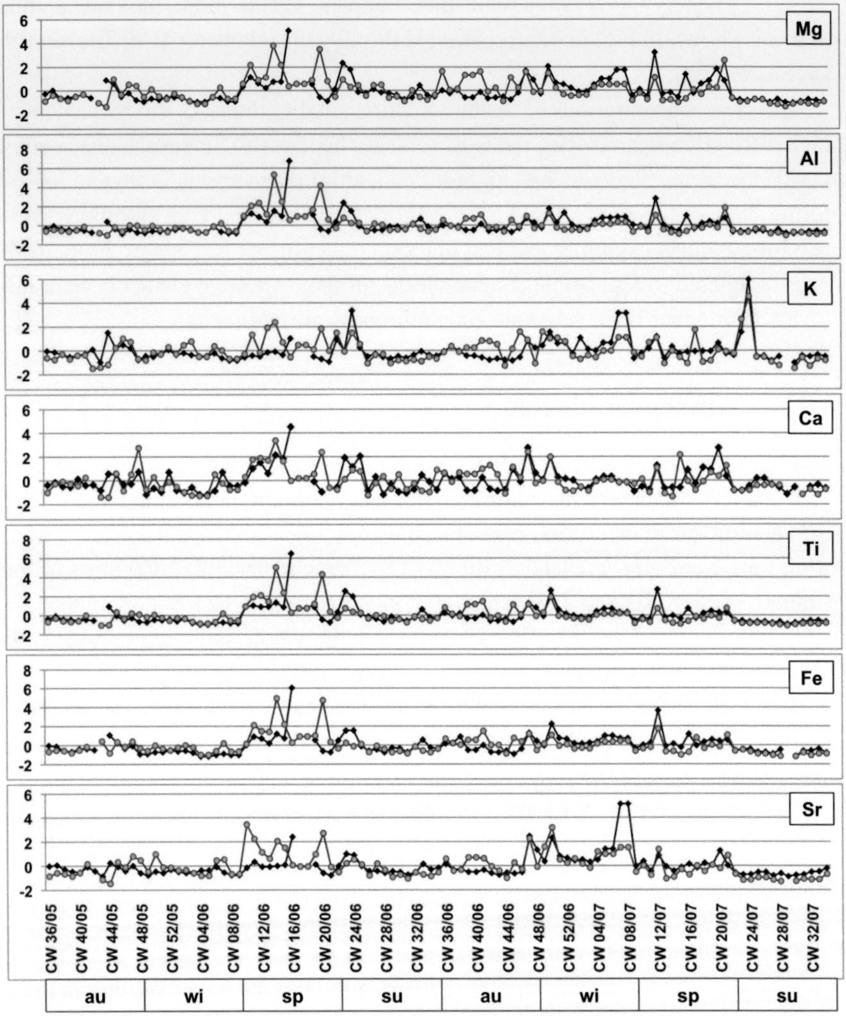

Figure 4.5: Standardized (z-transformation, Equation 3.4) element concentrations of seven selected elements from predominantly geogenic sources (Mg, Al, K, Ca, Ti, Fe, and Sr) of weekly $PM_{2.5}$ samples during period A at site 4 for day-time (light lines with dots) and night-time (dark lines with rhombi) samples, respectively.

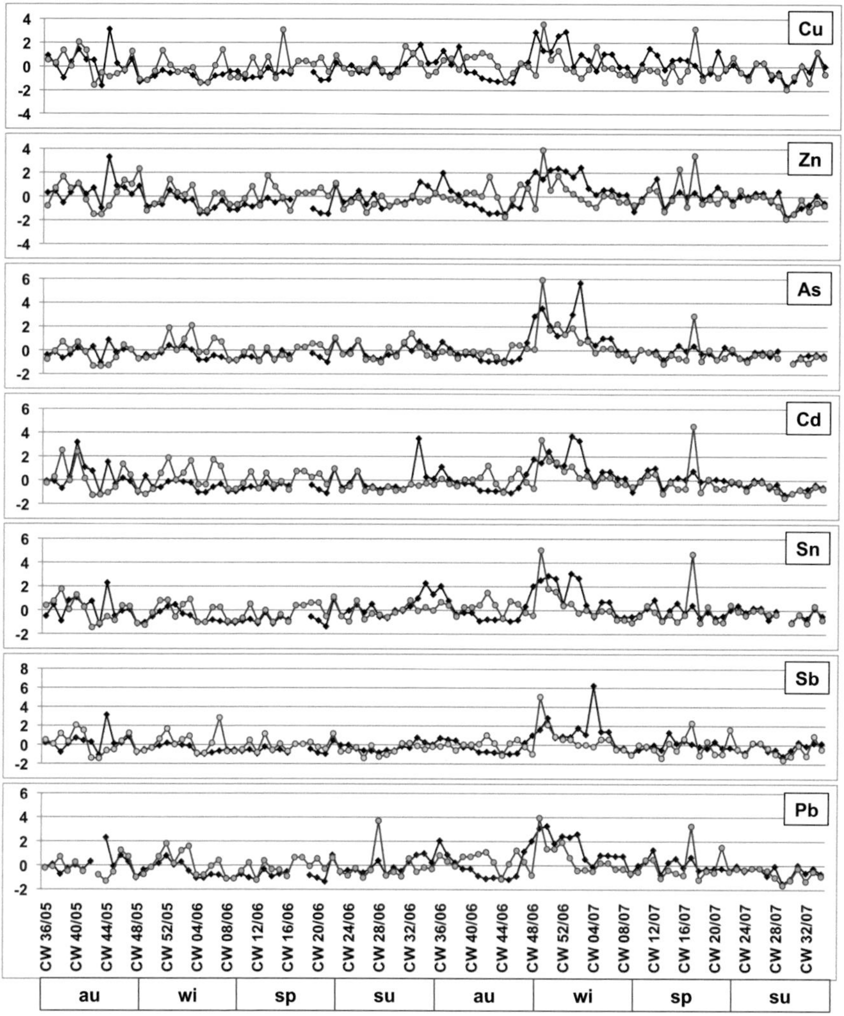

Figure 4.6: Standardized (z-transformation, Equation 3.4) element concentrations of seven selected elements from predominantly anthropogenic sources (Cu, Zn, As, Cd, Sn, Sb, Pb) of weekly $PM_{2.5}$ during period A at site 4 for day-time (light lines with dots) and night-time (dark lines with rhombi) samples, respectively.

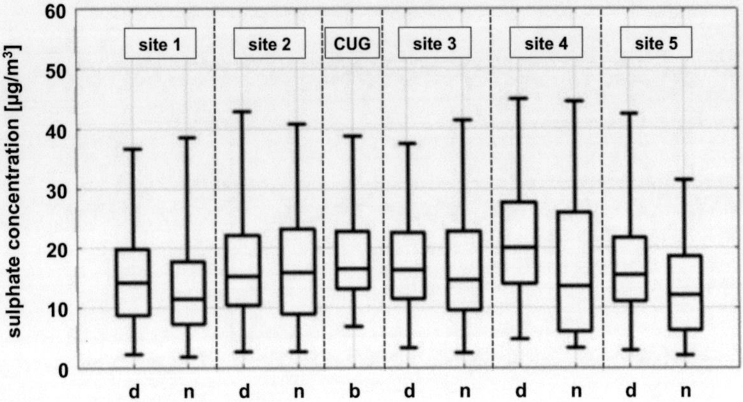

Figure 4.7: Boxplots of sulphate concentrations of the year 2006 from site 1 – site 5 for day- (d) and night-time (n) samples, respectively. At site CUG, no distinction between day- and night-time samples was made (b). The lower end of the box is represented by the lower quartile, the upper end by the upper quartile, and the median value is marked by the line within the box. The whiskers represent the complete data range (minimum to maximum).

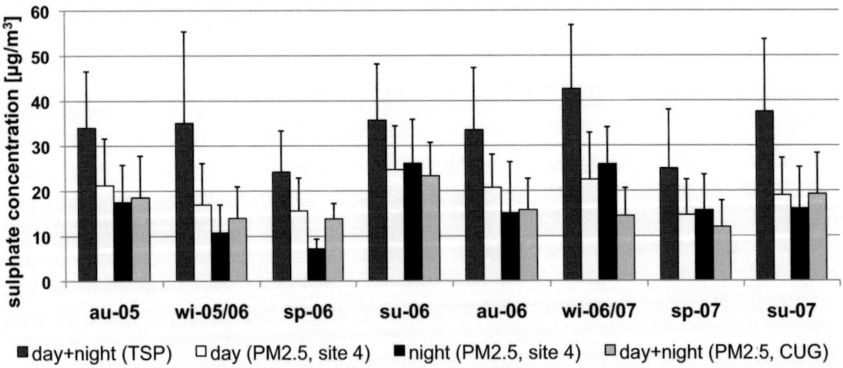

Figure 4.8: Average seasonal sulphate concentrations at site 4 for TSP and day- and night-time $PM_{2.5}$ samples as well as for $PM_{2.5}$ samples from site CUG. The whiskers represent the standard deviation. Seasons were defined meteorologically; au: autumn, wi: winter, sp: spring, su: summer.

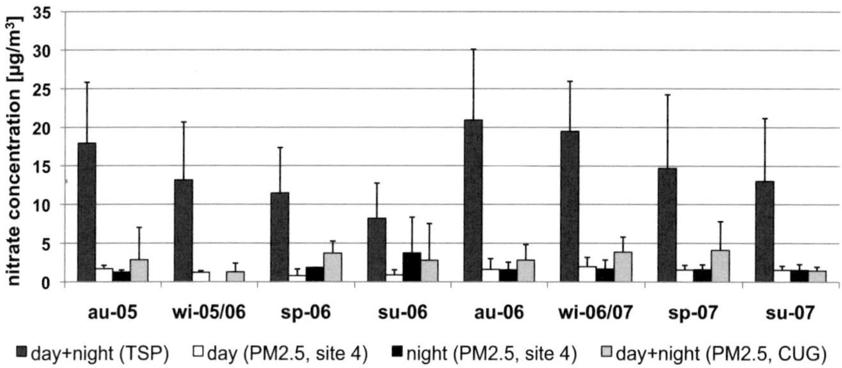

Figure 4.9: Average seasonal nitrate concentrations of TSP and day- and night-time $PM_{2.5}$ samples at site 4, as well as of $PM_{2.5}$ at site CUG for sampling period A. The whiskers represent the standard deviation. Seasons were defined meteorologically; au: autumn, wi: winter, sp: spring, su: summer.

Nitrate (NO_3^-), which was the anion with the second highest concentrations, showed a different seasonal trend compared to sulphate. Lowest nitrate concentrations occurred in summer (8.2 μg/m^3 for TSP), whereas winter and spring concentrations were high (winter: 16.2 μg/m^3, spring: 11.5 μg/m^3 for TSP). The average nitrate concentrations for the different seasons are displayed in Figure 4.9 for TSP and $PM_{2.5}$ samples from site 4 as well as for $PM_{2.5}$ from site CUG for the whole sampling period A. Nitrate concentrations in TSP samples correlated with total mass concentration ($r = 0.72, N = 53$).

Chloride (Cl^-) and **fluoride** (F^-) concentrations were often below LODs (LODs: 0.2 and 0.4 mg/L, respectively) in $PM_{2.5}$ samples and therefore not measured for the majority of $PM_{2.5}$ samples. For TSP samples, the average annual concentrations were 3.3 μg/m^3 for fluoride and 0.8 μg/m^3 for chloride.

Calcium (Ca^{2+}) was the most abundant water-soluble cation in TSP samples. Its concentrations varied from 5.25 to 27.9 μg/m^3 with an average value of 11.9 μg/m^3 for TSP samples from period A. In $PM_{2.5}$ samples, Ca^{2+} concentrations were much lower. Average concentrations were 1.46 μg/m^3 for day-time and 0.86 μg/m^3 for night-time $PM_{2.5}$ samples from 2006 at site 4. All other cations in TSP samples had lower concentrations compared to Ca^{2+}. Average TSP concentrations of **sodium** (Na^+) were 1.35 μg/m^3, of **ammonium** (NH_4^+) 2.51 μg/m^3, of **magnesium** (Mg^{2+}) 1.93 μg/m^3, and of **potassium** (K^+) 2.36 μg/m^3.

The average seasonal concentrations of ammonium, sodium, potassium, and calcium for the year 2006 are plotted in Figure 4.10 for TSP and day- and night-time $PM_{2.5}$ samples from site 4, as well as for $PM_{2.5}$ samples from site CUG. Water-soluble magnesium concentrations were not included in the Figure, since its concentrations were below LOD for many $PM_{2.5}$ samples. Summer and winter had higher NH_4^+ concentrations compared to autumn and spring concentrations (Figure 4.10). Spring was the season with lowest NH_4^+ concentrations for TSP and $PM_{2.5}$ samples. In all seasons, $PM_{2.5}$ samples had higher NH_4^+ concentrations compared to those of TSP samples. On the contrary, water-soluble sodium had higher concentrations in TSP than in $PM_{2.5}$ samples (Figure 4.10). For Na^+, summer was the month with lowest concentrations, whereas highest concentrations occurred in winter, and for TSP samples also in spring (Figure 4.10). Water-soluble potassium concentrations were highest in winter. Concentrations of K^+ were in a similar range for both, TSP and $PM_{2.5}$ samples (Figure 4.10).

4.2.4 Black carbon

Black carbon concentrations were measured only for $PM_{2.5}$ samples due to analytical reasons (see chapter 3.2.5). Figure 4.11 illustrates the BC concentrations during sampling Period A at four sites (site 1, site 3 – site 5) separately for day and night. Furthermore, BC concentrations from site CUG without distinction between day- and night-time samples are included in the Figure. Site 4 in the SE of Beijing had highest BC concentrations of all sites. Lowest BC concentrations were measured at site 1 in the NW of Beijing. With the exception of site 1, night-time BC concentrations were usually higher than corresponding day-time concentrations.

Black carbon concentrations of $PM_{2.5}$ samples had pronounced seasonal variations. The average concentrations of each season from spring 2005 to winter 2007/2008 at site CUG are shown in Figure 4.12. Spring and summer BC concentrations were considerably lower than those in autumn and winter. In 2005 and 2007, autumn was the season with maximum average BC concentrations. In 2006, however, concentrations of winter 2006/207 were higher than those of autumn. In 2005, the season with lowest BC concentrations was summer, whereas in the following two years spring concentrations were even lower than those of summer.

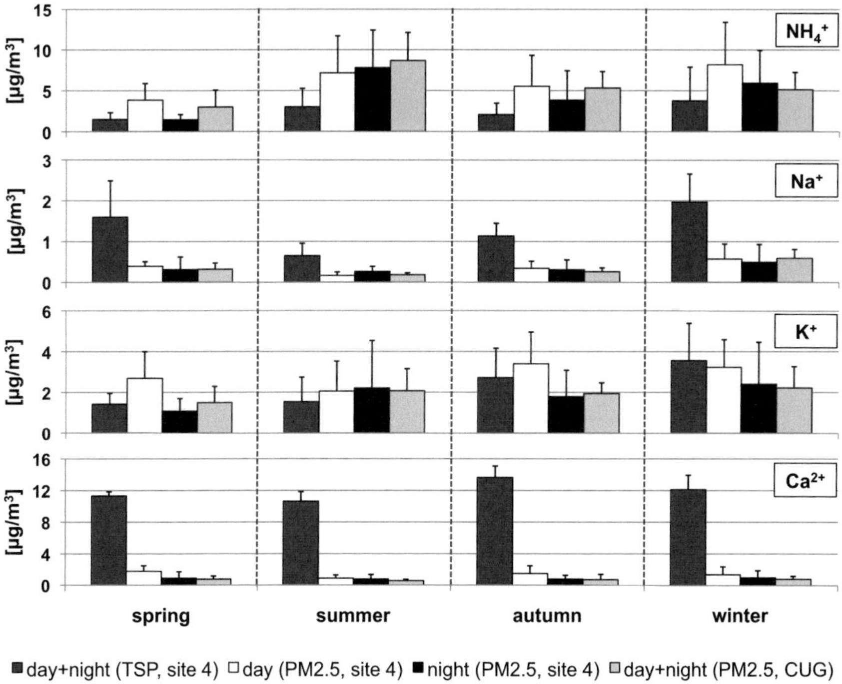

Figure 4.10: Average seasonal concentrations of water-soluble cations from TSP and day- and night-time $PM_{2.5}$ samples at site 4, as well as from $PM_{2.5}$ samples at site CUG for sampling period A. The cations are from top to bottom: ammonium, sodium, potassium, and calcium. The whiskers represent the standard deviation. Seasons were defined meteorologically.

4.2.5 Passively sampled atmospheric particles

Atmospheric particles ($APM_{2.5-80}$) between 2.5 and 80 μm d_g (geometrical diameter) were collected passively at site CUG. These particles were grouped in the following six size classes:

Size class one (SC1):	2.5 – 5.0 μm
Size class two (SC2):	5.0 – 10 μm
Size class three (SC3):	10 – 20 μm
Size class four (SC4):	20 – 40 μm
Size class five (SC5):	40 – 80 μm
Size class six (SC6):	80 – 160 μm

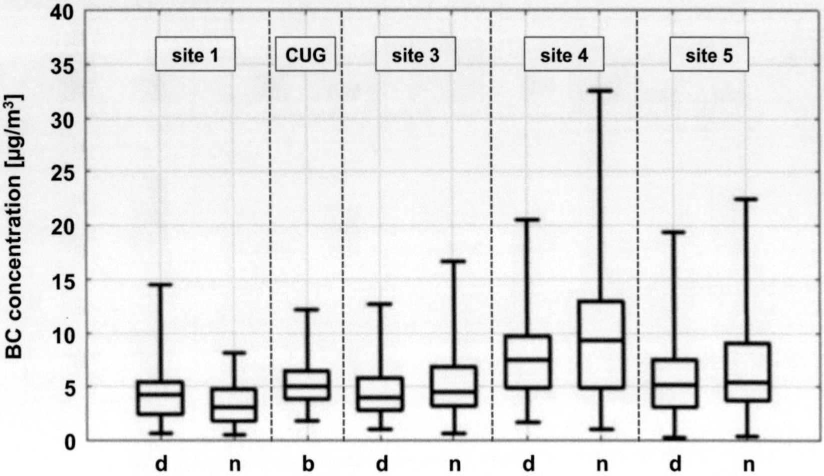

Figure 4.11: Boxplots of BC concentrations from site 1 and site 3 – site 5 for day (d) and night (n) samples, respectively. At site CUG, no distinction between day- and night-time samples was made (b). The lower end of the box is represented by the lower quartile, the upper end by the upper quartile, and the median value is illustrated by the line within the box. The whiskers represent the complete data range (minimum to maximum).

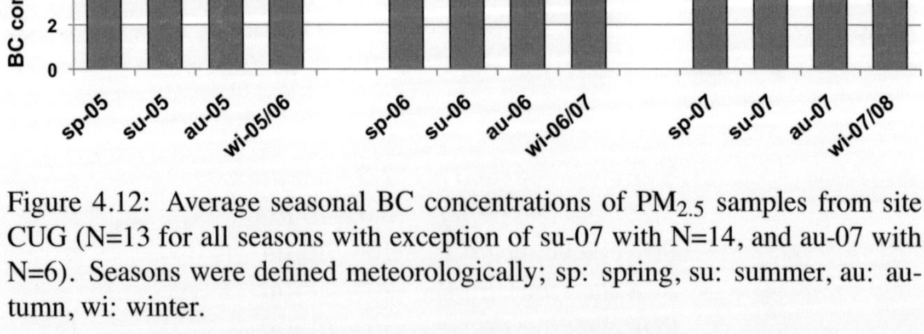

Figure 4.12: Average seasonal BC concentrations of $PM_{2.5}$ samples from site CUG (N=13 for all seasons with exception of su-07 with N=14, and au-07 with N=6). Seasons were defined meteorologically; sp: spring, su: summer, au: autumn, wi: winter.

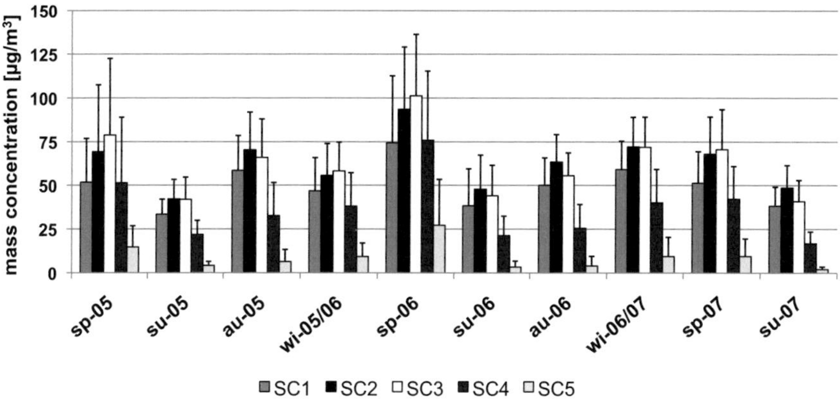

Figure 4.13: Average seasonal mass concentrations of total particles for the different size classes (SC1:2.5–5, SC2:5–10, SC3:10–20, SC4:20–40, SC5:40–80 μm). The whiskers represent the respective standard deviation. The seasons were defined meteorologically; sp: spring, su: summer, au: autumn, wi: winter.

Since only few of the collected particles belonged to SC6, only SC1–SC5 will be considered in the following sections. Furthermore, it was also distinguished between "transparent" and "opaque" particles. Both particle groups together are referred to as "total" particles. Descriptive statistics for total, transparent and opaque particles are summarised in Table A.23.

Seasonal variations were pronounced for all size classes. The average concentrations of the respective seasons from spring 2005 to summer 2007 are shown in Figure 4.13 for SC1–SC5. For particles of all size classes, spring 2006 was the season with maximum average concentrations and concentrations were lowest in summer. In spring, SC3 had highest concentrations during all three years. In winter 2005/2006, SC3 was also the size class with highest concentrations closely followed by SC2. Size class 2 had highest average concentrations for all other seasons. Consequently, particles between 5 and 20 μm constitute the most important size fractions of the total mass concentration in Beijing.

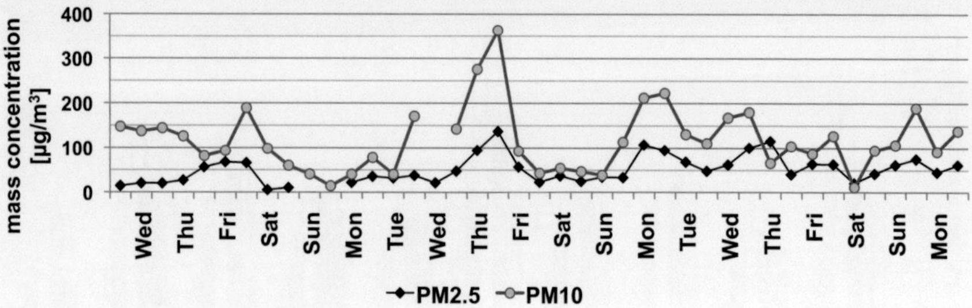

Figure 4.14: $PM_{2.5}$ and PM_{10} mass concentrations in $\mu g/m^3$ at site CUG from the 27th of March to the 14th of April 2007 (period B).

4.3 Results from short-term 24-hourly sampling

4.3.1 Mass and element concentrations

In spring 2007 (27th of March to the 16th of April 2007, sampling period B), $PM_{2.5}$ and PM_{10} samples were collected at site CUG in 12-hour intervals. During this time average $PM_{2.5}$ mass concentrations ranged from 5.54 to 137 $\mu g/m^3$ with an average concentration of 51.3 $\mu g/m^3$. PM_{10} mass concentrations varied from 11.4 to 363 $\mu g/m^3$ with an average concentration of 117 $\mu g/m^3$. Mass concentrations of particles of both size classes had a similar trend during these three weeks (Figure 4.14). On average, PM_{10} mass concentrations were 3-fold higher compared to those of $PM_{2.5}$ samples.

Beside the mass concentrations, also the element concentrations varied during this short-term sampling period. For geogenic elements, such as Mg, Al, K, Ca, Ti, Fe, and Sr, both particle size classes showed a similar trend. However, the PM_{10} concentrations were significantly higher compared to those of $PM_{2.5}$ samples (Figure 4.15). Anthropogenic elements, such as Cu, Zn, As, Cd, Sn, Sb, and Pb, on the contrary, had similar concentrations in both size fractions (Figure 4.16).

Descriptive statistics for mass and element concentrations of $PM_{2.5}$ and PM_{10} samples from period B are summarised in Table A.24 and Table A.25 as volume-related concentrations in ng/m^3. The same tables are provided for mass-related concentrations in $\mu g/g$ (Table A.26 and Table A.27).

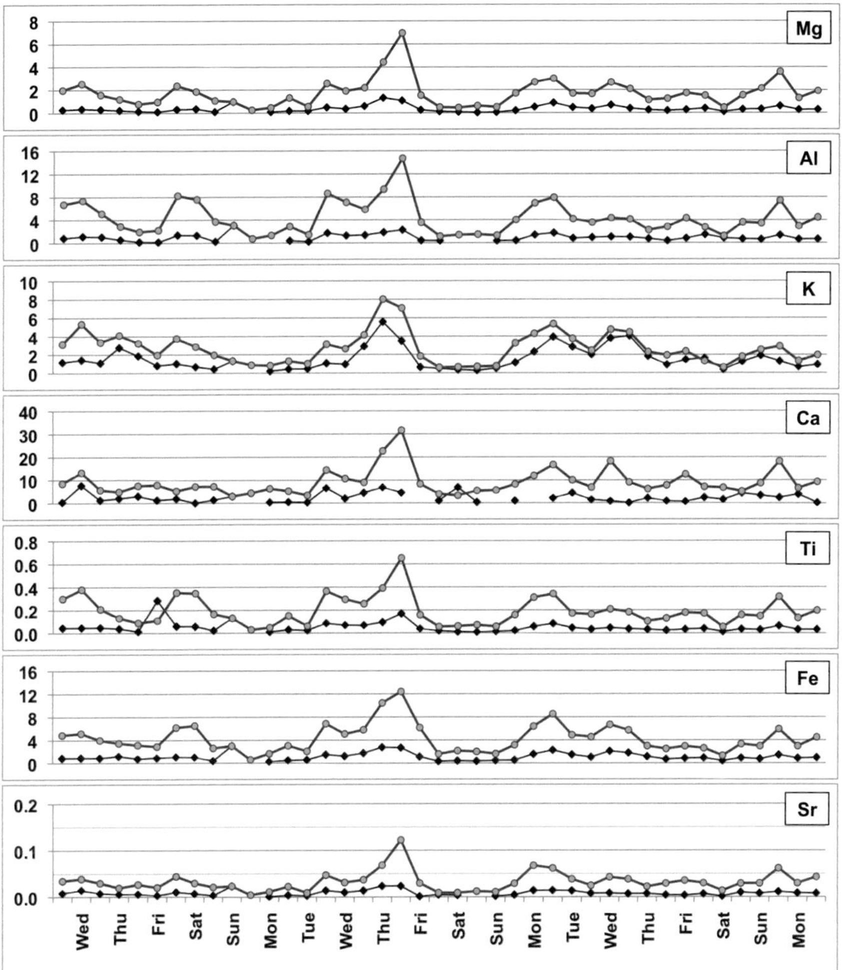

Figure 4.15: Element concentrations in $\mu g/m^3$ of seven selected elements from predominantly geogenic sources (Mg, Al, K, Ca, Ti, Fe, and Sr) of 12-hourly $PM_{2.5}$ (black line with rhombi) and PM_{10} (grey lines with circles) samples from the 27th of March to the 16th of April 2007 (period B) at site CUG.

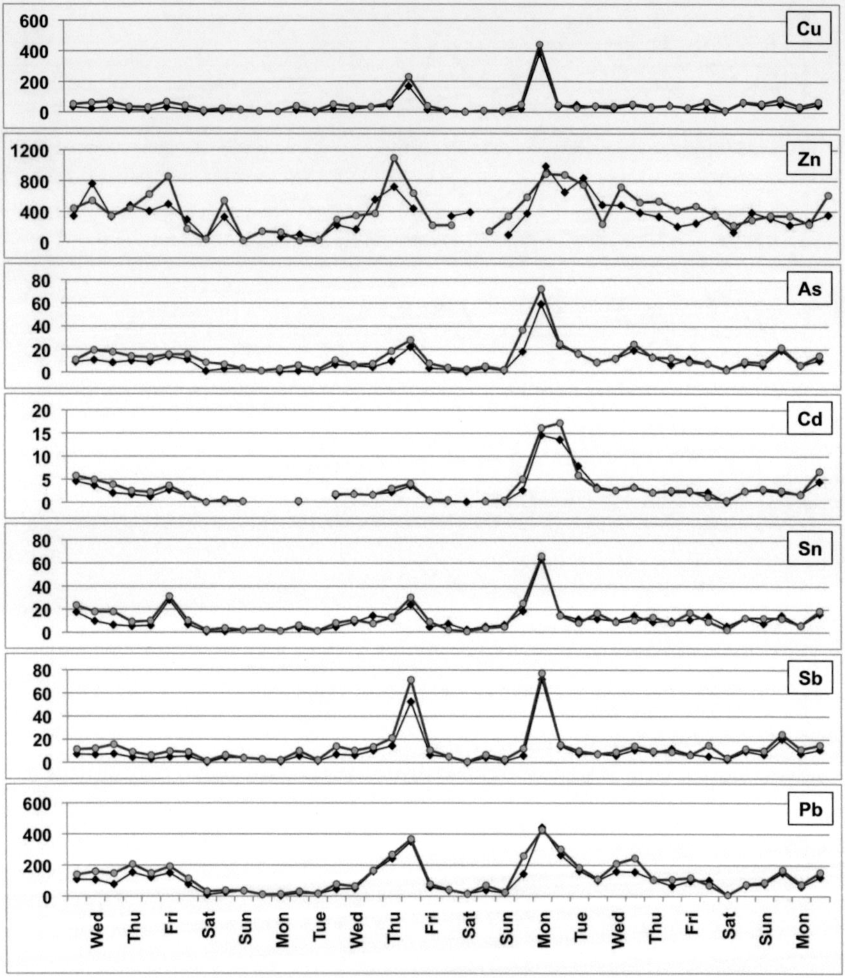

Figure 4.16: Element concentrations in ng/m^3 of seven selected elements from predominantly anthropogenic sources (Cu, Zn, As, Cd, Sn, Sb, and Pb) of 12-hourly $PM_{2.5}$ (black line with rhombi) and PM_{10} (grey lines with circles) samples from the 27th of March to the 16th of April 2007 (period B) at site CUG.

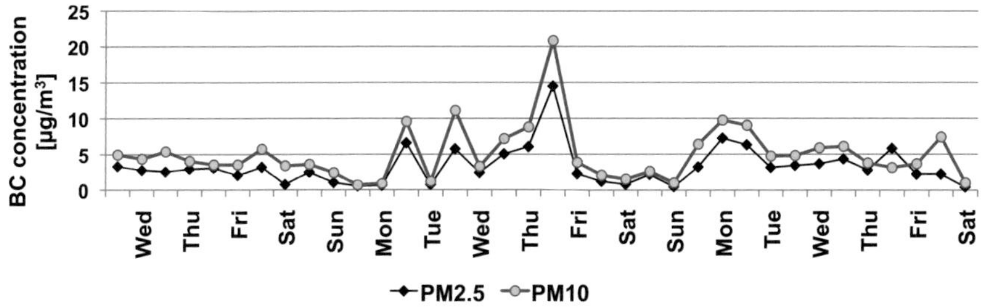

Figure 4.17: BC concentrations of $PM_{2.5}$ and PM_{10} samples in $\mu g/m^3$ at site CUG from the 27th of March to the 14th of April 2007 (Period B).

4.3.2 Black carbon

Black carbon concentrations at site CUG during sampling period B varied from 0.44 to 14.5 $\mu g/m^3$ with an average value of 3.29 $\mu g/m^3$ for $PM_{2.5}$ and from 0.68 to 20.8 $\mu g/m^3$ with an average value of 5.03 $\mu g/m^3$ for PM_{10} samples. The course from the 27th of March to the 14th of April 2007 (12-hour intervals) is illustrated in Figure 4.17 for both size fractions. On average, BC concentrations of $PM_{2.5}$ samples constituted 62% (ranging from 22 to 87%) of BC of PM_{10} samples.

4.3.3 Passively sampled atmospheric particles

During sampling period B (27th of March – 14th of April 2007), passively collected particles between 2.5 and 80 μm ($APM_{2.5-80}$) were also collected in 12-hour intervals at site CUG. Transparent particle concentrations varied from 84.5 to 694 $\mu g/m^3$ with an average concentration of 299 $\mu g/m^3$, whereas opaque particle concentrations ranged from 7.60 to 42.4 $\mu g/m^3$ with average value of 20.3 $\mu g/m^3$. The course for transparent and opaque particles, respectively, is shown in Figure 4.18 for the three sampling weeks in spring 2007.

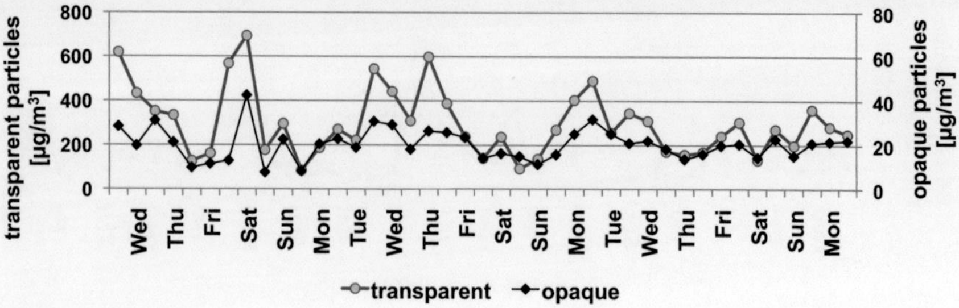

Figure 4.18: Mass concentrations of transparent and opaque particles between 2.5 and 80 μm d_g at site CUG from the 27th of March to the 14th of April 2007 (Period B).

4.4 Discussion

4.4.1 Spatial variations and their relation to different land-use patterns

$PM_{2.5}$ mass concentrations strongly varied between the five sites (site 1 – site 5) along the transect (Figure 3.1) with lowest concentrations at site 1 north-west of Beijing and highest at site 4 in the south-eastern parts of the city (Figure 4.4). Moreover, not only the mass but also the element concentrations differed strongly between the sites. In order to investigate the spatial pattern of various elements, the deviation of the average element concentrations of each site from the average value of all five sites were calculated. Firstly, four elements from predominantly geogenic sources (Mg, Ca, Ti, and Fe) were chosen (Figure 4.19) in order to investigate the spatial pattern of typical geogenic elements. Secondly, the same was done for four elements from predominantly anthropogenic sources (Zn, As, Cd, and Pb; Figure 4.20) in order to obtain information about the spatial distribution of typical anthropogenic elements.

Site 1 had lowest concentrations for both, geogenic and anthropogenic, elements. On average, element concentrations were 20–40% lower at this site compared to the average of all sites. The relatively lower particulate air pollution in this area can be explained by the location of site 1 in the NW of Beijing. Here, in the outskirts of Beijing the land-use is agricultural, residential and recreational. The site is located on the area of a hotel owned by the State Environmental Protection Administration (SEPA). Moreover, the traffic volume in the area of site 1 is lower and there are only few tall buildings, which could hinder the dispersion

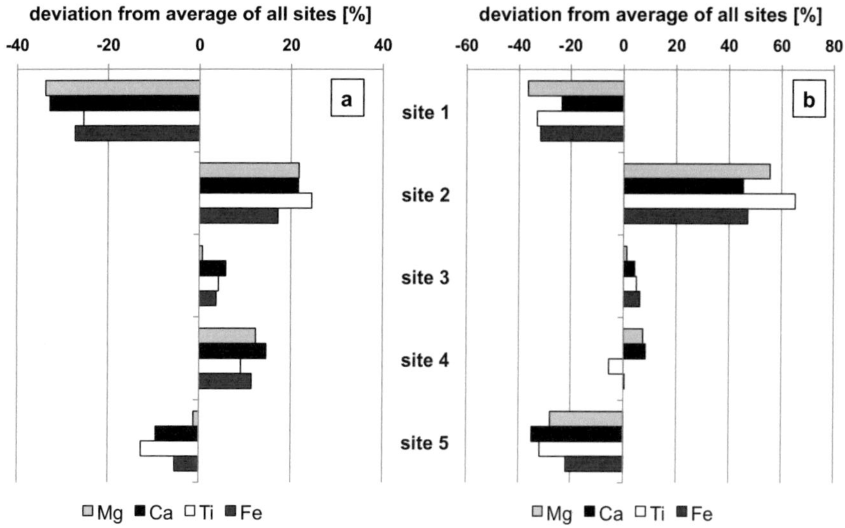

Figure 4.19: Deviation of average volume-related element concentrations (Mg, Ca, Ti, and Fe in ng/m^3) from site 1 – site 5 compared to the respective average value of all five sites; (a) day-time and (b) night-time concentrations.

(Liu *et al.*, 2009). Air masses from the mountainous region N and NW of Beijing reach this site without travelling over large industrial areas first and are, consequently less loaded with anthropogenic particles. For anthropogenic elements, it is worth mentioning that night-time concentrations were lower than those during daytime (Figure 4.20). Another special feature of site 1 is that it was the only site with lower BC concentrations during night- than daytime. The generally lower anthropogenic pollution during night time in the NW of Beijing can further be explained by the topography with the Yanshan Mountains in the N and the Taihang Mountains in the W of Beijing governing the wind fields. Hu *et al.* (2004) reported southern anabatic wind fields during day and northern katabatic wind during night time in Beijing. These katabatic winds, which transport relatively clean air masses towards the city, are probably the main reason for better air quality during night time in this area.

At **site 2** large differences between geogenic and anthropogenic elements were observed. Whereas this site had highest concentrations of geogenic elements compared to the other sites (Figure 4.19), the concentrations of anthropogenic elements were significantly lower (Figure 4.20). The specific charac-

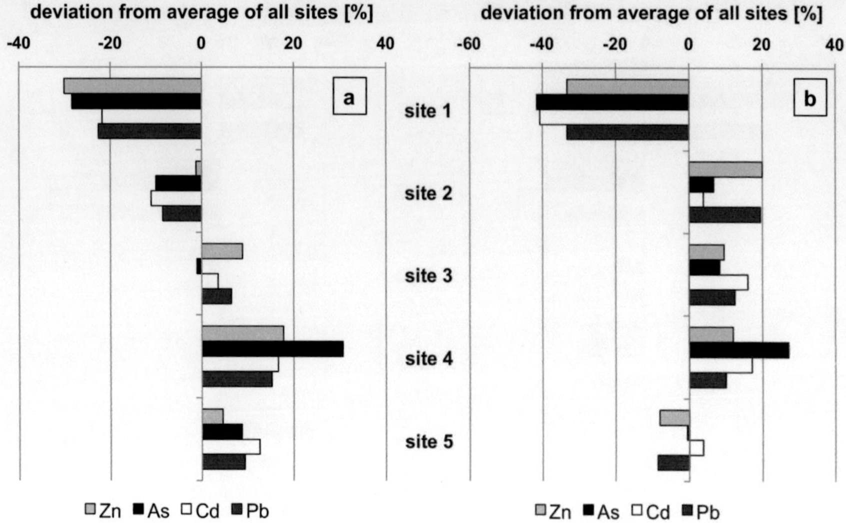

Figure 4.20: Deviation of average volume-related element concentrations (Zn, As, Cd, and Pb in ng/m^3) from site 1 – site 5 compared to the respective average value of all five sites; (a) day-time and (b) night-time concentrations.

teristic of sampling site 2 is that the sampling devices are located higher above ground (about 40 m) than it is the case for the other sites (1.5 m for all other sites during period A with exception of site 3 with about 6 m). At this height, the influence of geogenic particles transported over long distances from regional and supra-regional sources plays a relatively more important role. This assumption is further supported by the observation, that concentrations of geogenic elements were significantly higher at this site because of the dominance of minerals from semi-arid and arid areas due to long-range transport, whereas anthropogenic particles from local sources had lower concentrations compared to other sites.

Site 3 had slightly higher concentrations of geogenic elements and also of anthropogenic elements compared to the average of all sites (Figure 4.19 and 4.20). The enrichment at this site was higher for anthropogenic than for geogenic elements. Site 3 is located in the center of Beijing only about 500 m SE of the Tiananmen Square in an old "Hutong" quarter at the roof of a hotel (about 6 m above ground). The surrounding buildings are mostly the same size or smaller, and, thus, do not shield this sampling site. The roads in this area are narrow with only low traffic volume. Local sources are especially coal combustion for heating and cooking as well as resuspension processes of particles from unpaved streets, or construction and demolition waste.

At **site 4**, especially the anthropogenic element concentrations were high compared to the other sites (Figure 4.20). Arsenic had the highest enrichment at this site (about 30% above average). The high concentrations of anthropogenic elements can be explained by a high density of industry in the south-eastern parts of Beijing. Geogenic element concentrations were higher compared to the average of all sites during daytime but were very close to the average concentration during night time (Figure 4.19). The traffic influence at site 4, which is located in the south-eastern bend of the fourth ring road and is not sheltered from this road by vegetation or buildings, is high. Consequently, resuspension processes from the road during daytime with higher traffic volume and congestions (stop-and-go traffic) might explain the higher geogenic element concentrations of day- compared to night-time samples.

Compared to the average of all five sites, **site 5** had lower concentrations of geogenic elements (Figure 4.19) during daytime and especially during night time. Site 5 is located in the SE of Beijing in the outskirts of the city in the inner courtyard of a private house. Therefore, this site is more sheltered compared to the other sites and, thus, less affected by resuspended particles from bare soils and street surfaces. During night-time, especially As and Cd concentrations were enriched at this site. This observation highlights the stronger anthropogenic pollution in the south-eastern part of the city compared to the north-western areas.

Additional information about the spatial differences of particulate air pollution in Beijing can be gained by comparing also the mass related element concentrations in μg/g of the different sites. When considering the toxicity of elements in urban areas and their health impacts due to air pollution, it is important to considered also these concentrations in μg/g (equal to mg/kg as it is used in many sediment or soil studies), because anthropogenic elements are often enriched in the fine particle fraction and, thus, their proportion might be diluted by coarser particles. The spatial distribution of mass-related concentrations of Zn, As, Cd, and Pb is illustrated in Figure 4.21.

It is important to highlight the observation that the spatial variations are very different if the concentrations expressed in μg/g are considered for the comparison. In this context it should be noted that site 1 had relatively lowest concentrations of anthropogenic elements if expressed in $\mu g/m^3$ (Figure 4.20); this was the case for site 2 if the concentrations were calculated in μg/g (Figure 4.21). Furthermore, the situation at site 3 located just near to the central Tienamen Square should be pointed out. At this site the Zn, As, Cd, and Pb concentrations are

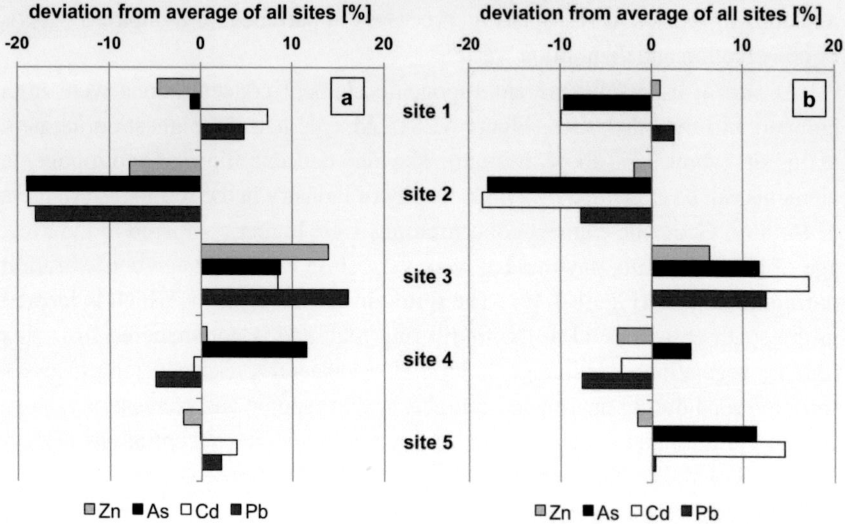

Figure 4.21: Deviation of average mass-related element concentrations (Zn, As, Cd, and Pb in μg/g) from site 1 – site 5 compared to the respective average value of all five sites; (a) day-time and (b) night-time concentrations.

lower than at the sites northwest and southeast if expressed in $\mu g/m^3$, but highest of all sites if expressed in μg/g. Those high concentrations might contribute to the health harming potential of aerosols although no standards are set for harmful metal concentrations in μg/g, yet. Consequently, for the monitoring of toxic element concentrations in urban areas, the values expressed in μg/g would provide further valuable information for the assessment of potential health effects in different parts of the city and should be considered in future.

4.4.2 Influence of meteorological conditions and seasonal sources on temporal variations

Seasonal variations and the influencing factors

The results from long-term sampling period A (Sep-05–Aug-07) showed that the amount of total particles varied with the seasonal course, but that also the concentrations of different elements had strong seasonal differences. A good proxy for toxicity of atmospheric particles are element concentrations from predominantly anthropogenic sources, such as Cd, As, or Pb. Consequently, seasons with high concentrations of these toxic anthropogenic elements can be considered as especially dangerous for the health of the affected inhabitants. The varying toxicity over the seasonal course shall be inspected in detail in the following.

This was done by calculating the deviation of the average TSP element concentrations over each season compared to the average value over all seasons, which is depicted in Figure 4.22. Four elements representing geogenic sources (Mg, Ca, Ti, Fe) were chosen for Figure 4.22a, whereas Figure 4.22b illustrates the seasonal variations for four anthropogenic elements (Zn, As, Cd, and Pb).

During the two-year sampling period (period A), **summer** was always the season with lowest concentrations for both, geogenic and anthropogenic elements (Figure 4.22). Consequently, during this time of the year, the population in Beijing is exposed to the relatively least polluted air. For anthropogenic elements, **winter** 2006/2007 was the season with relatively highest enrichments (Figure 4.22b). Zinc, Cd, and Pb were about 50 to 60% above average. Arsenic had even higher concentrations with an increase of about 100% compared to the overall average. In the previous winter, the enrichment of anthropogenic elements was considerably lower but still above average except for Zn. Arsenic was again the element with the highest enrichment. Consequently, winter is not only a season with high concentrations of atmospheric particles but also the time when the concentrations of toxic elements from anthropogenic sources are especially high. Therefore, winter can be regarded as the worst season with regard to negative health impacts. In **autumn** 2005, geogenic elements were 40 to over 90% higher compared to the average concentrations of the whole period. Magnesium and Ca had their relatively highest concentrations during this season. Titanium and Fe were only higher in spring 2006. In autumn 2006, Ca was again the element with the relatively highest enrichment of all geogenic elements. However, the deviation from the overall average (about 10% for all geogenic elements) was significantly lower in autumn 2006 compared to the previous year. For both years, all anthropogenic elements except As were about 20 to 30% above average during autumn. In 2005, Cd was even more strongly enriched with over 60% compared to the average of the whole sampling period A. Even though mass concentrations were sometimes higher in **spring** than in autumn, anthropogenic element concentrations were below average (Figure 4.22b). Hence, autumn must be considered as more dangerous with regard to negative health impacts than spring. The effect of the meteorological conditions on these seasonal variations will be evaluated in the following sections.

Figure 4.23 shows the concentrations of passively sampled transparent particles between 2.5 to 80 μm diameter. The amount of **precipitation** during the

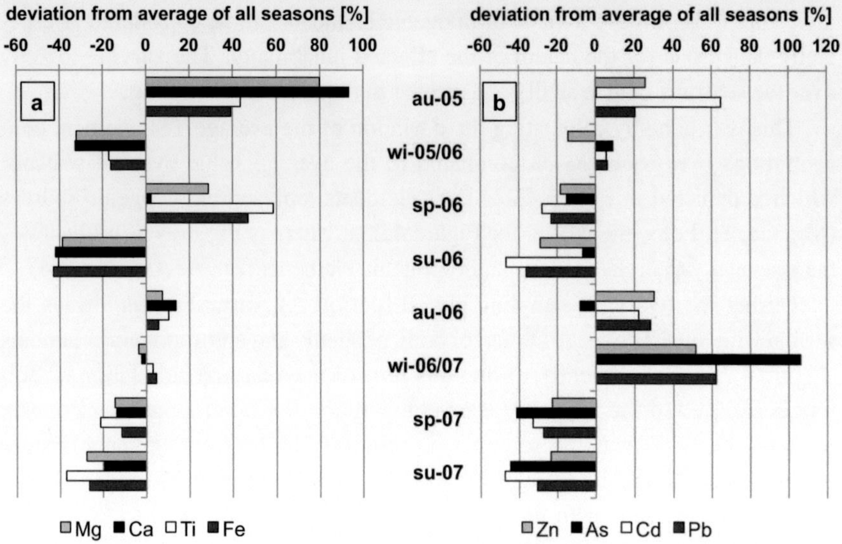

Figure 4.22: Deviation of the average volume-related element concentrations (in ng/m^3) of TSP samples of each distinct season compared to the average of all seasons during period A; (a) typical geogenic elements (Mg, Ca, Ti, Fe) and (b) typical anthropogenic elements (Zn, As, Cd, Pb). The seasons were defined meteorologically; au: autumn, wi: winter, sp: spring, su: summer.

respective sampling intervals is plotted in the same figure. It can be observed that especially in summer, when precipitation is highest in Beijing, the particle concentrations are lower compared to periods without precipitation. Wet deposition is an efficient mechanism to reduce particle concentration in the atmosphere (Seinfeld & Pandis, 2006). In order to illustrate, which particle size classes are reduced most significantly here, size distributions for two representative time intervals during summer 2005 are considered. In both intervals, 3 sampling days without rain were followed by four days with heavy rainfall. Figure 4.24 shows the size distributions of transparent particles for these two samples without rain (solid lines in Figure 4.24) and with heavy rainfall (dashed lines in Figure 4.24). The amount of reduction (in %) for the respective size classes for the two examples was:

	SC1	SC2	SC3	SC4	SC5
Jun-05	41	43	47	44	76
Jul-05	37	38	47	40	46

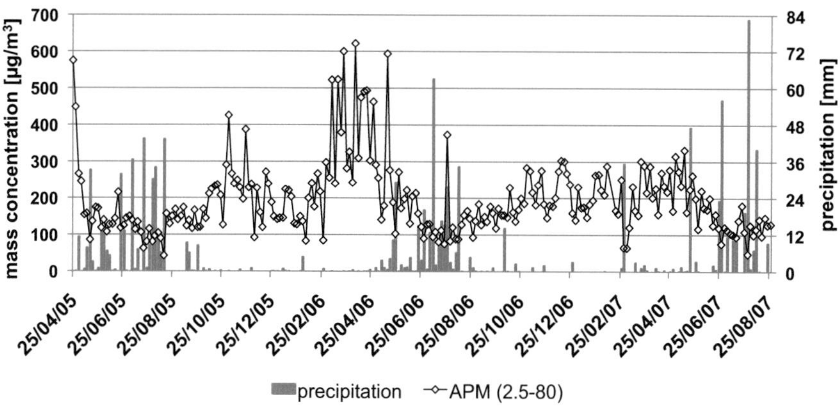

Figure 4.23: Transparent particle concentrations (2.5–80 μm d_g) at site CUG from Apr-05 to Aug-07 together with the amount of precipitation during the respective sampling intervals (3- and 4-daily sampling).

It could be observed that all size classes were reduced strongly and that the amount of reduction was relative uniform for all particles. The higher reduction of particles larger than 40 μm d_g (SC5) might be due to lower mass concentration of this size class and the respective higher analytical error. Particles between 10 and 20 μm d_g (SC3) are probably the size class with the strongest reduction by wet deposition.

Figure 4.25 shows the course of passively collected opaque particles and temperature. A negative correlation in winter between opaque particles and temperature can be seen. Several effects may be responsible for this. **Temperature** had an strong indirect effect on atmospheric pollution in Beijing due to heating activities.

Beside the heating activities, also stagnant wind conditions in Beijing during winter cause a the higher atmospheric pollution during this season. Air masses remain longer over the urban area and particles from local sources are consequently more enriched. Temporal variations of the urban **mixing height** might further contribute to the high winter concentrations in Beijing. A lower mixing height can lead to a relative accumulation of local particles in the smaller air volume of the boundary layer (Seinfeld & Pandis, 2006). Hao *et al.* (2000) studied the mixing height in Beijing and reported that, during daytime in winter, the mixing height could reach 800 m, whereas during daytime in other seasons, it could extend to 1000 m. Consequently, the lower mixing height during winter

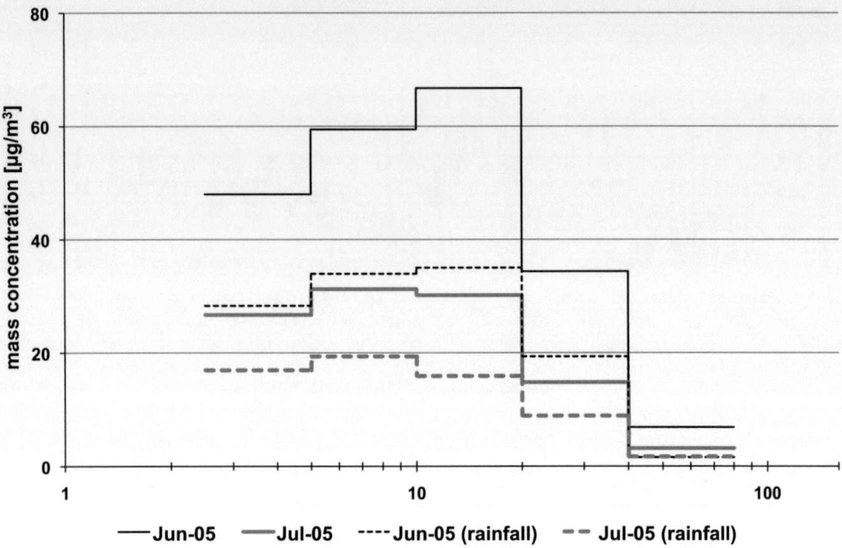

Figure 4.24: Size distribution of transparent particle concentrations (2.5–80 μm d_g) at site CUG for two samples in summer 2005 (20th to 23rd of June, 18th to 21st of July) without rainfall and two subsequent samples (23rd to 27th of June, 21st to 25th of July) during days with heavy rainfall (> 30 and > 40 mm, respectively).

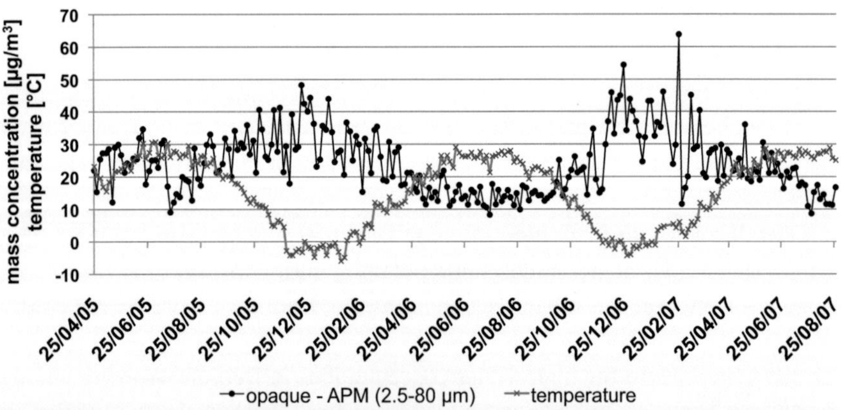

Figure 4.25: Opaque particle concentrations (2.5–80 μm d_g) at site CUG from Apr-05 to Aug-07 together with the average temperature during the respective sampling intervals (3- and 4-daily sampling).

in Beijing contributes to a relative accumulation of particles from local pollution sources. Since anthropogenic sources are dominant within the city of Beijing, this assumption would further explain the relatively higher concentrations of anthropogenic compared to geogenic particles during winter.

Wind direction is certainly an important factor influencing the particle concentration and composition. Determining the effect of wind direction on measured concentrations of atmospheric particles is very challenging. Firstly, during the weekly sampling period, the wind direction may change several times. For this study, it was decided that it is only meaningful to use the most abundant wind direction of each week, which is defined as the dominant wind direction. Unfortunately, only a few dominant wind directions occurred more than 7 times during sampling period A, and only these should be used for data interpretation. These wind directions were 30 (N=9), 150 (N=7), 210 (N=63), and 330 (N=11) degree for day-time samples and 30 (N=54), 60 (N=11), 210 (N=21), and 330 (N=9) degree for night-time samples and are marked in black in Figure 4.26. In Figure 4.26a the average day-time $PM_{2.5}$ mass concentrations for each wind directions are plotted for all five sampling sites whereas the same was done for night-time samples in Figure 4.26b. Variation of mass concentrations with respect to wind direction is small for all sites during daytime (Figure 4.26a). During night, concentrations were slightly lower for wind conditions from NNW (330°) for all sites except site 5, where particle concentrations were lower under NNE conditions (30°). Site 2 to site 5 had highest mass concentrations during westerly winds (60°) during night-time in Beijing.

Day-night variations and the influencing factors

As mentioned before, many element concentrations, and hereby especially those from anthropogenic sources, as well as BC concentrations, were often higher during night compared to day-time concentrations. Consequently, no relief of air pollution can be expected during night time in Beijing. There are various factors contributing to this serious night-time air pollution.

On the one hand, many emission sources operate also during night such as coal combustion or domestic heating and traffic. Furthermore, heavy duty traffic only is allowed to enter Beijing during night to prevent traffic congestions at daytime and, thus, these heavy duty vehicles constitute an additional source of atmospheric particles, especially of BC and traffic related elements.

On the other hand, the atmospheric mixing height might play a very important role for the observed variations of the daily concentration pattern in Beijing.

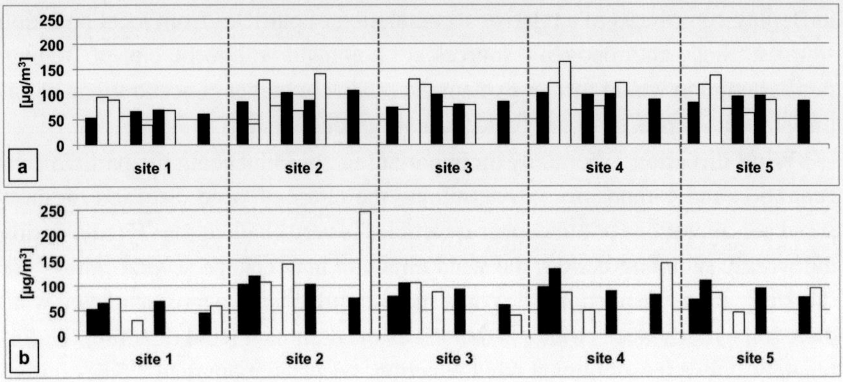

Figure 4.26: Average $PM_{2.5}$ mass concentration versus dominant wind direction for five sampling sites of sampling period A. (a) daytime, (b) night-time samples. The wind-angle bins for each site cover the range from 30 to 360° (N) with step size 30°.

Especially particles with strong local sources are determined by the strength of atmospheric mixing and the height of the mixing layer tends to be lower during night time (Seinfeld & Pandis, 2006). Guinot *et al.* (2006) investigated the urban boundary layer in Beijing and reported a strong diurnal pattern with low height values at night (80±50 m) and much higher values (up to 3000 m) at daytime. This finding explains the lower day-time concentrations due to a higher dilution of particles because of the higher boundary-layer volume of air. Consequently, it can be concluded these strong variation of the mixing layer height is the most important factor influencing the daily concentration pattern of atmospheric particles in Beijing. This conclusion is further supported by the finding that especially anthropogenic elements from the abundant sources in Beijing reached higher pollution levels during night time whereas geogenic particles originating predominantly from regional and supra-regional sources had a relatively higher importance during daytime.

Beside the observed differences between day- and night-time samples, also variations between the different weekdays and especially between workdays and weekends could be observed. Due to the weekly sampling intervals during the long-term measurement period A, only the samples collected in spring 2007 during the short-term sampling campaign with 12-hourly samples (active and passive samples) could be used for the investigation of diurnal variations of air pollution in Beijing.

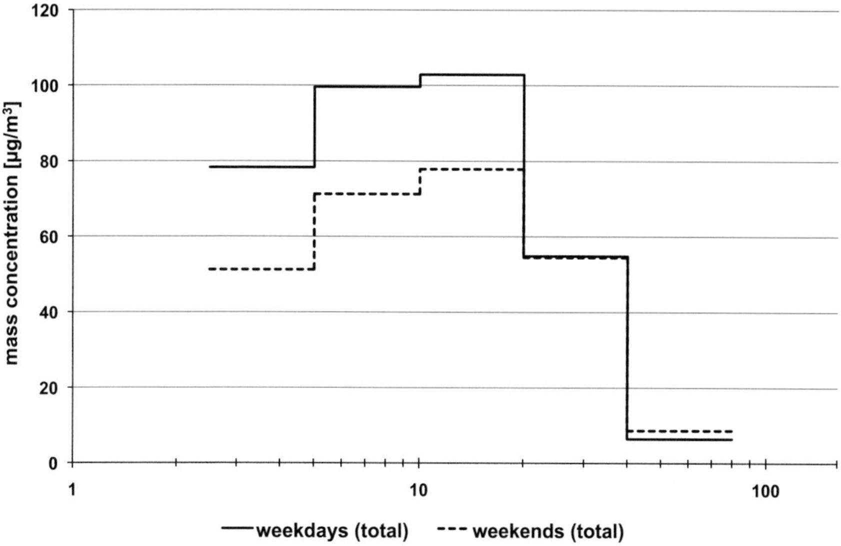

Figure 4.27: Size distribution of total particle concentrations (2.5–80 μm d_g) at site CUG for all 12-hourly samples of period B. Weekdays were defined from Monday morning to Friday night (N=29), weekends from Saturday morning to Sunday night (N=12).

In order to investigate, which particle size class was reduced how much during the weekend, the size distribution of the average of all samples during workdays (Monday morning to Friday night, N=29) were compared with the average concentrations during weekends (Saturday morning to Sunday night, N=12). A reduction was observed for all size classes (Figure 4.27). That indicates that sources for both, finer and coarser particles, were reduced during the weekends in Beijing. Less traffic volume and traffic congestions, and, thus, fewer particles from vehicle exhaust, but also fewer particles from resuspension and abrasion processes, is probably the most important reason for these observed lower particle concentrations. Furthermore, also a reduction of industrial sources is assumable.

In order to gain more information about the sources, which were reduced during weekends in Beijing, also the element concentrations during weekends were investigated in more detail. Figure 4.28 shows the average concentrations of selected anthropogenic elements (As, Cd, Cu, Sb, Sn, and Pb) during weekdays (workdays, Mon–Fri) and weekends (Sat+Sun) for both, $PM_{2.5}$ and PM_{10} samples at site CUG. For all these elements, average concentrations during weekends

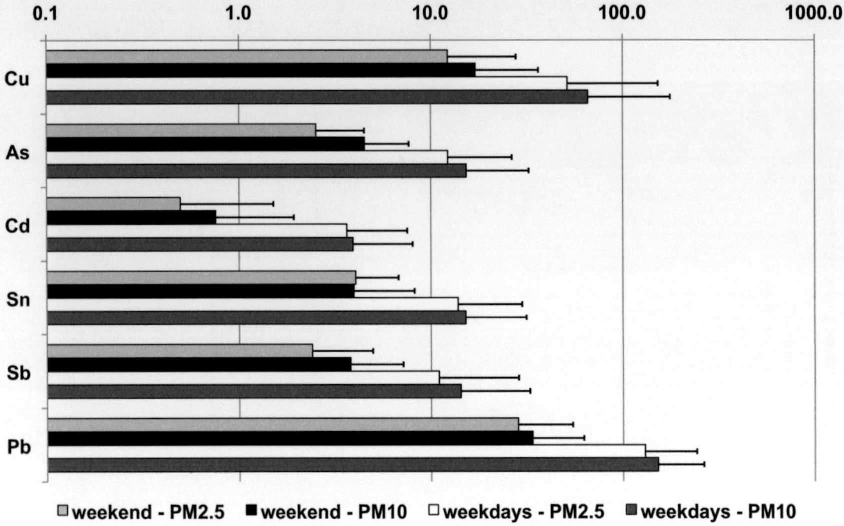

Figure 4.28: Element concentrations (Cu, As, Cd, Sn, Sb, Pb) of $PM_{2.5}$ and PM_{10} samples during sampling period B at site CUG (concentrations in ng/m^3). Only day values (7 a.m to 7 p.m) were considered and it was distinguished between weekends (Sat+Sun, $N = 6$) and weekdays (Mon–Fri, $N = 14$).

were considerably lower than during the rest of the week. In this respect it is worth mentioning that the PM_{10} concentrations of the anthropogenic elements during weekends were also considerably lower than those of $PM_{2.5}$ samples during working days. The lower concentrations of traffic-related elements, such as Cu, Sn, and Sb support the assumption that the lower traffic volume during weekends is one of the most important reasons for air pollution reductions. Lower As, Cd, and Pb concentrations indicate that also industrial sources are fewer during weekends. In this respect it is worth mentioning that the PM_{10} concentrations of the anthropogenic elements during weekends were also considerably lower than those of $PM_{2.5}$ samples during working days.

4.4.3 Acidity of atmospheric particles

The ion balance is a good indicator to study the acidity in the environment and amongst others also of aerosols. For the calculation of the the ion balance of TSP samples from site 4, the ions' mass concentrations were converted into anion (A)

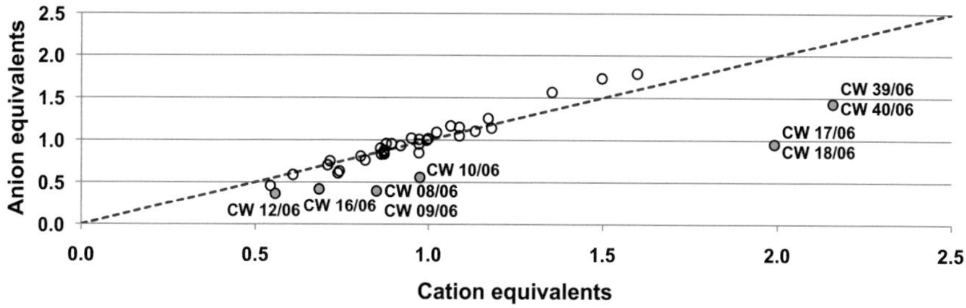

Figure 4.29: Anion versus cation micro-equivalents for water-soluble ions from TSP samples (circles) during period A. The 1:1 relation is marked with the dashed line.

and cation (C) micro-equivalents according to equation 3.1 and 3.2. The A/C ratios varied from 0.5 to 1.2 for TSP samples of the year 2006.

Figure 4.29 illustrates that the ion balance for most TSP samples was close to a 1:1 relation, which indicates that acidity of atmospheric particles seems not to be high in Beijing. The calendar weeks (CW ww/yy) during which the water-soluble ions deviated more from the 1:1 relation are additionally marked in this Figure.

In this context, it should be highlighted that the samples with a A/C ratio significantly <1, and consequently with relatively higher cation concentrations, were collected mostly in spring (CW 08/06–10/06, CW 12/06, CW 16/06–18/06) and also during two autumn weeks (CW 39/06–40/06). This observation indicates that the acidity seems to be lower during dust storm (DS) periods in Beijing. Furthermore, this conclusion confirms findings from a study by Sun *et al.* (2010) in which the authors interpreted DSs to play an important role in buffering and neutralising the acidity of aerosol over northern China. From Xi'an, a large city located about 1000 km south-west of Beijing, Shen *et al.* (2009) also reported lower A/C ratios (average of 0.6 for TSP and 0.8 for $PM_{2.5}$ samples) during DS days. Consequently, the capacity of DSs to buffer acidity could be regarded as an positive effect to reduce acidic rainfall over northern China.

For TSP samples, a significant correlation between ammonium and sulphate ($r = 0.77$, $N = 53$) was found, whereas ammonium and nitrate showed no correlation ($r = 0.10$, $N = 53$). Therefore, ammonium in Beijing's aerosols seems to be mainly present in form of ammonium sulphate. For $PM_{2.5}$ from site CUG, the correlation between ammonium and sulphate is also highly significant ($r = 0.86$,

N = 40) and no correlation was found between ammonium and nitrate (r = 0.23, N = 42).

4.4.4 Source identification with special focus on geogenic versus anthropogenic pollution

Enrichment factors

As already mentioned, elements with high EFs were Ni, Cu, Zn, As, Cd, and Pb (Table 4.1) and, consequently, can be interpreted as elements originating from predominantly anthropogenic sources. $PM_{2.5}$ samples mostly had higher EFs than it was the case TSP samples (Table 4.1). The differences between EFs of fine and total particles were especially large for anthropogenic elements (Figure 4.30). This indicates that anthropogenic sources contribute relatively more to the fine particle load. Since finer particles are more harmful to human health and many of the predominantly anthropogenic elements can be regarded as toxic, the combination of fine and toxic anthropogenic particles poses a serious threat to the inhabitants of Beijing.

Moreover, the EFs for anthropogenic elements were mostly higher during night than during daytime. This is exemplary shown for site 4 in Figure 4.31. Consequently, night time can be considered as more critical with regard to negative health effects. The reasons for the higher night-time concentrations of anthropogenic elements were already discussed in section 4.4.2.

Indicator elements and element ratios

Concentrations of indicator elements of APM can help to estimate the toxicity of urban dust for humans and to understand the reasons for seasonal variations, especially due to varying source contributions, in more detail. Indicator elements are chemical elements, which mainly are associated to certain sources or source groups, such as specific anthropogenic and geogenic sources. However, elements purely indicating geogenic sources are hardly to be found, because almost any chemical element is used for some technical material and purpose (Nath *et al.*, 2007).

An example for such indicator elements are **Pb** and **Ti**. Lead mainly indicates anthropogenic sources, such as fuel combustion and industrial metal processing activities, which are frequently present in Beijing. Nevertheless, Pb also can occur in geogenic particles (e.g. galena, cerussite or anglesite) but in most cases to

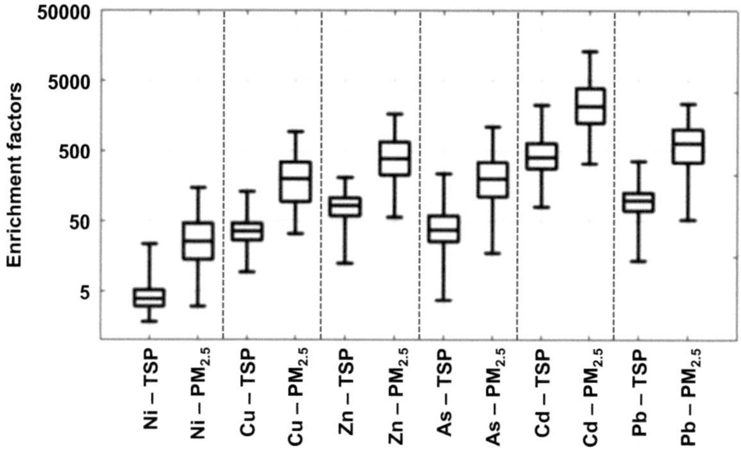

Figure 4.30: Boxplots of EFs from element concentrations at site 4 for TSP and $PM_{2.5}$ samples, respectively. The lower end of the box is represented by the lower quartile, the upper end by the upper quartile, and the median value is illustrated by the line within the box. The whiskers represent the complete data range (minimum to maximum).

a lesser degree as in anthropogenic particles, especially in urban areas. Titanium indicates mainly geogenic sources, although also used in some paints (Reimann & Caritat, 1998). The ratio of **Pb/Ti** indicates the variable importance of anthropogenic or geogenic sources of aerosol particles over time (Figure 4.32). High TSP mass concentrations during autumn 2005 only showed relative low ratios of Pb/Ti, indicating a predominance of geogenic particles whereas during winter 2005/2006 high ratios indicate the peak of the heating time. Again, in spring 2006, during dust storms, the ratio decreased to its minimum over the period of observation because of the predominating geogenic origin of the particles. For the rest of 2006, both, Pb/Ti and TSP mass were more or less constantly increasing indicating predominant anthropogenic emissions as cause for increasing TSP mass concentrations. Winter 2006/2007 again showed high Pb/Ti values due to intensive heating activities. In spring 2007, the dust storm season was not very pronounced, which was also reflected in the Pb/Ti ratio.

The Pb/Ti ratio showed a similar annual trend for both day- and night-time samples. This is illustrated exemplarily for concentrations of $PM_{2.5}$ samples from site 4 in Figure 4.33a. In this context, however, a distinctive feature was the

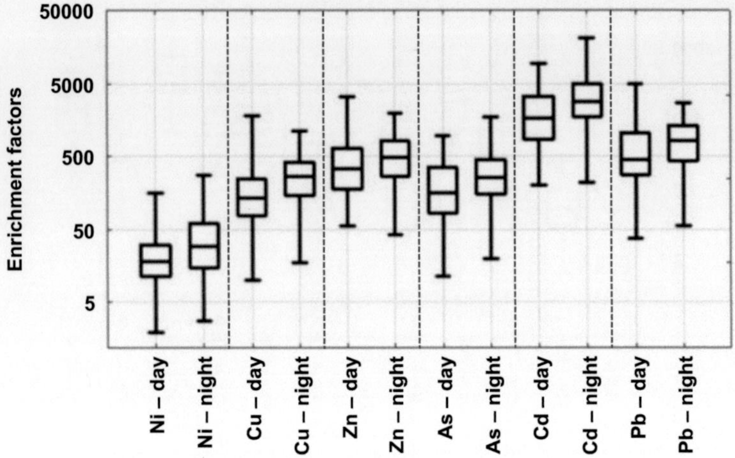

Figure 4.31: Boxplots of EFs from element concentrations at site 4 for $PM_{2.5}$ day- and night-time samples, respectively. The lower end of the box is represented by the lower quartile, the upper end by the upper quartile. The median value is illustrated by the line within the box. The whiskers represent the complete data range (minimum to maximum).

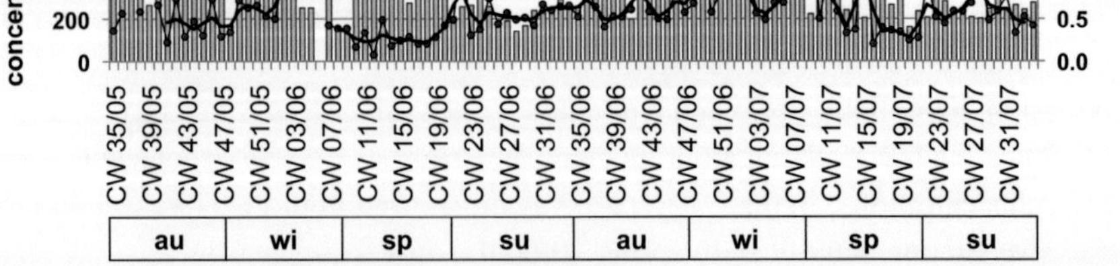

Figure 4.32: Pb/Ti ratio of weekly TSP samples from site 4 from September 2005 to August 2007 together with the TSP mass concentrations.

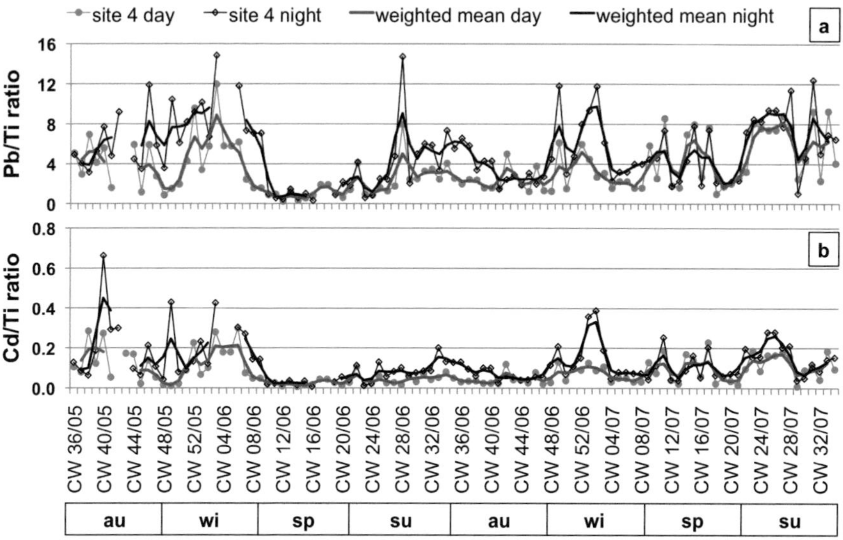

Figure 4.33: Pb/Ti ratio (a) and Cd/Ti ratio (b) for weekly $PM_{2.5}$ samples from site 4 from September 2005 to August 2007.

Table 4.2: Percentage of sample weeks, when Pb/Ti and Cd/Ti ratios, respectively, were higher during night than day time.

ratio	site 1	site 2	site 3	site 4	site 5
Pb/Ti	68	78	76	74	73
Cd/Ti	66	75	78	80	70

observation that, for the majority of samples, the ratio is higher for night- than for day-time samples (Table 4.2). The same annual trend with higher ratios during night time can also be observed for the **Cd/Ti** ratio (Figure 4.33b, Table 4.2). Cadmium, which had highest EFs of all elements within this study (Table 4.1), can also be used as an well suited indicator element for anthropogenic pollution in Beijing. From these observations, it can be concluded that the anthropogenic sources seem to play a relatively more important role during night time in Beijing, which was also indicated by higher EFs of night-time samples (see section 4.4.4). Reasons for these differences, such as the lowering of the atmospheric mixing height during night time or night-time traffic of heavy duty vehicles, were already discussed in section 4.4.2.

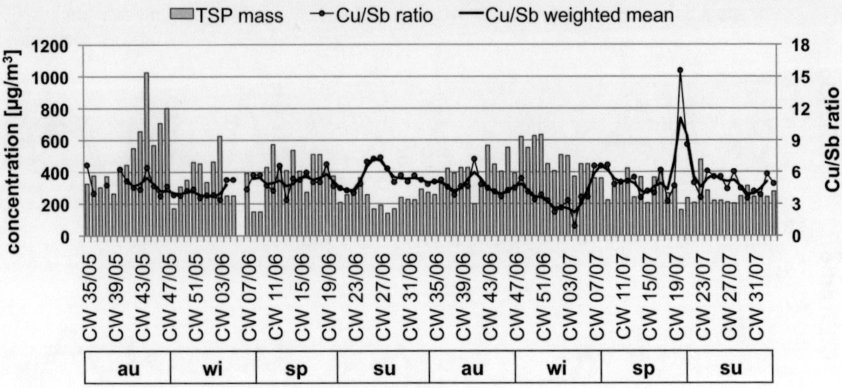

Figure 4.34: Cu/Sb ratio for weekly TSP samples from site 4 from September 2005 to August 2007 together with the TSP mass concentrations.

The **Cu/Sb** ratio of atmospheric particles was proposed to be a good indicator for brake wear (Sternbeck *et al.*, 2002) and, consequently, a good tracer for traffic sources. Weckwerth (2001) found that a Cu/Sb ratio of 4.6±1.2 for brake linings compared well with measured enrichments of these elements in Cologne, Germany. Sternbeck *et al.* (2002) conducted a tunnel study in Gothenburg, Sweden, and proposed a Cu/Sb ratio of 4.6±2.3 as diagnostic criteria for brake wear particles in ambient air. Traffic related Sb in Buenos Aires, Argentina, was investigated by Gómez *et al.* (2005) and the authors reported a Cu/Sb ratio between 5.6 and 8.0. Higher ratios varying from 13.2 to 17.4 were published in a study by Voutsa *et al.* (2002) about PM_{10} concentrations from Thessaloniki, Greece. The Cu/Sb ratio of TSP samples from site 4 varied from 0.9 to 15.6 with an average ratio of 4.9±1.6. The ratio for $PM_{2.5}$ samples at the same site was 5.6±4.3 for day-time and 4.8±2.3 for night-time samples. Over the annual course, the Cu/Sb ratio showed no clear seasonal pattern (Figure 4.34).

The **nitrate/sulphate ratio** was proposed to be an indicator of the relative importance of stationary versus mobile sources of S and N in the atmosphere (Arimoto *et al.*, 1996) and also investigated in different studies in China (Yao *et al.*, 2002; Shen *et al.*, 2009). Sulphate mainly originates from stationary sources (industrial or household burning of fossil fuels/coal), whereas the main source for nitrate are mobile sources (especially traffic emissions). Figure 4.35 illustrates the nitrate/sulphate ratio of TSP samples at site 4 during period A. The high sulphate concentrations in winter can mainly be associated with coal burning emis-

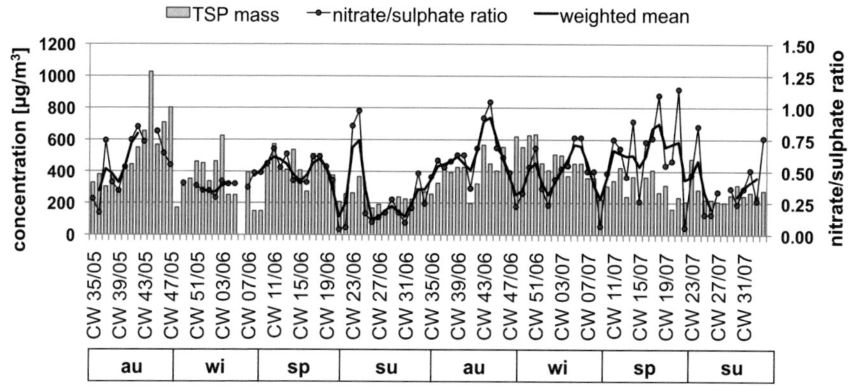

Figure 4.35: Nitrate/sulphate ratio for weekly TSP samples from site 4 from September 2005 to August 2007 together with the TSP mass concentrations.

sions. In summer, accelerated secondary formation of sulphate is of importance in Beijing due to high humidity and strong solar radiation (Yao *et al.*, 2002).

Multivariate statistics (e.g. factor analysis) is another useful tool to identify the most important sources of atmospheric particles in Beijing. It will be applied in the next chapter (chapter 5, section 5.3.4) in the context of discrimination between sources responsible for mobile and immobile element concentrations. The mixing of particles from anthropogenic and geogenic sources will be further discussed on a single particle level in chapter 7.

4.4.5 Comparison of air pollution in Beijing with other cities worldwide

Comparison of mass and element concentrations of Beijing with other cities worldwide

Table 4.3 summarizes the TSP mass and element concentrations of selected elements (Cr, Mn, Ni, Cu, Zn, As, Cd, Pb) of different cities and megacities in Asia in order to compare them with the results of this study. Most element concentrations of TSP samples from sampling period A from Sep-05 to Aug-07 were more or less constant or slightly increased compared to the concentrations during previous years (2001–2003) in Beijing (Okuda *et al.*, 2004). Lead concentrations, on the contrary, decreased clearly, which probably still reflects a long-term effect of phasing out leaded gasoline, which was done in Beijing already in 1997

(Sun *et al.*, 2006). However, compared with Pb concentrations of TSP samples from other Chinese cities, such as Guangzhou (Lee *et al.*, 2007), Hong Kong (Lee *et al.*, 2007), or Qingdao (Hao *et al.*, 2007), the concentrations in Beijing were still higher (Table 4.3). Only Guiyang (Wu *et al.*, 2008) had even higher Pb concentrations than Beijing. Manganese concentrations in Beijing were 3.5- to almost 8-fold higher compared to other Asian cities (Table 4.3) with the exception of Qingdao (Hao *et al.*, 2007). In Qingdao, where Mn concentrations were 3-fold higher than in Beijing, the authors assumed soil to be the dominant source of Mn. Unfortunately, most of the mentioned studies did not report As concentrations. Compared to Ho Chi Minh City (Hien *et al.*, 2001), As concentrations in Beijing were about 40-fold higher.

In order to evaluate the air pollution in Beijing in a broader global context, TSP mass and element concentrations were also compared with those of other large cities outside Asia. Table 4.4 summarises TSP mass and element concentrations from studies in North America (Mexico, USA), South America (Argentina, Brazil), and Europe (France, Italy). Cadmium and Pb were higher in Beijing compared to all other studies (Table 4.4). Chromium, Mn, and Zn were only higher in Rio de Janeiro (Quiterio *et al.*, 2004). Nickel concentrations in Beijing were in the same range than those measured in Mexico City (García *et al.*, 2008). Generally, this comparison illustrates that Beijing still is one of the most polluted cities worldwide and that measures to improve air quality should be the central focus of future urban planning.

4.5 Summary and conclusions

From the results of this study it could be shown that the spatio-temporal variability of aerosol concentration and composition was pronounced in the megacity Beijing. Furthermore, it was indicated that the factors influencing these variations were manifold and often interfered with each other.

With regard to spatial patterns, the north-western parts of the city were less polluted than regions in the south-eastern areas. Generally, the air pollution situation is worse in the central parts of the city compared to the outer regions and, thus, people living there have to cope with higher concentrations of polluting particles. With respect to the varying spatial burden of the inhabitants it is noteworthy to emphasize the importance of calculating the concentrations of toxic elements also as mass-related concentrations in μg/g in order to assess the potential negative health impact more comprehensive.

Table 4.3: Average TSP mass and element concentrations of different Asian cities. Mass concentrations in $\mu g/m^3$, element concentrations in ng/m^3.

City (Country)	Reference	Period	N	Mass	Cr	Mn	Ni	Cu	Zn	As	Cd	Pb
Bejing (CN)	this study	09/05–08/07	100–103	369	33	294	20	130	880	58	10	353
Bejing (CN)	Okuda *et al*. (2004)	03/01–08/03	618–731	–	19	240	22	110	770	48	6.8	430
Busan (KR)	Kim *et al*. (2009)	2004	12	–	25.9	48.1	15.2	–	–	–	1.72	53
Guangzhou (CN)	Lee *et al*. (2007)	12/03–01/05	22	–	20.9	84.7	–	82.3	1190	–	7.85	269
Guiyang (CN)	Wu *et al*. (2008)	04/06–01/07	40	–	26.4	–	–	31.2	1161	–	14.5	393
Ho Chi Minh City (VN)	Hien *et al*. (2001)	08/96–05/98	61	73.6	8.63	38	–	1.28	203	1.5	–	146
Hong Kong (CN)	Lee *et al*. (2007)	12/03–01/05	25	–	15.3	48.3	–	70.8	298	–	1.61	56.5
Qingdao (CN)	Hao *et al*. (2007)	06/01–05/02	23	–	–	910	43.6	57.3	341	–	1.4	185
Seoul (KR)	Kim *et al*. (2009)	2004	12	–	11.5	61.4	8.46	–	–	–	1.7	78.7
Taichung (TW)	Fang *et al*. (2003)	07/01–04/02	43	114	29.3	83.7	15.8	199	395	–	8.5	574

Table 4.4: Average TSP mass and element concentrations of different cities worldwide. Mass concentrations in $\mu g/m^3$, element concentrations in ng/m^3.

City (Country)	Reference	Period	N	Mass	Cr	Mn	Ni	Cu	Zn	Cd	Pb
Bejing (CN)	this study	09/05–08/07	100–103	369	33	294	20	130	880	10	353
North America											
Mexico City (MX)	García *et al*. (2008)	2003–2004	100	–	23.8	19.2	23	–	–	2.64	50.8
Los Angeles (US)	Lim *et al*. (2006)	08/02–06/03	–	–	4.9	–	9.2	52	84	–	14
South America											
La Plata (AR)	Bilos *et al*. (2001)	1993	28–36	–	4.32	25.5	3.15	29.5	273	0.41	64.5
Niteri City (BR)	Sella *et al*. (2004)	07/01	5	248*	–	96.3*	–	–	442*	1.5*	86.6*
Rio de Janeiro (BR)	Quiterio *et al*. (2004)	03/01–02/02	43	87*	421	1216	0.5	335	2120	0.9	101
Salvador (BR)	Pereira *et al*. (2007)	10/04–11/04	7, 8	182, 169	–	19.2, 13.5	–	21, 121	4.5, 4.0	–	–
Europe											
Paris (FR)	Ayrault *et al*. (2010)	03/02–09/03	21	–	4.12	6.95	2.17	18.4	45.8	0.39	15.4
Tito Scalo (IT)	Ragosta *et al*. (2002)	1997–1999	558–559	60	13	27	5	58	304	2	60

Over the annual course, the lowest aerosol concentrations occurred during summer, due to meteorological conditions (e.g. most rainfall occurs during summer months in Beijing) and the lack of certain sources (especially emissions from heating processes). In spring, high aerosol concentrations are predominantly caused by geogenic particles, whereas the concentrations of toxic elements from anthropogenic sources were considerably lower. Conclusively it can be stated that the burden of air pollution for the inhabitants of Beijing are highest during winter due to stagnant meteorological concentrations on the one hand, and additional sources, especially coal combustion for heating purposes, on the other hand. During this time, especially the high concentrations of potentially toxic elements, such as Cd, As, and Pb, are of great concern for human health. In this context, also the relatively higher concentrations of these anthropogenic elements compared to less harmful geogenic elements if calculated as mass-related concentration in μg/g is alarming. In the next chapter (chapter 5) element concentrations in Beijing will further be investigated with respect to their varying mobility and, consequently, different bioavailability over the seasonal course in order to gain additional information about the health harming potential for the exposed inhabitants.

Moreover, this study showed that indicator elements help to understand the different roles of geogenic and anthropogenic sources contributing to the aerosol pollution in Beijing. In this study, the Pb/Ti and Cd/Ti ratio proved to be a good instrument to discriminate between periods of dominant geogenic and those of dominant anthropogenic pollution in Beijing. In this regard, it should further be pointed out that anthropogenic air pollution was relatively higher during night- than during daytime. On the one hand, the meteorological conditions, especially the lower urban boundary layer during night time, were responsible for these differences. On the other hand, special anthropogenic activities, such as the night-time traffic of heavy duty vehicles, contribute to this effect.

Generally, the rapid growth of megacities such as Beijing with regard to the number of inhabitants, the vehicle fleet and the amount of energy consumption, is a huge challenge for a sustainable urban planning for the future.

Chapter 5

Temporal variability of trace metal mobility determined by sequential extractions - a contribution to health impact assessment

5.1 Introduction

Aerosol particles are known to have negative health effects on humans (see chapter 1.3.1). For the estimation of potential health effects, it is crucial to know not only the total amounts of toxic substances, but also their chemical forms, which determine the respective behaviours in the environment. The bioavailability of elements depends on characteristic surface properties of atmospheric particles, the strength of the chemical bonds, and on the properties of solutions in contact with atmospheric particulate matter (APM) (Smichowski *et al.*, 2005). Sequential extractions are suitable tools for assessing the mobility of elements, because they provide additional information about the solubility of the elements in specific reagents, which can be associated to specific matrix components (Richter *et al.*, 2007). Generally, chemical elements mobilised by weak acids are much more available for organisms as elements, which can be mobilised only by very strong acids. Therefore, the chemical mobility of elements determined by leaching procedures is a good indicator for their bioavailability.

In different geochemical studies, sequential extractions have been applied for several decades, especially for soil and sediment samples (e.g. summarised by Hirner, 1992; Linge, 2008). For dust filter samples, the small amount of sampling material poses a special challenge. In the past 15 years, some fractionation schemes for APM were published and were summarized in a review article by Smichowski *et al.* (2005). The authors classified five main sequential chemical fractionation schemes, that are operationally defined: (a) based on Tessiers procedure (Tessier *et al.*, 1979), (b) based on Chesters procedure (Chester *et al.*, 1989), (c) based on Zatkas procedure (Zatka *et al.*, 1992), (d) based on BCR procedure (Ure *et al.*, 1993), and (e) other procedures. However, the lack of uniformity of the schemes raises problems when comparing the results of different studies.

For this work, a modified Tessier scheme after Fernández Espinosa *et al.* (2002) was applied and the concentrations of 18 selected elements were analysed for each of the four fractionation steps. The methodology is explained in more detail in chapter 3.2.4. Fraction f1 is considered to be highly mobile, fraction f2 mobile, fraction f3 less mobile, and fraction f4 immobile. This extraction scheme was chosen because it was optimised for filter-collected fine urban particles and provides conditions more similar to the deposition and solubilisation in the human lung (Fernández Espinosa *et al.*, 2002). Within this chapter, the total element concentrations, the element mobilities estimated by the sequential fractionation steps and the main sources of these elements are discussed with a particular focus on the toxicity and mobility of Zn, Cd, Mn, As, Cu, and Pb.

In Beijing already several studies focussed on total element concentrations and sources of atmospheric particles (He *et al.*, 2001; Okuda *et al.*, 2004; Sun *et al.*, 2004; Duan *et al.*, 2006; Norra *et al.*, 2007, 2010; Schleicher *et al.*, 2010*a*,*b*). All above mentioned studies on APM carried out in Beijing do not consider the speciation of the elements and their mobility. Only at few other locations, such as Veszprm, Hungary (Hlavay *et al.*, 1996), Karlsruhe, Germany (Heiser *et al.*, 1999), Sevilla, Spain (Fernández *et al.*, 2000; Fernández Espinosa *et al.*, 2002; Fernández Espinosa & Ternero Rodríguez, 2004), San Nicols, Argentina (Fujiwara *et al.*, 2006), Tartous and Darya, Syria (Al-Masri *et al.*, 2006), Santiago, Chile (Richter *et al.*, 2007), Rome, Italy (Canepari *et al.*, 2008), and Tirupati, India (Praveen Kumar *et al.*, 2008), investigations using leaching procedures for urban APM samples were carried out. To the author's knowledge, the study by Wu *et al.* (2008), which focussed on very few elements (Cd, Cr, Cu, Pb, and Zn) from TSP samples from Guiyang, is the only investigation that applied sequential extractions in China. Other studies solely determined the speciation of single elements, such as Al (Wang *et al.*, 2007*a*).

This study aims at elucidating the mobility of metals in atmospheric particles of different sizes (TSP and $PM_{2.5}$) from Beijing for the first time. The knowledge obtained by this approach helps to identify those elements, which are abundant and at the same time highly mobile and are, therefore, the most critical for possible negative effects on human health. Furthermore, the sources of these especially health-relevant elements are addressed within this chapter. This information further provides valuable information for a sustainable urban planning and the implementation of mitigation measures in urban areas and megacities.

5.2 Results

5.2.1 Mass concentrations of the samples selected for sequential extractions

TSP mass concentrations for the selected 35 weeks from July 2005 to May 2008 at site 4 and site CRAES varied from 168 to 634 $\mu g/m^3$ with a median concentration of 373 $\mu g/m^3$. Lowest mass concentrations were found in summer with a median concentration of 263 $\mu g/m^3$ whereas winter concentrations were highest with 462 $\mu g/m^3$. In spring and autumn median concentrations were 375 and 393 $\mu g/m^3$, respectively. Results for TSP samples from the whole period, including the weeks not selected for sequential extraction, are reported in chapter 4.2. The weekly TSP concentrations at site 4 for a the two-year period A are illustrated in Figure 5.1.

For sequential extractions, $PM_{2.5}$ samples from site CUG were selected because of a higher mass load on the filters due to continuous sampling for a full week. At site 4, on the contrary, the $PM_{2.5}$ concentrations are too low for the sequential extraction procedure because of separated day and night sampling. The weekly $PM_{2.5}$ concentrations from site CUG from February 2005 to August 2007 are illustrated in Figure 5.2. The $PM_{2.5}$ mass concentrations for the selected 32 weeks at site CUG ranged from 28 to 129 $\mu g/m^3$ with a median value of 63 $\mu g/m^3$. Total $PM_{2.5}$ concentrations showed a similar seasonal pattern with lowest concentrations in summer (median: 58 $\mu g/m^3$). Median concentrations for autumn and winter were 71 and 68 $\mu g/m^3$, respectively, while spring concentrations were highest with a median concentration of 80 $\mu g/m^3$.

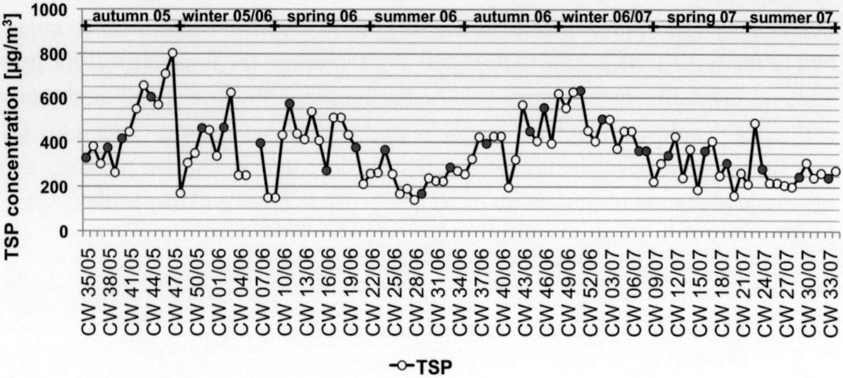

Figure 5.1: Annual course for weekly TSP mass concentrations at site 4 from September 2005 to August 2007. Filled symbols correspond to TSP samples selected for sequential extractions.

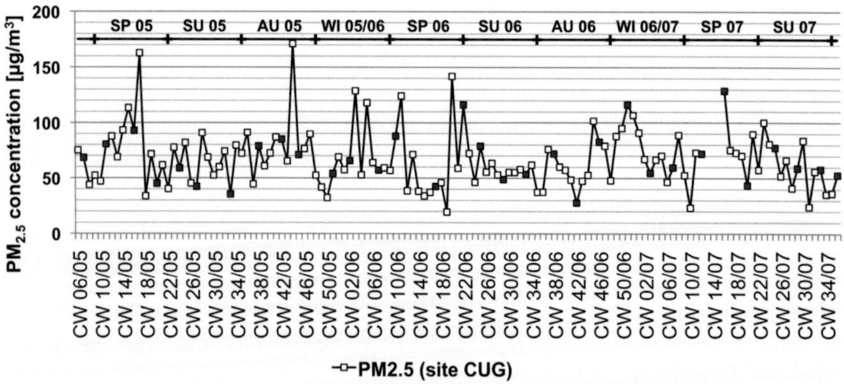

Figure 5.2: Annual course for weekly $PM_{2.5}$ mass concentrations at site CUG from February 2005 to September 2007. Filled symbols correspond to the $PM_{2.5}$ samples selected for sequential extractions.

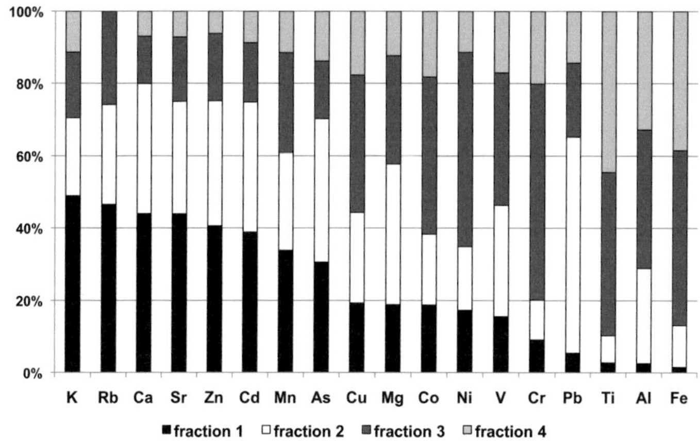

Figure 5.3: Distribution of the median element concentrations of TSP filter samples (N=35) in percentage for each fractionation step ((f1) water-extractable, (f2) bound to carbonates, oxides and reducible metals, (f3) bound to organic matter, oxidised metals and sulphides, and (f4) residual fraction). Elements are ordered by their percentage in fraction f1.

5.2.2 Element concentrations in the four different fractions

The median element concentrations in the four fractions are shown for all TSP samples (N=35) in Figure 5.3, and for all $PM_{2.5}$ samples (N=32) in Figure 5.4, expressed as percentage of each element in the considered fraction. The four fractions f1 – f4 can be interpreted as highly mobile (f1), mobile (f2), less mobile (f3), and immobile (f4). The detailed methodology is described in chapter 3.2.4. The elements in both figures are arranged according to their percentage in the water-soluble fraction (f1). Descriptive statistics for the results of TSP samples are summarised as volume related concentrations in ng/m^3 in Table B.1 and as mass related concentrations in μg/g in Table B.2. The same results for $PM_{2.5}$ samples are listed in Table B.3 and Table B.4.

In the following paragraphs, the elements are classified in different groups according to their mobility determined by the sequential extraction, their seasonal pattern, as well as the amount of anthropogenic influence determined by their enrichment factors (EFs) (Table 5.1).

Enrichment factors were calculated according to equations 3.5 and 3.6, which are explained in chapter 3.4. The calculation of EFs provides a good first estimation of the degree of anthropogenic influence on the concentration of each

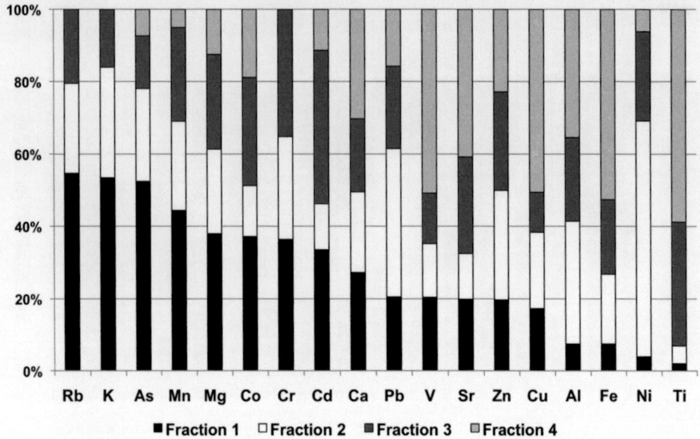

Figure 5.4: Distribution of the median element concentrations of $PM_{2.5}$ filter samples (N=32) in percentage for each fractionation step ((f1) water-extractable, (f2) bound to carbonates, oxides and reducible metals, (f3) bound to organic matter, oxidised metals and sulphides, and (f4) residual fraction). Elements are ordered by their percentage in fraction f1.

element and, thus, supports the distinction between elements originating predominantly from geogenic and those from anthropogenic sources. The EFs of the mobile elements K, Rb, Ca, Sr, and Mg are low (Table 5.1), which provides a good indication for their origin from predominantly geogenic sources. Titanium, Al, and Fe, which were very immobile, also show very low EFs (EFs ≤ 5, Table 5.1). This supports the interpretation that for those elements geogenic sources are dominant, too. Cadmium is the element with the highest EFs for both TSP and $PM_{2.5}$ samples (Table 5.1). Lead, Zn, As, and Cu also show high EFs. For all these elements, the EFs for the fine particles ($PM_{2.5}$) are considerably higher (about twice as high for As and Pb, 3-fold for Cu, 4-fold for Cd and 5-fold for Zn) than for TSP samples. Nickel is the only element with high EFs solely in the fine fraction. The higher EFs for $PM_{2.5}$ samples indicate that the influence from anthropogenic sources is especially high for the fine particles. For, As, Cd, Pb, and Zn, the EFs calculated using average concentrations from Chinese topsoils are lower than those calculated with the average crustal concentrations (Table 5.1). From the elements with high EFs, Cu is the only one, which showed slightly higher EFs when calculated with average soil concentrations. It is believed that the use of local soil concentrations is more representative compared to concen-

Table 5.1: Calculated enrichment factors (EF) referring to Ti for all analysed elements. EFs were calculated using the median concentrations for the sum of all extraction steps of all TSP (N=35) and $PM_{2.5}$ (N=32) samples, respectively. Crustal concentrations (A) were taken from Reimann & Caritat (1998) and soil concentrations (B) from Chinese topsoils from Chen *et al.* (2008). For some elements (-), no soil data were available from literature.

	(A) TSP EF(crust)	(B) TSP EF(soil)	(A) $PM_{2.5}$ EF(crust)	(B) $PM_{2.5}$ EF(soil)
Mg	6	16	2	6
Al	1	-	1	-
K	5	-	15	-
Ca	16	-	9	-
Ti	1	1	1	1
V	3	4	4	5
Cr	3	6	4	8
Mn	10	12	12	14
Fe	4	5	2	2
Co	4	8	4	7
Ni	5	10	47	88
Cu	122	133	403	439
Zn	437	418	2320	2220
As	731	155	1640	349
Rb	6	-	19	-
Sr	8	-	10	-
Cd	2670	2080	9500	7400
Pb	589	391	1380	917

trations of average earth crust because the metal concentrations in topsoils better represent the background pollution situation in China.

Mobile, less toxic, and predominantly geogenic elements

The EFs of K, Rb, Ca, Sr, and Mg are low (Table 5.1). Concurrently, their relative abundance in the mobile fractions f1 and f2 is high (Figure 5.3 and Figure 5.4). Potassium and Rb are the most mobile elements for TSP as well as for $PM_{2.5}$ samples within this study. All five elements have lowest concentrations in summer (median values of 3.98 (K), 0.02 (Rb), 19.88 (Ca), 0.08 (Sr), and 4.26 (Mg) $\mu g/m^3$ in TSP samples, respectively). Highest concentrations were measured in winter for K, Rb, and Sr (median values of 6.80, 0.03, and 0.23 $\mu g/m^3$, respectively) and in autumn for Ca and Mg (median values of 40.00 and 8.40 $\mu g/m^3$, respectively). The annual course of each fraction is illustrated for K, Rb,

Ca, Sr, and Mg concentrations in TSP samples in Figure 5.5a-e and their relative abundance in the mobile fractions is displayed in Figure 5.5f-j.

Immobile, less toxic, and predominantly geogenic elements

Titanium, Al, and Fe are most abundant in the residual fraction f4 for TSP as well as for $PM_{2.5}$ samples and are, consequently, the most immobile elements within this study (Figure 5.3 and Figure 5.4). Moreover, all of them show very low EFs (EFs $\leq$ 5, Table 5.1). Their seasonal course is similar to each other (Figure 5.6 a-c) with highest concentrations in spring (median values of 0.29 (Ti), 7.22 (Al), and 10.26 (Fe) $\mu g/m^3$, respectively). Concentrations in autumn and winter are also elevated (median values of 0.21 (Ti), 5.51 (Al), and 8.96 (Fe) $\mu g/m^3$, respectively, for autumn and 0.23, 6.33, and 9.80 $\mu g/m^3$, respectively, for winter). Lowest Ti, Al, and Fe concentrations occur in summer (median values of 0.18 (Ti), 3.17 (Al), and 6.43 (Fe) $\mu g/m^3$, respectively). The seasonal variability of the relative occurrence in the different fractions is less pronounced for Fe and a little more variable for Ti and Al (Figure 5.6 d-f).

Highly mobile, toxic, and predominantly anthropogenic elements

Cadmium is the element with the highest EFs for both TSP and $PM_{2.5}$ samples (Table 5.1). Lead, Zn, As, and Cu also show high EFs. The relative abundance in the mobile fractions f1 and f2 is high for all those elements. In samples including the coarser fraction (TSP), Pb occurs predominantly in fraction f2, while this is the case for Cd in the fine fraction ($PM_{2.5}$). Within the seasonal course, the relative abundance of all elements in the mobile fractions is lowest in spring and autumn and highest in summer (Figure 5.8). Zinc, Cd, As, Pb, Cu, and Ni concentrations are highest in winter (median values of 1940 (Zn), 23.5 (Cd), 101 (As), 692 (Pb), 194 (Cu), and 21.7 (Ni) ng/m^3, respectively). Autumn concentrations are also elevated (median values of 1690 (Zn), 20.0 (Cd), 60.9 (As), 539 (Pb), 178 (Cu), and 18.3 (Ni) ng/m^3, respectively). For all of these elements, concentrations are lowest in summer (median values of 1060 (Zn), 7.9 (Cd), 44.6 (As), 281 (Pb), 119 (Cu), and 10.1 (Ni) ng/m^3, respectively), while spring concentrations are also relatively low (median values of 1280 (Zn), 8.3 (Cd), 52.6 (As), 408 (Pb), 133 (Cu), and 17.0 (Ni) ng/m^3, respectively). The annual courses for the element concentrations for all four fractions are illustrated in Figure 5.7.

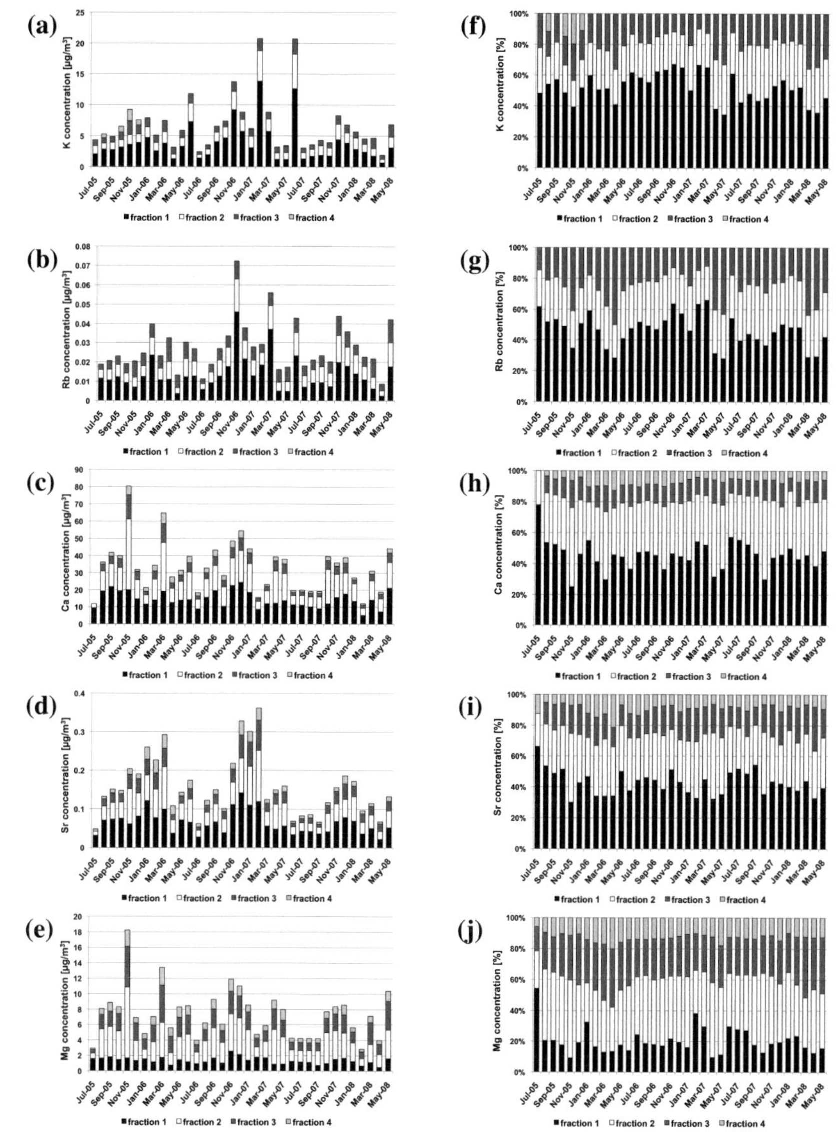

Figure 5.5: Annual courses of element concentrations in TSP filter samples in $\mu g/m^3$ (a–e) and in percentage for each of the four fractions (f-j) from July 2005 to May 2008. The selected elements are (a, f) K, (b, g) Rb, (c, h) Sr, and (i, j) Mg. Fraction (1) water-extractable, (2) bound to carbonates, oxides and reducible metals, (3) bound to organic matter, oxidisable and sulfidic metals, and (4) residual fraction.

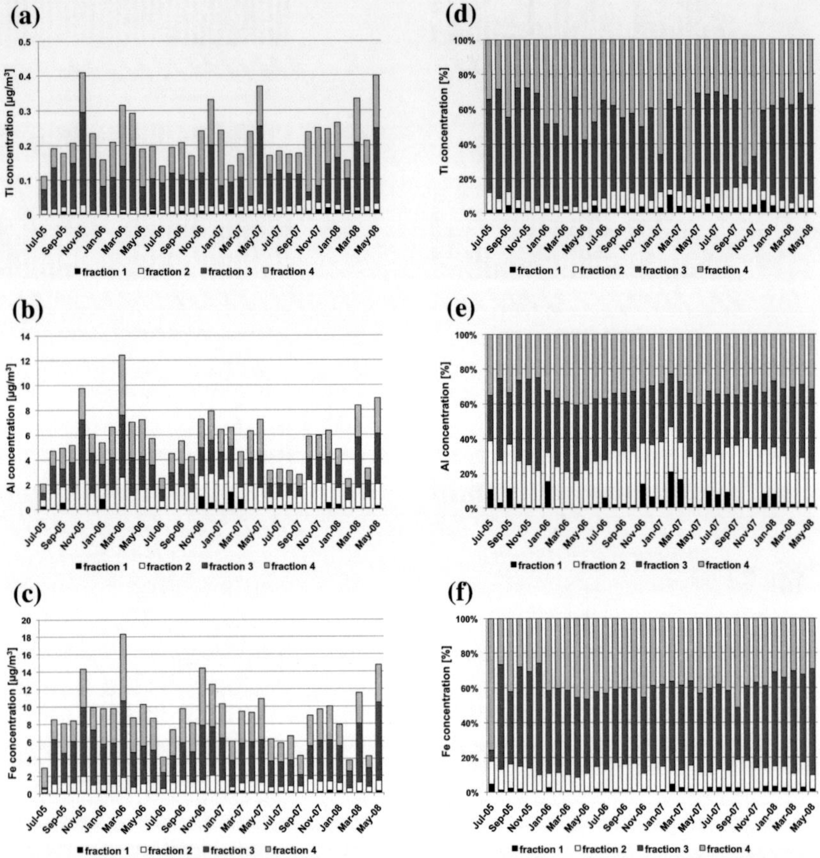

Figure 5.6: Annual courses of element concentrations in TSP filter samples in $\mu g/m^3$ (a–c) and in percentage for each of the four fractions (d-f) from July 2005 to May 2008. The selected elements are (a, d) Ti, (b, e) Al, and (c, f) Fe. Fraction (1) water-extractable, (2) bound to carbonates, oxides and reducible metals, (3) bound to organic matter, oxidisable and sulfidic metals, and (4) residual fraction.

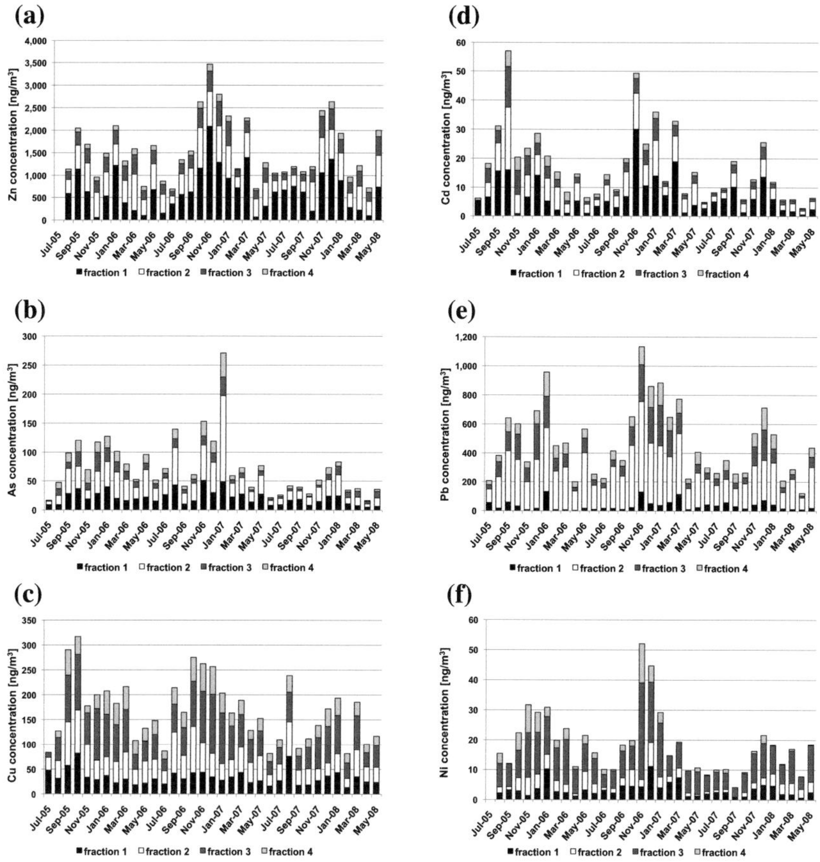

Figure 5.7: Annual courses of element concentrations in TSP filter samples in ng/m^3 from July 2005 to May 2008. The selected elements are (a) Zn, (b) Cd, c) As, (d) Pb, (e) Cu, and (f) Ni. Fraction (1) water-extractable, (2) bound to carbonates, oxides and reducible metals, (3) bound to organic matter, oxidisable and sulfidic metals, and (4) residual fraction.

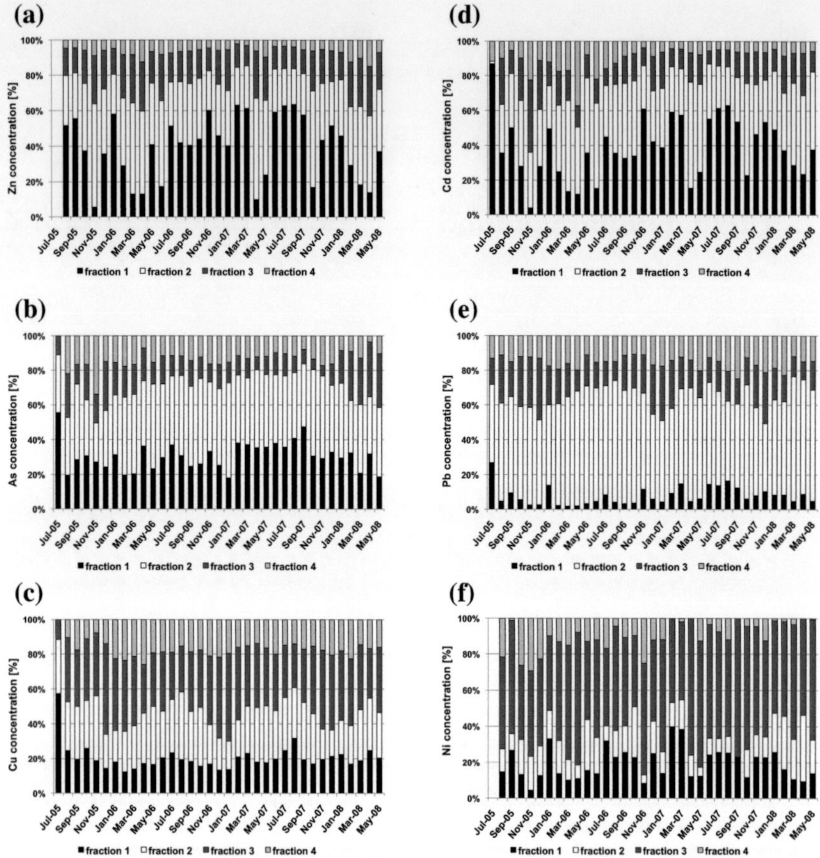

Figure 5.8: Annual courses of element concentrations in TSP filter samples in percentage for each of the four fractions from July 2005 to May 2008. The selected elements are (a) Zn, (b) Cd, c) As, (d) Pb, (e) Cu, and (f) Ni. Fraction (1) water-extractable, (2) bound to carbonates, oxides and reducible metals, (3) bound to organic matter, oxidisable and sulfidic metals, and (4) residual fraction.

Table 5.2: Range of deviation in % of element concentrations of TSP samples from sequential extractions (sum of all four fractions f1–f4) compared to element concentrations from total digestions.

Element	N	Min (%)	Max (%)
Mg	32	-20	69
Al	32	-74	-38
K	32	-46	52
Ca	32	-19	104
Ti	32	-80	-48
V	32	-50	0
Cr	31	-47	21
Mn	32	-14	80
Fe	32	-52	18
Co	32	-33	35
Ni	31	-60	56
Cu	32	-26	80
Zn	31	13	200
As	32	3	84
Rb	32	-53	60
Sr	32	-28	71
Cd	32	13	137
Pb	32	5	88

5.2.3 Comparison with results of total digestion

The results of element concentrations from the sequential extractions (sum of all four fractions) were compared to the results of total acid digestion of the same samples. However, the deviation between the results of both methods was quite large. Some elements, such as Al, Ti, and V, had higher concentrations after total digestion than after summing up the four extraction steps. Other elements, however, such as Zn, As, Cd, and Pb, had lower concentrations after the total digestion. Some element concentrations, such as Mg, K, Ca, Cr, Mn, Fe, Co, Ni, Cu, Rb, and Sr, were sometimes lower and sometimes higher after the total digestion. The range of the deviation (minima and maxima) of TSP samples is shown in Table 5.2.

There are various reasons for the wide differences. First, different acids are applied for the two methods. For total digestion, hydrofluoric acid (HF) is used. With this acid, it is possible to digest elements incorporated in residual mineral phases, including silicates. On the contrary, in the fourth step of the applied sequential extractions procedure, no HF is used, leaving silicates undissolved. This explains why the concentrations measured after total digestion were higher for

Al, Ti, and V. Those elements often occur in silicates (e.g. feldspars, pyroxenes, amphibols) and, thus, in very stable forms. Consequently, the silicates are dissolved during total digestion with HF, which leads to higher Al, Ti, and V concentrations after this digestion method than after the four digestion steps.

Since the total element concentrations for the extractions are calculated from four independent steps, the error of each fraction results in a larger overall error and, therefore, the uncertainty is higher for the sequential extraction method. Moreover, the dilution factor applied for the ICP-MS measurements was higher for some of the extractions steps than for the solution obtained after total digestion. Consequently, a dilution error might additionally contribute to the larger deviation.

The most important reason for the different results and, therefore, the most important reason why it is difficult to compare both results, is probably caused by inhomogeneities of the filter load. For each analysis, a different share of the filter was used. A single, comparatively large, particle with a high concentration of a distinct element, could already change the overall result. Since the particle size is very variable for the TSP samples, elements, which are very frequently incorporated in larger particles, are probably less homogeneously distributed over the whole filter area. The inhomogeneity of single particles is discussed in chapter 7 in more detail. There are only very few studies, which investigated the homogeneity of element distribution on filter samples. Pöykiö *et al.* (2003) studied the deposition of Cd, Cr, Cu, Fe, and Ni collected with a high-volume sampler on glass fibre filters. Their results indicated that Cr, Cu, Fe, and Ni were not necessarily uniformly distributed over the filter. Another study by Marrero *et al.* (2005) tested the uniformity of deposition of 12 elements of PM_{10} samples collected with a HV-sampler on glass fibre filters. The authors reported the largest variations for Cd $>$ Ni $>$ Fe $>$ Sb. The most homogeneous pattern was found for As $>$ Ti $>$ V $>$ Al, while variations within an intermediate range were observed for Cr, Cu, Mn and Pb. In contrast to the mentioned studies, QF filters were used for this study. Their surface area is a lot rougher than it is the case for glass filters. The quartz fibres vary in size and diameters and are arranged irregularly resulting in a even more inhomogeneous pattern of the distributed elements.

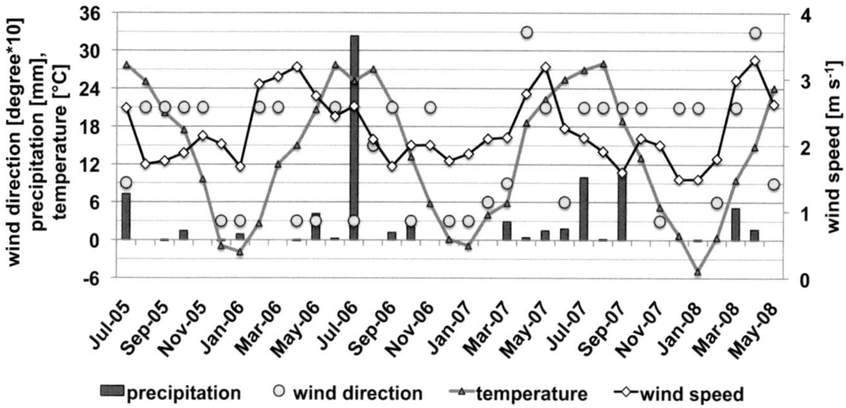

Figure 5.9: Precipitation (mm), average temperature (°C), average wind speed (m s^{-1}), and predominant wind direction (°) for the selected TSP sampling weeks (one week of each month) in Beijing, China.

5.3 Discussion

5.3.1 Temporal variability of trace metal mobility due to meteorological conditions and seasonal sources

Particle concentrations in the atmosphere in Beijing and in general are controlled by source emissions and meteorological factors. The influence of wind speed, wind direction, air temperature, relative humidity, precipitation, and atmospheric stability on particle number and mass concentrations, as well as size distributions were investigated in different APM studies worldwide (e.g. Davidson, 1994; Whiteaker *et al.*, 2002; Gietl & Klemm, 2009; Nicolás *et al.*, 2009). Usually, particle mass concentrations decrease with less stable atmospheric layering as indicated by higher wind speeds, unless resuspension processes play a dominant role at very high wind speeds. Wet deposition due to precipitation leads to a reduction of particles in the atmosphere. Air mass origin is highly responsible for the composition of the particles because it determines the source areas. The meteorological parameters for Beijing for the weeks in which the samples were analysed by sequential extractions (one sampling week each month) are illustrated in Figure 5.9.

Summer is the season with highest precipitation leading to the lowest particle mass concentration in Beijing due to wet deposition. Average TSP concentrations in summer 2006 were about 40% lower than the average concentrations of spring and autumn 2006, respectively, and more than 50% lower than TSP concentrations in winter 2006/2007. In winter, wind speed is usually very low in Beijing, which leads to stagnant wind conditions and a relative accumulation of APM. Average TSP concentrations in winter 2005/2006 and 2006/2007 were 351 and 485 $\mu g/m^3$, respectively. Moreover, temperature has an indirect effect in winter. Winters in Beijing are cold (average temperature of -1.7°C for winter 2005/2006 and 0.4°C for winter 2006/2007) and, thus, extensive heating activities (usually from Mid-November to Mid-March, He *et al.*, 2001) contribute as an additional source to higher particulate air pollution. The heating period is associated with high concentrations of potentially mobile Cd, As, and Pb (Figure 5.7b-d). In spring, geogenic sources are predominant, especially during so-called dust storm days (Xie *et al.*, 2005; Wang *et al.*, 2006*a*, 2007*b*). However, while contributing widely to total mass concentrations, geogenic dust is less problematic concerning potential bioavailable metal loads, which is illustrated by predominantly low Cd, As, and Pb total concentrations (Figure 5.7b-d), as well as lower percentages of toxic elements in the highly mobile fraction (f1), which can exemplary be shown for the toxic Cd (Figure 5.8b).

Due to meteorological conditions and varying influences of different sources, the air pollution situation in Beijing is different for the four seasons. In the following paragraph, four characteristic sampling weeks are discussed in detail representing the distinct seasons. In addition to the meteorological conditions illustrated in Figure 5.9, also the respective air mass origins for those selected weeks are shown in Figure 5.10 (NOAA HYSPLIT Model).

For **autumn**, the sampling week from the 6th to the 12th of October 2005 was chosen (CW 40/05, Figure 5.10a). Here, the air masses reach Beijing from southern directions and remain over densely populated and industrialized areas for several days. As a consequence, elements typically associated with anthropogenic activities, such as Cd and Cu, have their maximum concentrations during this week (Figure 5.7b+e), and also Pb concentration is above the WHO guideline value of 500 ng/m^3 (Figure 5.7d).

Figure 5.10b illustrates the air masses in the week from the 15th to the 22nd of March 2006 (CW 11/06) during a dust storm situation, as it often occurs in **spring** (Xie *et al.*, 2005; Wang *et al.*, 2006*a*, 2007*b*). The air masses originate from the semi-arid and arid areas northwest of Beijing and transport geogenic minerals over long distances. This special situation could explain why predominantly typical geogenic elements, such as Fe, Ti, and Al (Figure 5.6a-c) or Ca, Sr, and Mg (Figure 5.5b-d) have high concentrations in spring. On the other hand,

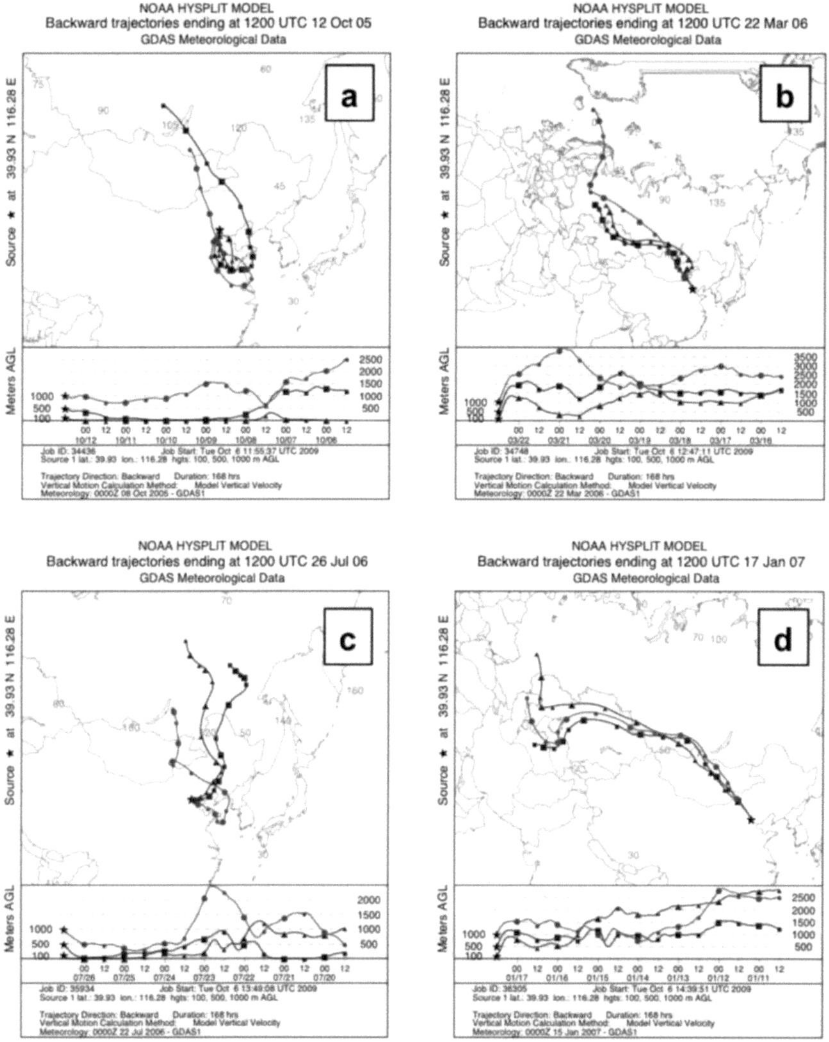

Figure 5.10: Backward trajectories for three different heights (100, 500, and 1000 m AGL) calculated with the NOAA Hysplit Model. Four sampling weeks were chosen, which stand for the four seasons: a) autumn (CW 40/05), b) spring (CW 11/06), c) summer (CW 29/06), d) winter (CW 02/07).

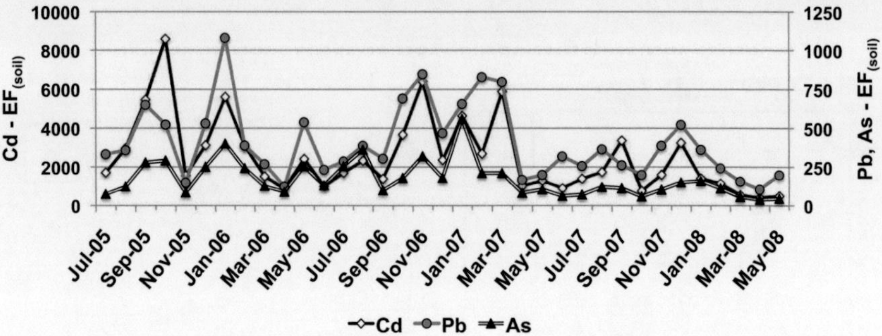

Figure 5.11: Enrichment factors (EF) for Cd, Pb, and As from TSP samples (sum of all four fractions) calculated with average values of Chinese soils (Chen *et al.*, 2008) according to equation 3.6.

concentrations of toxic elements, such as Cd, Pb, and As, are lower during this time (Figure 5.7b-d), and also their relative amount in the mobile fraction is low (Figure 5.8b-d). As a consequence, it can be concluded that although spring is a season with high particle mass concentrations, it has at the same time lower shares of critical elements with respect to toxicity and bioavailability than other seasons. Therefore, it is important to bear in mind that during the annual course, times with highest atmospheric mass concentrations are not necessarily the periods most threatening for human health and the environment.

The sampling week from the 19th to the 26th of July 2006 (CW 29/06, Figure 5.10c) was the week with highest precipitation within this study and represents a typical **summer** situation in Beijing. As expected, the TSP and $PM_{2.5}$ mass concentrations are very low due to wet deposition (168 and 45 $\mu g/m^3$, respectively, Figure 5.1 and 5.2). Many elements, such as Fe, Ti, Al, Ca, Sr, and Mg also show minimum or at least very low absolute concentrations during this week. However, As concentrations are still quite high (72 ng/m^3) and the percentage of As in the highly mobile fraction f1 is also high (37%). Consequently, negative health implications from distinct metals can also occur during summer.

The sampling week from the 10th to the 17th of January 2007 (CW 02/07, Figure 5.10d) represents **winter** conditions in Beijing with lowest temperatures and, consequently, a high amount of heating activities. Moreover, wind speed is very low and, therefore, mass concentrations are high due to stagnant meteorological conditions. Arsenic has its maximum concentration (271 ng/m^3) during this sampling week and the WHO guideline values for TSP in air of 5 ng Cd/m^3

and 500 ng Pb/m^3 in air are exceeded by factor 10 and 2, respectively. With respect to overall concentrations of toxic metals, as well as a high bioavailability of those elements, in Beijing, winter is the season most problematic regarding potential negative health effects.

The varying importance of anthropogenic sources during the annual course can also be illustrated by the EFs of elements from predominantly anthropogenic origin, such as Cd, Pb, and As (Figure 5.11). The high EFs of those elements in winter and autumn indicate that anthropogenic sources dominate during this time of the year. On the contrary, in spring and some autumn weeks (e.g. the sampling week in November 2005), the EFs are lower and, consequently, the relative importance of geogenic sources is higher.

5.3.2 Assessment of health impacts based on guideline values and the estimated mobility of potential toxic metals

The possible health effects of certain elements and their compounds is discussed in more detail in chapter 1.3.1. Within this chapter, the health effects of the mobile and, thus, bioavailable elements Mn, Ni, Zn, As, Cd, and Pb are of special interest. Manganese is considered as especially toxic if taken up through inhalation causing various psychiatric and movement disorders, respiratory effects such as pneumonitis and pneumonia, as well as reproductive dysfunction (WHO, 2000). Lippmann *et al.* (2006) found Ni and V to be significantly associated with daily mortality rates in 60 U.S. cities included in their study. Although Zn is essential for all organisms, toxicological studies proved that elevated Zn concentrations have negative health effects (Adamson *et al.*, 2000). Arsenic, Cd and their chemical compounds are considered as highly carcinogenic by the International Agency for Research on Cancer, while Pb and inorganic Pb compounds are classified as less carcinogenic (IARC, 2008).

Guideline and threshold values for element concentrations in the atmosphere do not exist for all potential toxic elements. For some metals considered within this chapter, the WHO guideline values for total concentrations in air of 5 ng Cd/m^3, 150 ng Mn/m^3, and 500 ng Pb/m^3 were regularly exceeded. For Mn, this was the case for all samples. The minimum value of Mn (sum of all fractions for TSP samples) within this study of 151 ng/m^3 was equal to the recommended threshold level and the maximum Mn concentration of 812 ng/m^3 exceeds this value by more than factor 5. On the contrary, the WHO guideline value for V in air of 1000 ng V/m^3 was never reached within this study. Highest V concentrations (sum of all four fractions) in Beijing were 33 ng/m^3 for TSP and 2.7 ng/m^3 for $PM_{2.5}$ samples. In addition to the elevated total Cd concentrations in Beijing (median of 14 ng/m^3 and maximum of 57 ng/m^3 for the sum of all four fractions), its high proportion in the two mobile fractions (f1+f2 $>$ 70%) indicates a high mobility and, thus, a potential high bioavailability. In this context, it is notably that even when considering only the Cd concentrations in the most bioavailable fraction, 60% of all samples exceed the WHO guideline value. Therefore, Cd can be considered as a key element when assessing the toxicity of urban dust in Beijing. In $PM_{2.5}$ samples, As was detected to a even higher percentage in the highly mobile fraction (median: 50%) than in TSP samples. Since smaller particles can penetrate deeper into the human respiratory tract, the high concentration of very mobile As in the fine particles is of particular concern.

While the volume related metal concentrations expressed in ng/m^3 are higher for most elements in TSP samples, some concentrations are higher in $PM_{2.5}$ samples, when expressed in mass related values in μg/g (for the different fractions as well as for the sum of all fractions). Based on this observation, it should be highlighted that the particle size fraction above 2.5 μm considerable contributes to the overall particle mass, but dilutes the metal concentrations within the particle mass. This is the case for many toxic elements predominantly from anthropogenic sources, such as Cd, As, Cu, or Cr (Figure 5.12). Only Ni and Zn show higher concentrations in $PM_{2.5}$ as in TSP when expressed in ng/m^3 as well as expressed in μg/g, a strong indication for a very high anthropogenic share of these elements in aerosols. For other elements, which are less toxic and mostly originated from geogenic sources, such as Mg, Al, Ca, and Fe, TSP concentrations are higher expressed in ng/m^3 as expressed in μg/g. Consequently, when considering the toxicity of elements in urban areas and their health impacts due to air pollution, it is important to know also their total concentrations in μg/g (equal to mg/kg as it is used in many sediment or soil studies) as well as the concentrations in the bioavailable fractions in μg/g.

5.3.3 Particulate air pollution and associated metal mobility in Beijing in a global context

Compared to other large cities, the air pollution situation in Beijing is serious. In chapter 4.4.5 it was already shown that the total TSP and metal concentrations in Beijing were often elevated in comparison to other cities worldwide. Since there are only few studies applying sequential extractions for atmospheric particles and

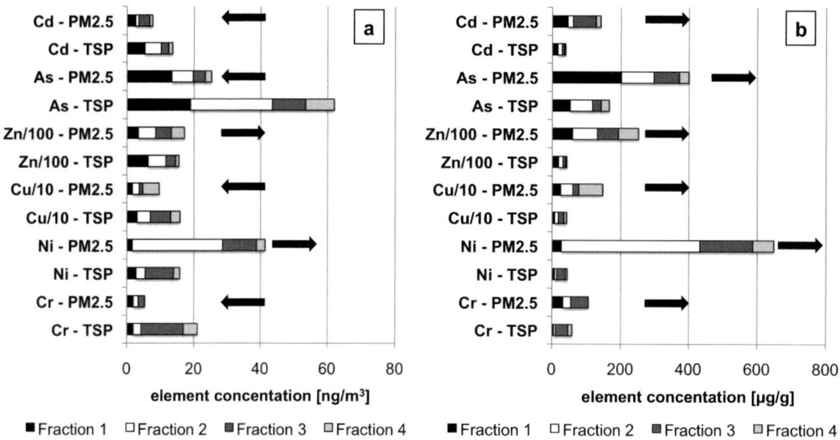

Figure 5.12: Selected metal concentrations for $PM_{2.5}$ and TSP samples in ng/m^3 (a) and μg/g (b). The arrows indicate those elements with higher concentrations in $PM_{2.5}$ relative to TSP samples when expressed in μg/g.

the leaching schemes and samples particle fractions are often different, a comparison of metal mobilities with other cities is difficult. In Table 5.3 the results for TSP and $PM_{2.5}$ samples from this study were compared to those from studies using similar leaching procedures for samples in Seville, Spain (Fernández Espinosa & Ternero Rodríguez, 2004) and Santiago, Chile (Richter *et al.*, 2007). The total metal concentrations (sum of all fractions) of all metals included in the comparison are higher in Beijing than in Seville or Santiago. That means that the overall atmospheric pollution is even more serious for the inhabitants of Beijing. Besides the total concentrations, also the distribution in the distinct fractions varies between the different studies, which will be discussed for a few selected elements in the following paragraphs. The mobile metal concentrations are the crucial key-points for health impact assessment in the different cities.

The toxic elements Cd and Pb are present to an even higher percentage in the easily mobile fraction in the study of PM_{10} from Santiago (55 and 43%, respectively) compared to $PM_{2.5}$ from Beijing (34 and 20%, respectively). In this regard, it is important to note that the study from Santiago comprised the years 1997 to 2003 and that leaded fuel was allowed in Chile until March 2001 (Richter *et al.*, 2007). In Beijing, on the contrary, leaded fuel was already banned in 1997 (Sun *et al.*, 2006), followed by other Chinese megacities, such as Shanghai, Guangzhou, and Tianjin, and, finally, in 2000 all over China. This explains

Table 5.3: Comparison of average APM concentrations for the distinct fractions for different cities in China, Spain and Chile. A and B: this study, median values of TSP and $PM_{2.5}$ quartz fibre filter samples from Beijing, China (N=35 and N=32, respectively); C: Fernández Espinosa & Ternero Rodríguez (2004), $PM_{0.6}$ quartz filter samples from Seville, Spain; D: Richter *et al.* (2007), PM_{10} quartz fibre filter samples from Santiago, Chile (average values of the reported years 1997-2003 during winter).

	Study	f1 ng/m^3 (%)	f2 ng/m^3 (%)	f3 ng/m^3 (%)	f4 [ng/m^3] (%)	sum ng/m^3
As	A	18.9 (31)	24.6 (40)	9.8 (16)	8.6 (14)	60.9
	B	13.3 (52)	6.5 (26)	3.7 (15)	1.9 (7)	25.3
	C	– (–)	– (–)	– (–)	– (–)	–
	D	– (56)	– (14)	– (25)	– (9)	110
Cd	A	5.3 (39)	4.9 (36)	2.2 (16)	1.2 (9)	13.5
	B	2.5 (34)	1 (13)	3.2 (42)	0.9 (11)	7.5
	C	0.1 (20.5)	0.1 (31.3)	0.1 (28.8)	0.1 (19.5)	0.4
	D	– (83)	– (9)	– (5)	– (8)	3.6
Cu	A	30.4 (19)	39.4 (25)	60 (38)	27.9 (18)	158
	B	16.5 (17)	20.1 (21)	10.5 (11)	48.3 (51)	95.3
	C	1.6 (15.6)	1.4 (13.4)	5.3 (51.9)	2 (19)	10.2
	D	– (55)	– (4)	– (12)	– (30)	176
Fe	A	137 (2)	1060 (12)	4380 (48)	3490 (38)	9060
	B	55.3 (7)	145 (19)	154 (21)	394 (53)	749
	C	1.5 (0.5)	1.2 (6.2)	140 (45.3)	149 (48)	310
	D	– (–)	– (–)	– (–)	– (–)	–
Mn	A	129 (34)	103 (27)	105 (28)	44 (12)	382
	B	33.3 (44)	18.6 (25)	19.5 (26)	3.8 (5)	75.2
	C	2.4 (31)	1.7 (21.1)	2.5 (32.5)	1.2 (15.4)	7.8
	D	– (64)	– (5)	– (10)	– (21)	81
Ni	A	2.7 (17)	2.8 (18)	8.4 (54)	1.8 (11)	15.7
	B	1.6 (4)	27 (65)	10.2 (25)	2.6 (6)	41.4
	C	1 (30.9)	0.3 (11)	1.4 (44.2)	0.4 (13.8)	3.1
	D	– (30)	– (ND)	– (27)	– (49)	27
Pb	A	23.9 (5)	265 (60)	90.1 (20)	63.6 (14)	443
	B	31.7 (20)	63.5 (41)	35.2 (23)	24.4 (16)	155
	C	5 (3.5)	49.3 (34.6)	71.2 (49.9)	17 (12)	143
	D	– (43)	– (37)	– (14)	– (5)	394
V	A	2.4 (16)	4.8 (31)	5.8 (37)	2.7 (17)	15.7
	B	1.3 (20)	0.9 (15)	0.9 (14)	3.2 (51)	6.2
	C	1.7 (39.2)	1.1 (24.9)	1.1 (26.1)	0.4 (9.8)	4.2
	D	– (4)	– (ND)	– (93)	– (2)	681
Zn	A	625 (41)	533 (35)	287 (19)	96.5 (6)	1540
	B	336 (20)	515 (30)	466 (27)	390 (23)	1710
	C	14.4 (36.3)	10.5 (26.4)	13.5 (34)	1.3 (3.2)	39.7
	D	– (70)	– (7)	– (4)	– (21)	756

why the percentages of Pb from Beijing and Santiago in the highly mobile and the mobile fraction are more similar, if only the year 2003 of the study from Santiago is chosen for comparison. In 2003, the percentage of Pb in Santiago was 18% in the highly mobile fraction and 52% in the mobile fraction. Iron was found to be one of the most immobile metals in all studies and can, consequently, be considered as not bioavailable. As a consequence, such immobile elements do not need to be focussed in health impact studies and air quality guideline values.

As already mentioned in the previous chapter 5.3.2, the mobile As concentrations in Beijing are problematic with regard to negative health impacts. In a study carried out in an industrial city in Argentina, Fujiwara *et al.* (2006) also observed high concentrations of mobile As. The authors distinguished between summer and winter samples and concluded that the higher mobile As concentrations in winter originate from coal combustion processes. The situation in Beijing is quite similar and the important sources, such as coal combustion, are discussed in the following sections (see chapter 5.3.4).

5.3.4 Source apportionment and varying influences of geogenic and anthropogenic sources

For the successful implementation of mitigation measures, it is essential to know the main sources of those elements considered to be most toxic and at the same time very mobile and, therefore, easily bioavailable. Factor analysis (FA) is a frequently used tool to group related elements and, thus, provides indications of possible sources. Factor analysis was carried out separately for the sum of the two mobile fractions f1+f2 and the sum of the two immobile fractions f3+f4 in order to distinguish between sources contributing to the bioavailable element concentrations and those, which represent the less bioavailable shares of element concentrations.

Mobile fractions of TSP samples

First, FA was carried out for the concentrations of 18 elements (concentrations in μg/g) of the two mobile fractions (sum of f1+f2) of TSP samples together with some meteorological parameters such as temperature, precipitation, wind direction, and wind speed. Here, six factors were extracted (Table 5.4). These factors accounted for 82% of the total explained variance (Expl. Var.) of the whole data set.

Table 5.4: Factor (Fa) loadings of the seven extracted factors (Principal Component Analysis, case by case exclusion of missing data, Varimax standardized rotation) for sum of the two mobile fractions f1+f2 of TSP samples (concentrations in μg/g), together with the respective communalities (comm). Meteorological parameters (predominant wind direction, wind speed, precipitation, and temperature) are included. Loadings $\geq \pm 0.55$ are marked in bold.

	Fa1	Fa2	Fa3	Fa4	Fa5	Fa6	Fa7	comm
Mg	**0.89**	0.06	-0.29	0.07	0.01	-0.06	0.06	0.94
Al	0.32	-0.04	0.11	-0.13	**0.84**	0.15	-0.06	0.91
K	-0.29	0.02	-0.29	**-0.81**	0.29	0.05	0.02	0.90
Ca	**0.87**	-0.15	-0.22	0.31	-0.09	0.00	0.06	0.97
Ti	**0.62**	0.02	0.25	-0.13	0.15	**0.61**	-0.04	0.92
V	-0.13	0.10	0.54	**-0.69**	0.00	-0.01	-0.11	0.83
Cr	0.05	0.27	0.21	-0.38	**0.55**	0.53	-0.02	0.89
Mn	-0.07	-0.07	**0.87**	-0.05	-0.02	0.05	0.10	0.79
Fe	**0.81**	0.04	0.15	0.17	0.03	0.33	-0.02	0.92
Co	-0.16	0.18	**0.86**	-0.02	0.18	0.02	-0.14	0.89
Ni	**-0.56**	0.34	0.30	-0.01	0.00	-0.02	0.25	0.86
Cu	0.07	0.49	-0.24	0.30	-0.05	**0.68**	-0.10	0.80
Zn	-0.02	**0.84**	0.23	-0.27	-0.10	0.19	0.09	0.92
As	0.00	0.49	0.02	0.34	0.27	0.01	**0.59**	0.80
Rb	-0.18	0.47	0.16	**-0.79**	-0.10	0.11	0.08	0.95
Sr	-0.31	0.04	0.04	0.06	**0.89**	-0.17	0.02	0.92
Cd	0.00	**0.86**	-0.12	0.08	0.01	-0.02	0.04	0.80
Pb	-0.27	**0.82**	0.02	-0.13	0.11	0.08	0.26	0.94
mass	-0.07	0.09	-0.04	0.11	-0.01	**-0.88**	-0.24	0.78
wind dir	0.50	-0.14	-0.12	0.25	-0.01	0.24	**-0.59**	0.69
wind v	-0.04	**-0.68**	-0.18	0.28	-0.08	0.16	0.17	0.82
precip	0.09	0.00	-0.13	0.00	-0.15	0.32	**0.75**	0.68
temp	0.40	-0.16	-0.54	-0.05	-0.24	0.50	0.12	0.80
% of Expl. Var.	16	16	12	11	9	11	7	

Table 5.5: Factor (Fa) scores of the seven extracted factors (Principal Component Analysis, case by case exclusion of missing data, Varimax standardized rotation) for sum of the two mobile fractions f1+f2 of TSP samples (concentrations in μg/g) together with some meteorological parameters (predominant wind direction, wind speed, precipitation, and temperature). Scores $> \pm 1$ are marked in bold.

	Fa1	Fa2	Fa3	Fa4	Fa5	Fa6	Fa7
Aug-05	0.43	-0.12	**-1.28**	0.00	-0.72	-0.56	-0.67
Sep-05	0.92	**1.67**	-0.82	0.81	0.05	-0.09	-0.54
Oct-05	-0.23	**1.45**	**-1.26**	**1.05**	-0.62	0.00	-0.87
Nov-05	**1.45**	-0.79	-0.98	0.11	-0.37	**-1.94**	-0.53
Dec-05	**-1.58**	-0.16	-0.54	0.85	-0.52	-0.97	0.01
Jan-06	**-2.20**	0.83	**1.04**	0.57	0.05	-0.59	0.12
Feb-06	**-1.02**	-0.54	-0.30	**1.03**	0.23	-0.38	-0.36
Mar-06	0.05	**-1.22**	0.34	0.71	0.03	**-1.11**	-0.28
Apr-06	-0.92	**-1.55**	-0.31	0.98	-0.36	0.27	0.48
May-06	-0.56	-0.16	-0.02	0.23	-0.53	-0.38	**1.50**
Jun-06	0.16	**-1.07**	**-1.05**	-0.28	0.11	-0.06	-0.19
Jul-06	0.43	-0.21	-0.63	0.33	-0.28	0.96	**4.45**
Aug-06	0.97	0.99	-0.16	0.87	0.85	**1.09**	0.75
Sep-06	0.92	-0.11	-0.28	-0.28	0.06	-0.01	-0.52
Oct-06	-0.36	**1.88**	-0.45	0.05	-0.85	0.82	0.42
Nov-06	0.48	**1.90**	-0.58	-0.86	-0.01	**-1.31**	-0.24
Dec-06	-0.31	0.34	0.18	0.25	0.31	**-1.76**	0.12
Jan-07	-0.01	0.70	0.49	0.96	**1.15**	**-1.57**	**1.23**
Feb-07	**-1.77**	-0.38	**-1.21**	**-1.41**	**4.55**	0.37	-0.28
Mar-07	-0.72	**1.38**	0.54	**-1.24**	-0.83	0.10	0.10
Apr-07	0.99	**-1.19**	-0.45	0.68	0.34	0.00	-0.84
May-07	**1.12**	-0.50	0.15	0.75	0.91	0.58	0.37
Jun-07	-0.57	-0.79	**-1.63**	**-3.79**	**-1.12**	0.17	0.09
Jul-07	-0.15	-0.27	-0.47	-0.14	-0.68	**1.27**	-0.07
Aug-07	-0.49	0.91	-0.39	0.50	-0.24	**2.60**	**-1.53**
Sep-07	**1.44**	0.92	0.25	-0.81	0.12	**1.12**	0.19
Oct-07	**2.64**	-0.74	**1.28**	-0.49	**1.06**	0.04	-0.17
Nov-07	0.51	-0.11	**1.41**	**-1.07**	-0.24	-0.63	0.23
Dec-07	0.14	0.63	0.87	-0.31	-0.06	-0.53	-0.99
Jan-08	-0.13	0.60	**2.49**	0.06	0.24	-0.04	-0.77
Feb-08	-0.92	-0.45	**2.83**	**-1.08**	-0.10	0.78	0.27
Mar-08	**-1.06**	**-1.45**	0.16	0.90	**-1.07**	0.29	-0.52
Apr-08	0.16	**-1.35**	0.25	0.92	-0.16	**2.04**	-0.88
May-08	0.17	**-1.04**	0.54	-0.84	**-1.32**	-0.56	-0.08

Factor 1 has high positive loadings for Mg, Ca, Fe, and Ti, and a negative loading for Ni (Table 5.4). Wind direction is also included in this factor (loading of 0.50). This association comprises typical crustal elements representing predominantly geogenic sources, which is also supported by their low EFs (Table 5.1). Therefore, this factor can be interpreted as a predominantly geogenic factor. However, a certain influence of anthropogenic sources cannot be excluded in a megacity like Beijing, especially for the soluble shares of those elements. Magnesium and Ca can originate from minerals such as calcite or dolomite. However, building materials, such as mortar and concrete, are also possible source materials. Gypsum is another possible source for Ca. As white color pigment, TiO_2 is widely used in paints, paper, and plastics (Reimann & Caritat, 1998). Nickel can originate from waste disposal and waste incineration or chemical industry (Reimann & Caritat, 1998).

Factor 2 comprises Cd, Zn, and Pb with high loadings, as well as As with a lower loading (Table 5.4). These elements are among the elements with highest EFs in this study and, thus, originate mostly from anthropogenic sources. Therefore, factor 2 can be seen as an anthropogenic factor. This interpretation is supported by the fact, that factor scores for factor 2 are most negative in April (Table 5.5), the time when geogenic sources are more important than during other seasons due to the transport of minerals from arid areas NW of Beijing, which leads in turn to a relative less importance of anthropogenic sources. Moreover, a high percentage of the total concentration of Cd, Zn, Pb, and As was determined to be very mobile. Therefore, this factor not only represents mostly anthropogenic sources, but also is of special importance with regard to negative health effects. Coal combustion can be regarded as the dominant source for Cd, Pb, and As in Beijing and combustion processes can also emit Zn. Gong *et al.* (2010) investigated coal ash dumps in China and found an enrichment of leachable Cr, Cd, Pb, Zn, and Co. A significant correlation between water-soluble sulphate and mobile Pb (Pearson correlation factor of 0.79, Spearman correlation factor of 0.64, N=25) reveals that coal combustion is the major source for these elements. This aspect will be discussed in the following paragraph in more detail. The negative correlation with wind speed indicates that these elements accumulate especially under stagnant atmospheric conditions with low wind speeds.

As already mentioned, water-soluble sulphate and the environmentally mobile Pb (sum of the fractions f1+f2) correlate significantly and show a similar annual trend. The annual courses are illustrated in Figure 5.13 for a time period of two years. Both show their maximum concentrations in winter and minimum concentrations in April. The high sulphate and Pb concentrations in winter can

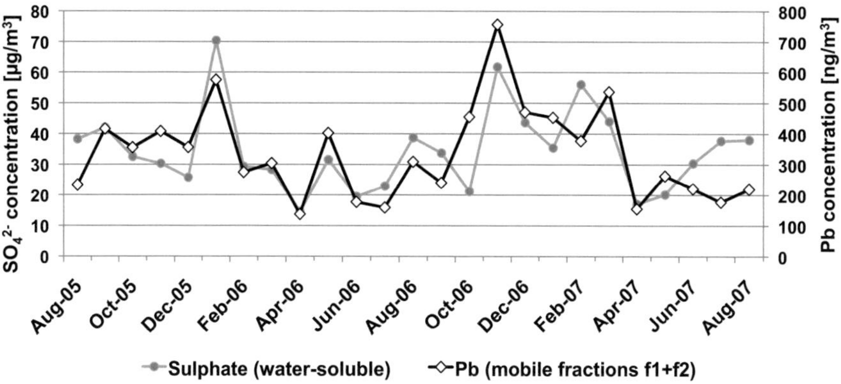

Figure 5.13: Annual courses for water-soluble sulphate and mobile Pb (sum of the fractions f1+f2) concentrations in TSP samples from Beijing.

mainly be associated with coal burning emissions. In summer, however, the sulphate concentration increases relatively more than the mobile Pb concentration. As an additional source, accelerated secondary formation of sulphate could be of importance in Beijing during summer due to high humidity and strong solar radiation (Yao *et al.*, 2002).

Manganese and Co have high loadings in **factor 3** (Table 5.4). Moreover, V is included with a lower loading (0.54) in this factor. These elements can originate from geogenic, as well as anthropogenic sources. However, all three had low EFs within this study (EF $\leq$ 5 for V, EF $\leq$ 8 for Co, EF $\leq$ 14 for Mn). Therefore, the geogenic sources seem to prevail for Mn, Co, and especially for V in Beijing. Possible host minerals for all three elements are pyroxenes, amphibols, micas, as well as garnets and olivins for Mn and Co. Anthropogenic sources are mining and smelting processes, the use as catalysts and Mn and Co are also used in fertilisers (Reimann & Caritat, 1998). Nevertheless, coal combustion is also an important Co source and Mn and V can also be emitted by combustion processes. The negative correlation of temperature (-0.54) is an additional indication that coal combustion for heating processes in winter is an important anthropogenic source. Consequently, factor 3 can be interpreted as an mixed geogenic-anthropogenic factor.

Factor 4 comprises K, Rb, and V with negative loadings (Table 5.4). Vanadium is higher correlated with this factor than with factor 3. In contrast to other elements from mostly geogenic origin, K and Rb showed a high mobility. In a study of Kuang *et al.* (2004) about street sediments in Beijing in 2004, which included mineral analysis, the authors concluded that K and Rb concentrations in Beijing originated mainly from clay minerals. Fertiliser and chemical industry are possible anthropogenic sources for K (Reimann & Caritat, 1998).

Factor 5 has very high loadings for Sr and Al (Table 5.4). Moreover, Cr has also a loading of 0.55 in this factor. Generally, geogenic sources are more important for Sr and Al, which is supported by their low EFs (average EFs of 1 for Al, and 8 for Sr). However, anthropogenic sources also contribute. Aluminium and Sr are both used in alloys (Reimann & Caritat, 1998), while Sr is also used as pigments against corrosion (Kuang *et al.*, 2004).

Copper and Ti are positively correlated with **factor 6** (Table 5.4). Moreover, mass concentration has a negative loading in this factor. It is important to emphasize that Ti concentrations were very low in the two mobile fractions and, thus, mainly the immobile sources, and especially geogenic minerals, such as oxides and silicates are probably important. Anthropogenic uses for Cu and Ti are pigments and alloys (Reimann & Caritat, 1998).

Precipitation and As have positive loadings and wind direction has a negative loading in **factor 7**. For the interpretation of wind direction results it is important to act with caution. Due to the continuous sampling over a whole week, only the predominant wind direction could be considered. Coal combustion, which is the main energy source in China, is probably the most important source for As in Beijing.

Immobile fractions of TSP samples

Furthermore, FA was also carried out for the two less mobile fractions f3+f4. Here, also the concentrations of 18 elements (concentrations in μg/g) of TSP samples were considered together with the same meteorological parameters (temperature, precipitation, wind direction, and wind speed). The factor loadings for the seven extracted factors for the immobile fractions are listed in Table 5.6, while the corresponding factor scores are shown in Table 5.7. All extracted factors accounted for 83% of the total explained variance (Expl. Var.) of the whole data set.

Factor 1 has high loadings for Al, Mg, Mn, Fe, Ca, Co, and Ti and accounts for 26% of the Expl. Var. of the data set. All these elements have low EFs (Table 5.1) and originate predominantly from geogenic sources. However, an additional possible anthropogenic influence should not be neglected, for example steel manufacturing and other industrial activities are probably additional anthro-

Table 5.6: Factor (Fa) loadings of the seven extracted factors (Principal Component Analysis, case by case exclusion of missing data, Varimax standardized rotation) for sum of the two immobile fractions f3+f4 of TSP samples (concentrations in μg/g), together with the respective communalities (comm). Meteorological parameters (predominant wind direction, wind speed, precipitation, and temperature) are included. Loadings $\geq \pm 0.55$ are marked in bold.

	Fa1	Fa2	Fa3	Fa4	Fa5	Fa6	Fa7	comm
Mg	**0.88**	-0.37	-0.08	0.05	-0.04	0.03	-0.02	0.97
Al	**0.92**	0.05	0.02	-0.07	-0.17	-0.10	0.02	0.96
K	-0.11	0.02	-0.21	0.03	**-0.84**	-0.16	-0.02	0.86
Ca	**0.76**	-0.34	-0.06	0.23	0.17	0.05	-0.06	0.97
Ti	**0.58**	-0.16	0.45	-0.13	-0.29	-0.19	0.34	0.87
V	0.51	-0.07	0.37	0.02	**-0.62**	-0.05	0.06	0.92
Cr	-0.01	-0.10	0.01	-0.04	0.18	**0.90**	0.06	0.93
Mn	**0.83**	0.08	0.38	-0.16	0.04	0.13	-0.11	0.92
Fe	**0.80**	-0.09	0.06	-0.24	-0.06	0.26	0.11	0.94
Co	**0.75**	0.29	0.39	-0.07	0.16	0.03	-0.20	0.92
Ni	0.07	0.16	0.02	0.35	-0.01	**0.84**	-0.21	0.95
Cu	-0.32	0.06	0.45	0.41	0.49	0.05	0.02	0.74
Zn	0.15	0.10	**0.88**	0.06	0.12	0.03	-0.05	0.85
As	-0.04	0.37	0.07	**0.80**	0.16	0.18	0.05	0.82
Rb	0.30	-0.02	0.42	**-0.55**	-0.48	0.04	0.21	0.94
Sr	0.10	**0.63**	0.00	0.08	0.29	-0.29	-0.27	0.88
Cd	-0.11	-0.08	0.02	**0.91**	-0.14	0.06	-0.06	0.78
Pb	**-0.55**	0.50	0.14	0.35	0.23	0.23	-0.20	0.92
mass	0.07	0.13	-0.39	0.19	0.04	0.33	**-0.73**	0.86
wind dir	0.15	**-0.80**	-0.01	-0.05	0.16	-0.11	-0.15	0.80
wind v	**0.81**	0.04	-0.05	-0.15	-0.12	-0.13	0.17	0.87
precip	0.03	0.10	-0.10	0.08	0.01	0.08	**0.85**	0.73
temp	0.19	-0.54	-0.25	-0.17	-0.05	-0.05	**0.60**	0.86
% of Expl. Var.	26	10	9	11	9	9	9	

Table 5.7: Factor (Fa) scores of the seven extracted factors (Principal Component Analysis, case by case exclusion of missing data, Varimax standardized rotation) for sum of the two immobile fractions f3+f4 of TSP samples (concentrations in μg/g) together with some meteorological parameters (predominant wind direction, wind speed, precipitation, and temperature). Scores $> \pm 1$ are marked in bold.

	Fa1	Fa2	Fa3	Fa4	Fa5	Fa6	Fa7
Aug-05	-0.83	**-1.16**	**-1.28**	0.46	-0.59	0.01	0.10
Sep-05	-0.73	**-1.04**	0.09	0.83	**1.49**	-0.62	0.02
Oct-05	-0.86	**-1.42**	0.13	**3.46**	**-1.26**	-0.30	-0.01
Nov-05	0.83	**-1.21**	**-1.77**	**1.55**	**-1.91**	-0.12	**-1.38**
Dec-05	-0.42	**1.19**	-0.11	**1.34**	**-1.06**	0.51	-0.47
Jan-06	-0.98	**1.70**	-0.22	0.34	0.22	0.02	-0.49
Feb-06	0.93	0.71	0.09	0.91	0.83	-0.63	-0.61
Mar-06	**2.39**	0.17	-0.48	-0.23	0.49	0.18	**-1.22**
Apr-06	**2.42**	**1.04**	0.40	-0.05	-0.70	0.07	0.45
May-06	0.81	0.79	-0.60	-0.26	-0.40	0.28	0.64
Jun-06	0.51	-0.69	**-1.17**	-0.38	0.57	-0.51	-0.07
Jul-06	0.65	**1.15**	-0.86	**1.04**	0.66	0.62	**4.18**
Aug-06	0.09	-0.52	-0.06	0.75	**1.40**	-0.50	0.82
Sep-06	-0.16	**-1.23**	-0.67	-0.69	**1.12**	0.38	-0.16
Oct-06	-0.65	-0.13	**1.52**	0.22	**1.03**	0.22	0.40
Nov-06	-0.36	-0.47	0.04	-0.22	0.35	**4.90**	-0.71
Dec-06	-0.35	**1.02**	-0.42	0.14	**1.09**	0.08	**-1.23**
Jan-07	-0.19	**1.79**	0.26	**1.34**	0.88	0.31	-0.62
Feb-07	**-1.19**	**2.36**	**-1.75**	-0.67	0.21	**-1.59**	-0.39
Mar-07	**-1.13**	0.37	-0.58	-0.62	0.23	-0.13	0.05
Apr-07	0.71	**-1.27**	**-1.07**	-0.87	0.94	-0.73	-0.36
May-07	**2.11**	-0.08	**1.20**	0.14	-0.29	0.01	0.38
Jun-07	**-1.32**	0.08	**-1.00**	**-1.55**	**-2.68**	-0.13	0.64
Jul-07	-0.64	-0.59	-0.32	**-1.12**	-0.09	0.32	**1.42**
Aug-07	-0.80	-0.83	0.66	-0.77	0.64	0.35	0.74
Sep-07	-0.85	-0.85	0.93	-0.13	-0.88	-0.45	**1.17**
Oct-07	0.42	-0.79	0.16	**-1.10**	0.67	-0.46	-0.44
Nov-07	-0.51	0.56	0.59	**-1.08**	-0.28	0.04	-0.57
Dec-07	-0.61	-0.20	0.54	-0.07	0.57	0.01	**-1.20**
Jan-08	**-1.05**	-0.02	**1.09**	-0.27	0.48	-0.57	-0.71
Feb-08	-0.31	0.89	**3.14**	-0.10	**-1.46**	-0.22	-0.24
Mar-08	0.65	-0.15	0.21	-0.87	-0.91	-0.09	-0.07
Apr-08	0.55	**-1.24**	**1.46**	-0.23	0.04	**-1.52**	0.19
May-08	0.86	0.07	-0.16	**-1.23**	**-1.35**	0.26	-0.26

pogenic sources for these elements. The predominant geogenic origin of the elements within this factor is supported by the factor scores (Table 5.7), which were highest in Mar-06, Apr-06, and May-07 (factor scores > 2). During these three months, severe dust storm episodes occurred in Beijing. Wind speed is also included in factor 1 (loading of 0.81). The average wind speed for the three sampling weeks in Mar-06, Apr-06, and May-07 was > 3 m·s^{-1}. At this high wind speeds, resuspension processes from street surfaces, fallow land, and dry soils play an important role for the transport of minerals and lead to a high concentration of geogenic elements.

Factor 2 comprises wind direction and Sr (Table 5.6). Moreover, Pb is included in this factor. A possible anthropogenic source for both elements are alloys (Reimann & Caritat, 1998). The EFs of Sr were low ($\leq$10) and therefore, the geogenic sources seem to prevail. Typical Sr-minerals are strontianite and celestite. During the sampling weeks in Dec-05, Jan-06, Apr-06, Jul-06, Dec-06, Jan-07, and Feb-07 factor scores were high ($>$+1) and the predominant wind direction was 30° (NNE), except for Feb-07 when the wind direction was 60° (ENE). Therefore, a source north-east of Beijing seems to prevail.

Only Zn has high loadings in **factor 3**. With lower loadings of 0.45, Cu and Ti probably also contribute somewhat to this factor. Zinc and Ba are often used as tracers for tyre wear (e.g. Monaci *et al.*, 2000), while Cu can originate from brake abrasion (e.g. Muschak, 1990; Sternbeck *et al.*, 2002). Therefore, traffic is an important source for this factor. However, Ba was not analysed within this study. Nevertheless, traffic abrasion products seem to be an important source. Another important anthropogenic source for Zn are smelting processes.

Factor 4 includes Cd and As with high positive loadings, as well as Rb with a negative loading (Table 5.6). All these elements were highly mobile and could be associated with coal combustion. Their less mobile shares can also originate from geogenic minerals. Sphalerite is a mineral, which can incorporate both elements, As and Cd.

Potassium and V have high negative loadings in **factor 5** (Table 5.6). Both elements had low EFs and geogenic sources seem to prevail. Many of the major silicate minerals contain K, e.g. K-feldspar or mica. Host minerals for V are e.g. pyroxens, amphiboles, micas, apatite, or magnetite.

Factor 6 has high positive loadings for Cr and Ni (Table 5.6). Both elements can originate from various anthropogenic sources including chemical industry, steel works and waste incineration (Reimann & Caritat, 1998). Chromium and Ni both had their maximum concentrations in Nov-06. Therefore, the factor scores for the November sampling week are extremely high (Table 5.7).

Factor 7 is negatively correlated with total mass. Furthermore, the two meteorological factors precipitation and temperature have positive loadings. Therefore, this factor represents the overall diffuse pollution with higher mass concentrations during colder episodes (winter) and lower concentrations due to wet deposition in summer. This is also reflected by high negative factor scores in winter and positive factor scores in summer (Table 5.7).

Factor analysis was also applied for $PM_{2.5}$ samples separately for the sum of the two mobile fractions f1+f2 and for the sum of the two immobile fractions f3+f4. For $PM_{2.5}$, also 18 elements, the total mass concentration and meteorological parameters were considered for FA.

Mobile fractions of $PM_{2.5}$ samples

For the mobile fractions of $PM_{2.5}$ samples, 6 factors were extracted and these factors accounted for 78% of the Expl. Var. of the whole data set. The factor loadings are listed in Table 5.8, while the corresponding factor scores are shown in Table 5.9.

Factor 1 has high loadings for K, Cr, Rb, and Fe (Table 5.8). Geogenic sources are dominant for these elements, which is reflected by their low EFs. While Cr, and Fe had very low EFs of 4 and 2, respectively, the EFs of K and Rb were 15 and 19, and, thus, the anthropogenic influence is higher for these two elements. Potassium, Cr, and Rb concentrations in $PM_{2.5}$ samples were high in the mobile fractions (Figure 5.4), whereas Fe was one of the most immobile elements within this study.

Factor 2 comprises Ti, Fe, and Ca with high loadings (Table 5.8). This factor can be interpreted a second geogenic factor, which is supported by the low enrichment factors of these elements (Table 5.1). Titanium and Fe were the two most immobile elements in $PM_{2.5}$ samples and, consequently, the mobile shares of these elements only play a minor role. As already mentioned before, the use of TiO_2 as a white color pigment, is a possible anthropogenic source for Ti.

Zinc and Ni have high positive loadings in **factor 3** (Table 5.8). Furthermore, precipitation and mass (negative loading) are included in this factor. Especially in $PM_{2.5}$ samples, anthropogenic sources play an important role for both elements. The observation, that the EFs of $PM_{2.5}$ results are approximately 9-fold higher for Ni, and 5-fold higher for Zn if compared to the respective EFs of the TSP samples supports this interpretation.

Factor 4 comprises Pb, Cd, As, and V with high loadings (Table 5.8). With the exception of V, all these elements had very high EFs (Table 5.1), which were

Table 5.8: Factor (Fa) loadings of the seven extracted factors (Principal Component Analysis, case by case exclusion of missing data, Varimax standardized rotation) for sum of the two mobile fractions f1+f2 of $PM_{2.5}$ samples (concentrations in μg/g), together with the respective communalities (comm). Meteorological parameters (predominant wind direction, wind speed, precipitation, and temperature) are included. Loadings $\geq \pm 0.55$ are marked in bold.

	Fa1	Fa2	Fa3	Fa4	Fa5	Fa6	comm
Mg	-0.16	0.50	0.22	-0.13	**0.77**	0.04	0.99
Al	0.11	0.29	-0.02	0.07	**0.86**	0.15	0.99
K	**0.91**	-0.16	-0.06	0.10	0.13	-0.05	0.93
Ca	-0.43	**0.58**	0.35	-0.18	0.48	-0.10	0.99
Ti	0.03	**0.91**	0.08	-0.09	0.17	-0.20	0.94
V	0.25	-0.16	-0.29	**0.58**	0.12	-0.12	0.78
Cr	**0.80**	0.21	0.30	-0.03	0.07	0.14	0.95
Mn	0.31	0.53	-0.05	0.25	0.21	**0.62**	0.95
Fe	**0.66**	**0.67**	-0.08	0.01	0.05	0.14	0.93
Co	0.06	-0.12	-0.13	0.08	0.30	**0.85**	0.96
Ni	-0.04	0.00	**0.76**	-0.10	0.37	-0.08	0.83
Cu	-0.01	-0.04	0.24	0.34	-0.06	**0.67**	0.86
Zn	0.04	0.18	**0.84**	-0.10	-0.08	0.21	0.95
As	0.19	0.18	-0.05	**0.61**	-0.12	0.53	0.88
Rb	**0.79**	0.12	0.05	0.40	-0.25	0.04	0.94
Sr	0.12	-0.11	0.12	-0.24	**0.87**	0.18	0.98
Cd	0.22	0.13	0.17	**0.71**	-0.20	0.22	0.91
Pb	-0.11	-0.16	0.13	**0.85**	-0.10	0.18	0.93
mass	-0.40	-0.44	**-0.58**	-0.14	-0.15	-0.19	0.87
wind dir	-0.29	0.01	-0.28	-0.06	0.19	-0.50	0.87
wind v	-0.48	0.10	0.00	0.00	0.49	-0.36	0.94
precip	0.07	-0.12	**0.77**	0.29	0.01	-0.18	0.93
temp	0.19	0.11	0.37	0.10	-0.43	**-0.66**	0.98
% of Expl. Var.	15	12	13	11	14	13	

Table 5.9: Factor (Fa) scores of the seven extracted factors (Principal Component Analysis, case by case exclusion of missing data, Varimax standardized rotation) for sum of the two mobile fractions f1+f2 of $PM_{2.5}$ samples (concentrations in μg/g) together with some meteorological parameters (predominant wind direction, wind speed, precipitation, and temperature). Scores > ±1 are marked in bold.

	Fa1	Fa2	Fa3	Fa4	Fa5	Fa6
Feb-05	-0.36	-1.27	-0.35	0.83	**1.28**	0.23
Mar-05	**-1.24**	0.09	0.06	**1.06**	0.38	0.49
Apr-05	**-1.80**	0.65	-0.63	-0.51	0.95	**-1.11**
May-05	-0.58	0.41	**2.03**	0.60	0.52	-0.39
Jun-05	0.46	-0.93	0.57	-0.49	-0.26	-0.96
Jul-05	0.43	-0.53	**2.15**	0.71	0.14	-0.86
Aug-05	0.35	-0.43	**3.07**	0.08	-0.27	0.05
Sep-05	-0.66	-0.60	0.49	0.36	-0.95	0.13
Oct-05	-0.76	-0.38	-0.02	0.24	-0.79	0.24
Nov-05	-0.82	0.12	0.31	**-1.15**	0.39	0.67
Dec-05	-0.80	-0.15	0.27	-0.20	0.06	**1.52**
Jan-06	-0.03	-0.77	-0.50	**1.78**	0.14	**3.26**
Feb-06	-0.29	0.23	-0.95	0.82	0.61	0.42
Mar-06	**-1.16**	-0.55	**-1.35**	-0.81	0.54	**-1.02**
Apr-06	-0.31	**1.15**	-0.50	0.14	0.96	-0.24
May-06	0.18	**-1.21**	**-1.32**	**2.11**	-0.14	**-1.68**
Jun-06	0.45	-0.64	-0.41	0.02	-0.16	**-1.22**
Jul-06	0.14	-0.10	0.49	**2.12**	-0.15	-0.16
Aug-06	0.31	**1.02**	-0.49	**1.19**	**-1.00**	-0.35
Sep-06	0.86	0.74	-0.84	0.88	-0.90	-0.53
Oct-06	**3.18**	**1.90**	-0.74	0.10	0.05	0.11
Nov-06	0.78	0.30	**-1.15**	-0.37	-0.03	0.01
Dec-06	-0.22	0.00	**-1.32**	**-1.03**	-0.17	**1.12**
Jan-07	-0.28	0.94	-0.01	-0.95	-0.32	**1.96**
Feb-07	**1.92**	**-1.65**	0.23	**-1.32**	**3.70**	0.34
Mar-07	0.26	0.05	-0.03	-0.75	**-1.06**	0.39
Apr-07	**-1.49**	**-1.18**	-0.62	**-1.50**	-0.80	-0.98
May-07	-0.93	**3.39**	0.68	-0.12	**1.55**	-0.99
Jun-07	**1.34**	-0.91	-0.39	-0.67	-0.74	-0.89
Jul-07	-0.16	-0.12	0.50	-0.75	**-1.12**	0.12
Aug-07	0.64	0.19	0.44	**-1.33**	**-1.27**	0.14
Sep-07	0.61	0.23	0.38	**-1.07**	**-1.13**	0.19

even higher for $PM_{2.5}$ than for TSP samples, and, consequently, anthropogenic sources seem to prevail for this factor. Furthermore, all these elements had high shares in the very mobile fractions, which is also important with regard to negative health effects. As already discussed for TSP samples, coal combustion is probably the most important source for the mobile shares of all these elements.

Strontium, Al, and Mg have high loadings in **factor 5** (Table 5.8). These elements mainly originate from geogenic sources and therefore, this factor represents a third geogenic factor. However, for the mobile shares of these elements, anthropogenic sources most likely also contribute. For the mobile fractions of TSP samples, Sr and Al were also combined in one factor.

Factor 6 comprises Co, Cu, and Mn with positive loadings (Table 5.8). Moreover, temperature has also a high loading (negative loading) in this factor. Coal combustion is an important source for Co. The negative correlation with temperature indicates that coal combustion for heating purposes in winter are an important Co source. This interpretation is further supported by high positive factor scores in Dec-05, Jan-06, Dec-06, and Jan-07, respectively, indicating the high importance of this factor during that time. Furthermore, Co, Cu, and Mn could be related to agriculture. Cobalt and Mn are used in fertilisers, while high Cu levels on farmland can arise from fertilizing with sludge from pig farms (Reimann & Caritat, 1998). The influence from farmland is probably lower, when the vegetation is growing in spring and summer, and higher when wind transport from fallow land without vegetation plays an more important role in winter. The factor scores, which are most negative in spring and summer, and, as mentioned before, most positive in winter, support this statement.

Immobile fractions of $PM_{2.5}$ samples

For the FA of the immobile fractions f3+f4 of $PM_{2.5}$ samples, seven factors were extracted (Table 5.10). These factors accounted for 83% of the Expl. Var. of the whole data set.

Factor 1 comprises Cu, Cr, Zn, and Ni with high loadings (Table 5.10). With the exception of Cr, all these elements had high EFs (Table 5.1). Compared to TSP samples, the EFs of $PM_{2.5}$ samples were more than 3-times higher for Cu, 5-times higher for Zn, and 9-times higher for Ni. This means that the anthropogenic influence is even higher if only the small size classes are considered. Consequently, those elements originate from predominantly anthropogenic sources and their association can be classified as a diffuse anthropogenic factor. The most important sources are probably smelting processes and steel works. This factor had

Table 5.10: Factor (Fa) loadings of the seven extracted factors (Principal Component Analysis, case by case exclusion of missing data, Varimax standardized rotation) for sum of the two immobile fractions f3+f4 of $PM_{2.5}$ samples (concentrations in μg/g), together with the respective communalities (comm). Meteorological parameters (predominant wind direction, wind speed, precipitation, and temperature) are included. Loadings $\geq \pm 0.55$ are marked in bold.

	Fa1	Fa2	Fa3	Fa4	Fa5	Fa6	Fa7	comm
Mg	0.28	**0.84**	0.00	0.19	0.22	0.21	-0.01	0.98
Al	0.03	**0.95**	-0.12	0.02	-0.02	-0.09	0.02	0.96
K	-0.05	-0.09	0.00	-0.07	-0.03	**0.85**	0.07	0.75
Ca	-0.06	-0.07	-0.02	-0.07	**0.93**	0.00	-0.22	0.87
Ti	0.21	**0.85**	0.14	0.03	0.00	-0.22	0.14	0.98
V	0.24	0.28	-0.01	**0.83**	0.03	-0.06	0.14	0.86
Cr	**0.81**	0.01	0.17	0.02	-0.07	0.22	-0.09	0.96
Mn	0.13	0.25	0.44	0.31	0.08	0.24	**0.62**	0.97
Fe	0.03	**0.89**	0.17	0.02	-0.13	0.07	0.12	0.99
Co	-0.13	**0.57**	0.45	0.05	-0.06	0.07	0.43	0.91
Ni	**0.72**	-0.04	-0.09	-0.03	0.03	-0.10	-0.03	0.90
Cu	**0.86**	0.22	0.15	0.24	0.09	-0.08	0.06	0.98
Zn	**0.81**	0.19	0.15	0.46	0.10	-0.06	-0.05	0.98
As	0.14	0.08	0.10	**0.94**	0.05	0.02	-0.04	0.93
Rb	0.34	0.04	**0.76**	0.29	-0.07	0.34	-0.15	0.99
Sr	0.47	0.13	0.18	0.34	**0.70**	-0.08	0.17	0.93
Cd	-0.17	-0.11	0.44	0.54	0.02	-0.10	0.47	0.91
Pb	-0.19	-0.22	**0.63**	0.45	-0.10	0.41	0.28	0.97
mass	-0.23	-0.14	**-0.77**	-0.04	-0.23	0.15	0.11	0.73
wind dir	-0.12	0.29	**-0.60**	0.29	0.15	0.29	-0.34	0.74
wind v	-0.07	**0.74**	-0.21	0.14	0.04	-0.11	-0.12	0.83
precip	**0.56**	0.06	0.33	-0.32	0.12	-0.18	-0.36	0.80
temp	0.13	-0.09	0.13	0.00	0.16	-0.02	**-0.88**	0.94
% of Expl. Var.	16	19	13	13	7	6	9	

Table 5.11: Factor (Fa) scores of the seven extracted factors (Principal Component Analysis, case by case exclusion of missing data, Varimax standardized rotation) for sum of the two immobile fractions f3+f4 of $PM_{2.5}$ samples (concentrations in μg/g) together with some meteorological parameters (predominant wind direction, wind speed, precipitation, and temperature). Scores $> \pm 1$ are marked in bold.

	Fa1	Fa2	Fa3	Fa4	Fa5	Fa6	Fa7
Feb-05	0.21	0.05	**-1.12**	-0.03	-0.07	-0.36	0.40
Mar-05	0.81	0.89	-0.70	-0.05	0.65	0.54	**1.04**
Apr-05	-0.09	**2.97**	-0.96	-0.22	-0.23	0.63	-0.22
May-05	**1.67**	**1.27**	0.46	-0.41	0.46	-0.12	-0.73
Jun-05	0.39	-0.45	-0.18	-0.39	-0.21	-0.74	-0.56
Jul-05	**2.09**	0.02	0.97	-0.98	-0.34	-0.41	-0.86
Aug-05	**2.68**	-0.01	**1.16**	**-1.05**	0.74	-0.42	-0.43
Sep-05	**1.73**	**-1.00**	**-1.26**	-0.46	-0.67	0.01	0.16
Oct-05	0.06	-0.56	-0.80	-0.38	-0.73	0.03	0.07
Nov-05	0.47	0.70	-0.08	**-1.19**	-0.42	-0.27	**1.08**
Dec-05	-0.05	-0.07	-0.07	-0.55	0.08	-0.84	**1.36**
Jan-06	**-1.06**	-0.30	0.89	-0.21	-0.62	**-1.31**	**1.42**
Feb-06	-0.94	-0.10	-0.09	-0.11	0.03	-0.71	0.57
Mar-06	-0.99	**1.49**	**-1.04**	-0.42	-0.61	0.05	-0.07
Apr-06	**-1.26**	**2.02**	0.76	**-1.07**	-0.71	-0.88	-0.27
May-06	-0.82	-0.90	**-1.13**	0.14	**-1.02**	-0.32	-0.79
Jun-06	-0.98	-0.27	-0.24	-0.91	**2.29**	-0.04	-0.91
Jul-06	-0.88	-0.52	0.97	-0.79	-0.78	**-1.15**	**-1.33**
Aug-06	**-1.19**	-0.76	-0.17	-0.81	**4.04**	-0.08	-0.49
Sep-06	-0.93	**-1.11**	-0.45	-0.45	-0.46	-0.36	-0.73
Oct-06	**-1.12**	0.01	**3.55**	-0.31	-0.67	**1.24**	0.48
Nov-06	-0.47	-0.38	-0.22	-0.85	-0.26	**4.53**	0.46
Dec-06	-0.35	-0.74	-0.63	-0.81	-0.74	0.14	**1.76**
Jan-07	0.37	-0.60	0.83	**1.63**	0.90	-0.48	**2.84**
Feb-07	0.95	-0.34	-0.42	**1.32**	0.65	0.32	0.91
Mar-07	0.02	-0.59	0.48	**1.63**	-0.26	-0.18	0.39
Apr-07	-0.48	-0.36	**-1.91**	0.74	-0.41	-0.48	-0.32
May-07	-0.25	**2.25**	0.20	**2.97**	0.71	-0.03	-0.56
Jun-07	0.39	-0.59	-0.13	0.28	-0.11	**1.02**	**-1.11**
Jul-07	-0.35	-0.62	0.65	0.89	-0.08	-0.01	**-1.25**
Aug-07	0.21	-0.65	0.49	**1.50**	-0.40	0.42	**-1.20**
Sep-07	0.18	-0.76	0.20	**1.33**	-0.75	0.27	**-1.11**

high factor scores mainly in 2005 (Table 5.11). This indicates, that steel industry is the most important source for the immobile shares of Cu, Cr, Zn, and Ni in Beijing $PM_{2.5}$ samples. A big steel factory ("Capital Steel Group Company") was located in the western part of Beijing (Shijingshan district). This plant was built in 1950 and was one of the biggest steel-producing factories in China (Kuang *et al.*, 2004). However, the production was reduced bit by bit starting in summer 2005 and the factory was further relocated in order to improve air quality in the run-up to the Olympic Games (see also chapter 6). By 2010, all production facilities have been moved to Tangshan, a city in the neighboring Hebei Province (XINHUA online, 2010). Traffic, and especially the involved abrasion processes from tyres, brakes, and asphalt, is another important source for the element association of this factor.

Aluminium, Fe, Ti, Mg, and Co have high loadings in **factor 2** (Table 5.10). Furthermore, wind speed is included in this factor. All these elements had low EFs and, thus, geogenic sources seem to prevail. This is supported by the factor scores, which are highest in Apr-05, May-05, Mar-06, Apr-06, and May-07. Similar to the geogenic factor 1 extracted from the immobile element concentrations of TSP samples, the average wind speed was high during these five sampling weeks ($>3\ m{\cdot}s^{-1}$, except for Apr-06 with $2.5\ m{\cdot}s^{-1}$) and, thus, resuspension processes of minerals played an important role during these sampling weeks. However, it has to be considered that steel works is a possible additional anthropogenic sources for Fe.

Rubidium and Pb are positively correlated to **factor 3**, while total mass and the predominant wind direction have negative loadings (Table 5.10). In fraction f3, Rb had only low concentrations and no Rb (concentrations were below the LOD) was found in the residual fraction f4. Therefore, the immobile concentrations of Rb do not play an important role for source apportionment. Lead occurred in higher concentration in the fractions f3 and f4 than Rb, but highest Pb concentrations occurred in the mobile fraction f2. Immobile Pb can be present in form of minerals, such as galena, cerussite or anglesite. Rubidium is a common element in many silicates (mainly by replacing K).

Factor 4 comprises As and V with high loadings, as well as Cd with a lower loading (Table 5.10). Coal combustion is probably the dominant source for these three elements. However, insoluble parts can also originate from alloys and other industrial products.

Calcium and Sr are positively correlated to **factor 5** (Table 5.10). Geogenic sources, such as carbonates, seem to prevail for this factor, which is supported by their low EFs (Table 5.1). In various samples from Beijing, Ca-bearing minerals, such as calcite, dolomite, and gypsum, were abundant (see also SEM analysis results in chapter 7.2.2).

Manganese is the only element with high loadings in **factor 7** (Table 5.10). Only Cd and Co also contribute slightly to this factor. Temperature shows a strong negative correlation with factor 7. Therefore, the factor scores are highest during winter (Nov-05 – Jan-06, Dec-06 – Feb-07). The high importance of Mn during winter leads to the interpretation that it is also associated to an additional source during winter, which is probably combustion of fossil fuels.

5.4 Summary and conclusions

Regularly, mass and metal concentrations in Beijing's atmosphere exceed international guideline values. Within this study, this was reported for Mn, Cd, Pb and partially also for As. With regard to negative health effects, it should be highlighted that sometimes even the sum of the two mobile fractions already exceeds the guideline values. In this context, it is notably that even when considering only the Cd concentrations in the most bioavailable fraction, 60% of all samples exceed the WHO guideline value. Moreover, some toxic elements, such as As, had higher percentages in the very mobile fractions for $PM_{2.5}$ than for TSP samples. This is health-relevant because the smaller particles can penetrate deeper into the human lung system. Seasonal variations over the annual course were pronounced. The seasonal trend observed for most elements showed lowest concentrations in summer, which is also the time with lowest TSP total mass concentrations caused by wet deposition due to the high precipitation. In spring, geogenic sources are predominant. However, while contributing widely to total mass concentrations, geogenic dust is less problematic with regard to potential bioavailable metal loads, because of low concentrations of highly toxic metals in the mobile fractions. Winter, on the contrary, is the season most problematic regarding potential negative health effects because of maximum total concentrations of toxic metals, as well as a high bioavailability of those elements.

The study showed that the mobility of the particular metals is indicated to be very different. In this regard, a cause of concern is the high mobility and, thus, assumed high bioavailability of toxic metals, such as Zn, Cd, Mn, As, Cu and Pb, in Beijing. Furthermore, the study showed that the bioavailable eco-toxicological critical elements Zn, Cd, As, Cu and Pb originate mainly from anthropogenic sources. Many of those sources, like industry or traffic, are more or less constant during the whole year and mainly meteorological parameters determine the level of pollution, while other sources, like combustion processes for heating activities, occur only at distinct seasons and superimpose the overall air pollution during this time. For the implementation of mitigation measures for air pollution in urban areas and megacities the focus should, therefore, be on the reduction of these sources, which are closely related to high emissions of potentially toxic elements in mobile and, thus, bioavailable speciation. With regard to seasonal anthropogenic sources, coal combustion seems to play the most important role in Beijing during autumn and winter. Coal combustion, which is still widely used in Beijing for heating and cooking purposes, seems to be a crucial source for airborne concentrations of toxic and easily mobile metals, especially for Cd, As, and Pb. Therefore, the implementation of alternative energy sources instead of coal combustion is crucial for a sustainable urban development in Beijing. In preparation of the Olympic Games, which were hosted in Beijing in August 2008, the percentage of natural gas used for domestic heating was already increased (Wang *et al.*, 2010*a*), which seems to mark a first effort in the right direction.

In general, the approach to use sequential extraction procedures for APM characterisation, which was applied for the first time for atmospheric particles in a Chinese megacity, proved to be a good tool in order to gain important additional information needed for a future sustainable urban planning and especially for selecting mitigation measures. With regard to health studies, especially the mobile and bioavailable fraction of toxic metals from anthropogenic sources, as well as the concentration of those element fractions not only in ng/m^3, but also in absolute concentrations in μg/g should be considered. For Beijing, the enormous growth rates of the urban population, the industry and the energy consumption, will demand considerable efforts to reduce air pollution in the future.

Chapter 6

Influence of mitigation measures on air quality: the example of the Olympic Summer Games 2008

6.1 Introduction

The period of the 2008 Olympic Summer Games in Beijing can be considered as a large field experiment and unique opportunity to study the influence of emission reduction measures on air quality improvement. Both within and outside China, the air quality in Beijing during the Olympic Games was of huge concern and discussed extensively not only in the scientific world, but also in the international press. The importance of the event was tremendous for the Chinese Government and, therefore, unprecedented efforts were undertaken to reduce air pollution. Half of the budget for the Games was allocated to making the Games "green" and in total, US$12.2 billion (85.4 billion RMB) have been provided so that Beijing could present green Olympics to the world (Fang *et al.*, 2009). The mitigation measures, which will be described in more detail in the following section 6.2, comprised traffic reductions, expansion of public transportation, planting of plants and parks, cleaner production techniques for industries, as well as industry closures in Beijing and also in the surrounding areas, such as in Tianjin.

Measuring atmospheric particulate matter (APM) at the time before, during, and after the Olympic (8th – 24th of August 2008) and Paralympic (6th – 17th of September 2008) Games offered the unique possibility to observe and assess the success of short-term measures to mitigate extreme urban aerosol pollution and also to investigate the increase of aerosol pollution that was expected after the Olympic Games when many of the sources operated again. Within this chapter, the results of TSP, $PM_{2.5}$ and PM_1 measurements at two sampling sites (labelled CRAES and CUG, see Figure 3.1 in chapter 3.1.1) from October 2007 to February 2009 will be discussed. While sampling site CRAES was operated for the first time in October 2007, site CUG has already been in use since 2005. At this long-term measurement site also the passive sampling of particles between 2.5 and 80 μm geometrical diameter (d_g) with the Sigma-2 device was continued for the Olympic Games sampling period. With this comprehensive approach, detailed information about mass and element concentrations, water-soluble ions, BC, and size distribution of APM was obtained. The central questions of this study, which will be discussed later in more detail, were:

- How fast and to what extend was the particulate air pollution reduced?
- Which particle size class was reduced most efficiently?
- To what extend did the weather conditions influence the particle decrease?
- What was the chemical composition of atmospheric particulate matter during the Olympic Games and how did the composition change?
- Which sources can be identified to contribute to the aerosol concentration during the Olympic Games?
- How fast and to what extend did the aerosol pollution increase again after the Olympic Games?

6.2 Mitigation measures

In order to fulfil the air quality commitment made in the Olympic bid, which Beijing won in 2001, Beijing and neighbouring Tianjin (both are Olympic cities) and the surrounding provinces of Hebei, Shanxi, Shandong and the Inner Mongolia Autonomous Region formed an air pollution abatement region to minimize the transport of air pollutants to Beijing. This abatement region geographically covers a distance of 500-1000 km (Fang *et al.*, 2009). In early 2008, the Fourteenth Phase Intensive Control Program went into effect. Fang *et al.* (2009) gave an English summary of this program and its "5655" strategy:

- **Five** schemes to control pollutants associated with *coal combustion*:
  1. Particulate removal, desulfurization and denitrification of coal-fired power plant plumes and decommissioning of hundreds of coal-fired boilers in manufacturing facilities and small power plants
  2. More stringent control regulations for boiler emissions
  3. Conversion of domestic coal use to cleaner fuels in some 50,000 homes
  4. Banning of the ad hoc burning of coal in the city/village boundary areas
  5. Strict control of illegal small boilers and open air burning including barbecues
- **Six** *vehicular emissions* control schemes:
  1. Issuance of new automobile emissions standards
  2. Decommissioning of high emissions vehicles, buses and taxis
  3. Recovery of fuel vapors at pumping stations and from tankers
  4. Restricted use of high emissions vehicles
  5. Banning of large polluting vehicles from the roads
  6. Control of emissions from small stationary diesel generators
- **Five** retrofitting and upgrading of *manufacturing facilities* schemes:
  1. Closing of polluting petrochemical, cement, and other facilities
  2. Control of emissions from facilities not closed
  3. Control of industrial emissions of volatile organic compounds
  4. Abatement of emissions from kitchens
  5. Implementation of more stringent emission regulations
- **Five** schemes to abate *resuspended particulates*:
  1. Strict control of dust emissions from construction sites
  2. Elimination of barren earth by planting of vegetation or paving
  3. Citywide sanitary and cleaning activities by the residents
  4. Management of dust from farms and planting fields in the countryside
  5. Citizen and community participation in environmental projects

The vehicular emission control scheme was implemented successively. During the first stage from the 1st of July to the 20th of July 2008, all vehicles not meeting the European Euro I standard for exhaust emissions were forbidden all-day from the roads (Wang *et al.*, 2010*a*). The "odd-even" ban to halve the vehicle volume was enforced from the 20th of July to the 20th of September 2008. During this time, private vehicles could only operate on odd or even days depending on the last digit of their license plates. Seventy percent of government vehicles were ordered off the road during the event (Zhou *et al.*, 2010). Moreover, since the beginning of 2008, the sulphur content of vehicle fuel has been reduced below 50 ppm (Zhou *et al.*, 2010). Road sweeping and washing were increased to reduce fugitive dust (Fan *et al.*, 2009). Public transportation was promoted by a new fleet of buses, many of them using compressed natural gas (Zhou *et al.*, 2010). Moreover, the subway network was expanded. Subway line 13, Batong line, line 5, line 10, the Olympic line, and, the airport line were put into use so that Beijing's total number of subway lines increased to 8 with a rise in length from 42 km when bidding for the host in 2001 to 200 km in 2008 (Liu *et al.*, 2008). Daily flow of light-duty vehicles (LDV) and heavy-duty trucks (HDT) was reduced by 30% and 54%, respectively, during the Games while the flow of taxis and buses was increased by 30% and 10%, respectively (Zhou *et al.*, 2010).

For industries, the emission standards were also raised. Many factories, especially those that used a lot of coal, were relocated to other provinces and 19 industries around Beijing were ordered in 2007 to cut 2008 emissions by 30% (Stone, 2008). The relocation of heavily polluting enterprises included the Capital Iron and Steel Company and the Yanshan Petrochemical Plant (Yang *et al.*, 2010). Most construction material manufacturers were shut down during the period and all construction activities were temporarily ceased (Wang *et al.*, 2010*a*).

6.3 Results

6.3.1 Total suspended particles

Mass and element concentrations

Results of mass and element concentrations of TSP samples from site CRAES are summarized in Table C.1, Table C.2, and Table C.3, for the sampling periods before (period C1, weekly sampling from Oct-07 – Jul-08), during (period C2, daily sampling from Jul-08 – Sep-08), and after (period C3, weekly sampling from Oct-08 – Feb-09) the Olympic and Paralympic Games. Descriptive statis-

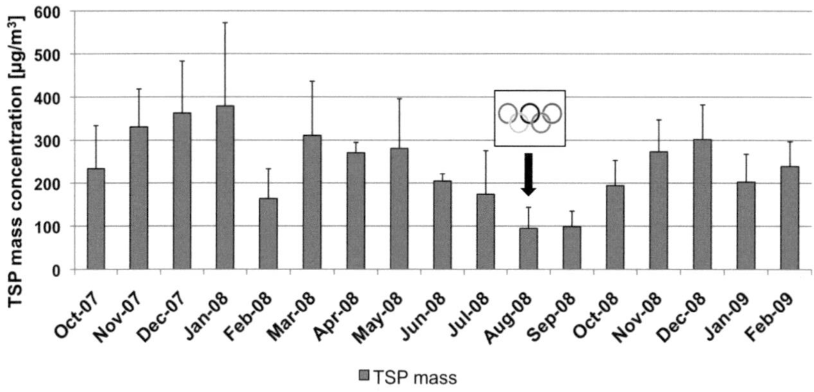

Figure 6.1: Average TSP concentration of each month of sampling period C from October 2007 to February 2009 at site CRAES. Whiskers represent the standard deviation. The Olympic Rings mark the month with the Olympic Summer Games.

tics for the whole Period C are summarized in Table C.4. Besides these data in volume air, the same tables are provided for mass related concentrations in μg/g (Table C.5, Table C.6, Table C.7, and Table C.8).

The average monthly TSP mass concentrations for the whole period C at site CRAES are shown in Figure 6.1. The annual course is characteristic for Beijing with highest concentrations in winter and lowest in summer and is similar to the results of the previous 2-year-long sampling period at site 4 (period A from August 2005 to August 2007, see chapter 4.2.2).

As expected and hoped from the implementation of the mitigation measures, August was the month with lowest TSP concentrations (93$\pm$49 μg/m^3). To what extent meteorological conditions were responsible for the low August values and how much the mitigation measures contributed will be discussed later. The winter concentrations before the Olympic Games were distinctly higher (average of 296$\pm$156 μg/m^3 (N=13) for Dec-07, Jan-08, and Feb-08), compared to the winter concentrations after the Olympic Games (average of 245$\pm$77 μg/m^3 (N=12) for Dec-08, Jan-09, and Feb-09). The difference is even higher if only the averages of the respective December and January are considered (371$\pm$149 μg/m^3 (N=8) for Dec-07 and Jan-08 compared to 247$\pm$85 μg/m^3 (N=9) for Dec-08 and Jan-09).

The monthly average of four selected elements from predominantly geogenic sources (Mg, Ca, Ti, and Fe) and four representative elements from predominantly anthropogenic sources (Zn, As, Cd, and Pb) is shown in Figure 6.2. For all those elements, Aug-08 was the month with lowest concentrations, followed by Sep-08 and Jul-08. In December 2007 highest concentrations for both, geogenic and anthropogenic, elements were measured. However, the deviation from the overall average of all months of the respective elements is higher for the anthropogenic than for geogenic elements (150–175 % compared to 95–110 %, respectively). A wide difference between geogenic and anthropogenic elements becomes obvious in spring 2008. Dust storm events occurred in March and May 2008, which is reflected by relative high concentrations of typical geogenic elements.

TSP concentrations at site CRAES during the Olympic Games (8th – 24th of August 2008) varied from 37 to 194 μg/m^3 with an average value of 78 μg/m^3. For the daily samples, also the course of different elements was considered and for better comparison the element concentrations were standardized using z-transformation (Equation 3.4, see chapter 3.3.2). In Figure 6.3 seven elements (Li, Al, K, Ti, Fe, Rb, and Sr), which had low EFs (Table 6.1) and, therefore, can be interpreted to originate from predominantly geogenic sources, were chosen to illustrate the diurnal course during period C2 with strictly enforced mitigation measures. The same course is shown in Figure 6.4 for seven elements, which had high EFs and, thus, can be interpreted to originate mainly from anthropogenic sources (S, Cu, Zn, As, Cd, Sn, and Pb).

All those element concentrations drop strongly during the first four days after the opening ceremony at the 8th of August and remain quite constant during the following days. Elements from predominantly anthropogenic sources, such as S, Cu, As, or Cd, had a less variable course (Figure 6.4) than elements, which originate mainly from geogenic sources, such as Fe, Al, or Rb (Figure 6.3). Element concentrations increase again at the 20th of August 2008. This concentration peak is visible for geogenic as well as for anthropogenic elements. Few days before, at the 13th of August, there is also a small concentration peak, which is pronounced only for the geogenic elements. During the Paralympic Games, when the mitigation measures were probably not enforced as strictly any more, daily differences of element concentrations are more pronounced than during the Olympic period. Only Cu and Sb concentrations remain quite low and do not vary significantly also during the Paralympic Games in September 2008.

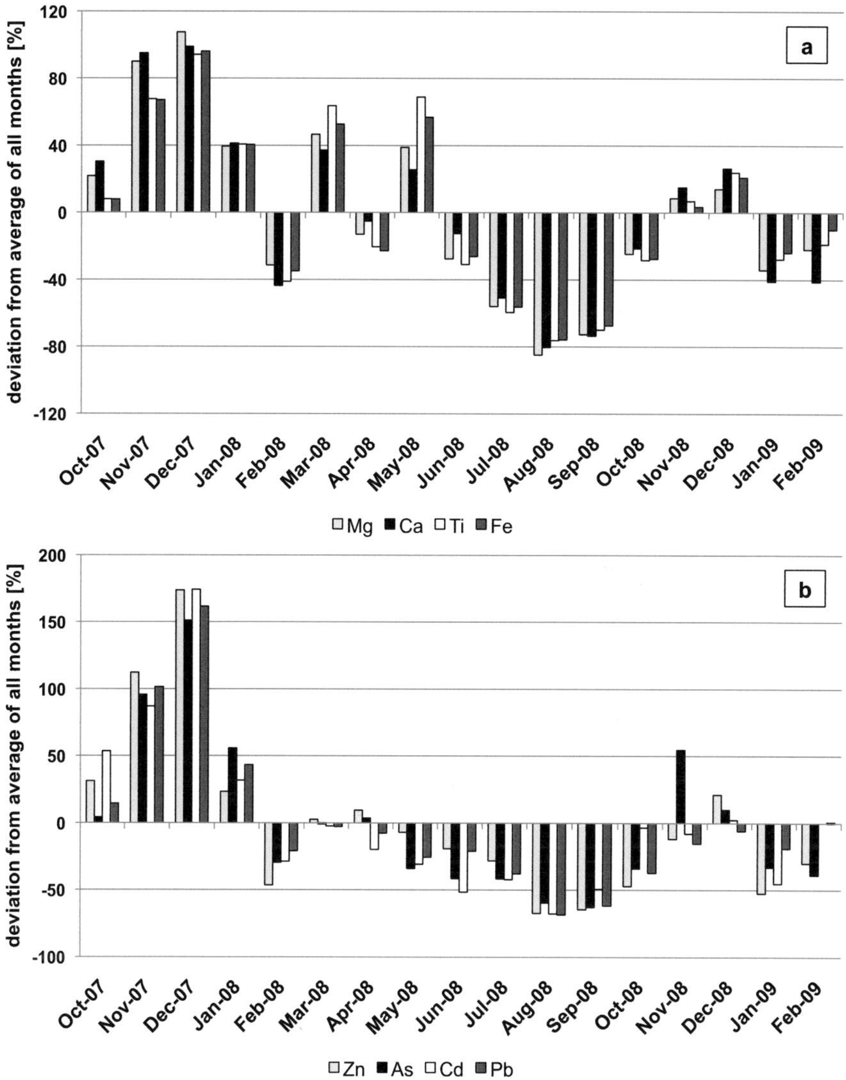

Figure 6.2: Deviation of monthly average concentrations from overall average value of all months of period C for element concentrations of TSP samples from site CRAES. (a) Elements from predominantly geogenic sources (Mg, Ca, Ti, and Fe); (b) elements from predominantly anthropogenic sources (Zn, As, Cd, and Pb).

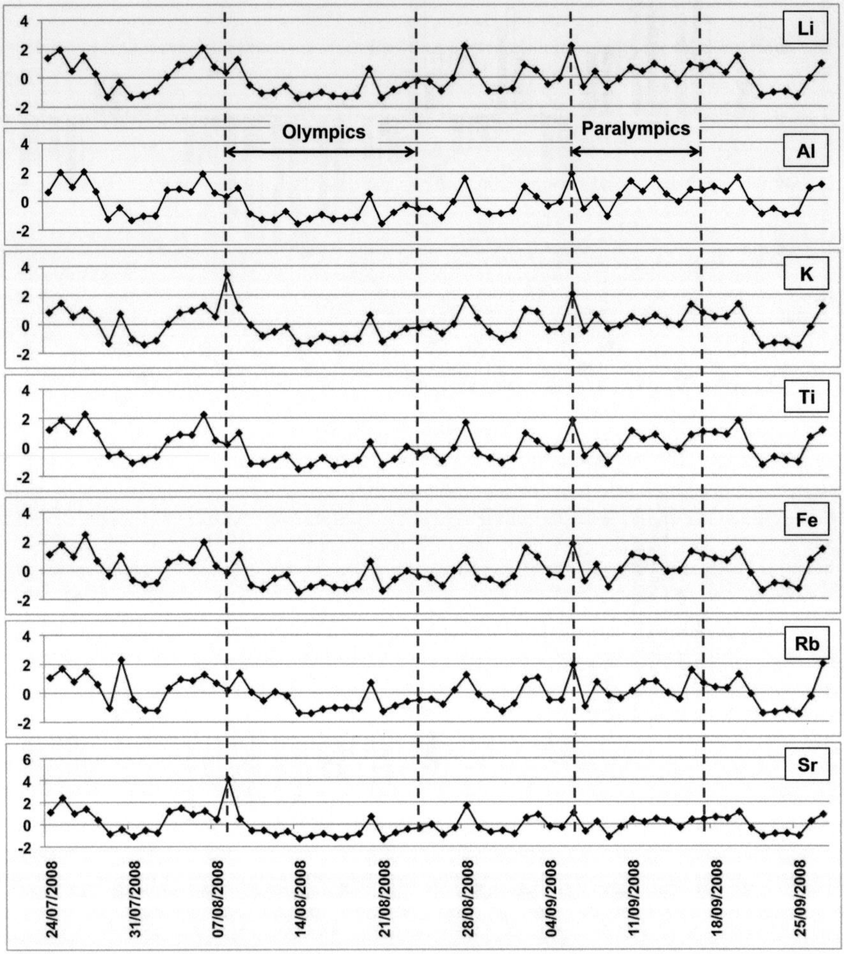

Figure 6.3: Standardized (Equation 3.4) element concentrations of seven selected elements from predominantly geogenic sources (Li, Al, K, Ti, Fe, Rb, and Sr) of daily TSP samples during period C2 at site CRAES.

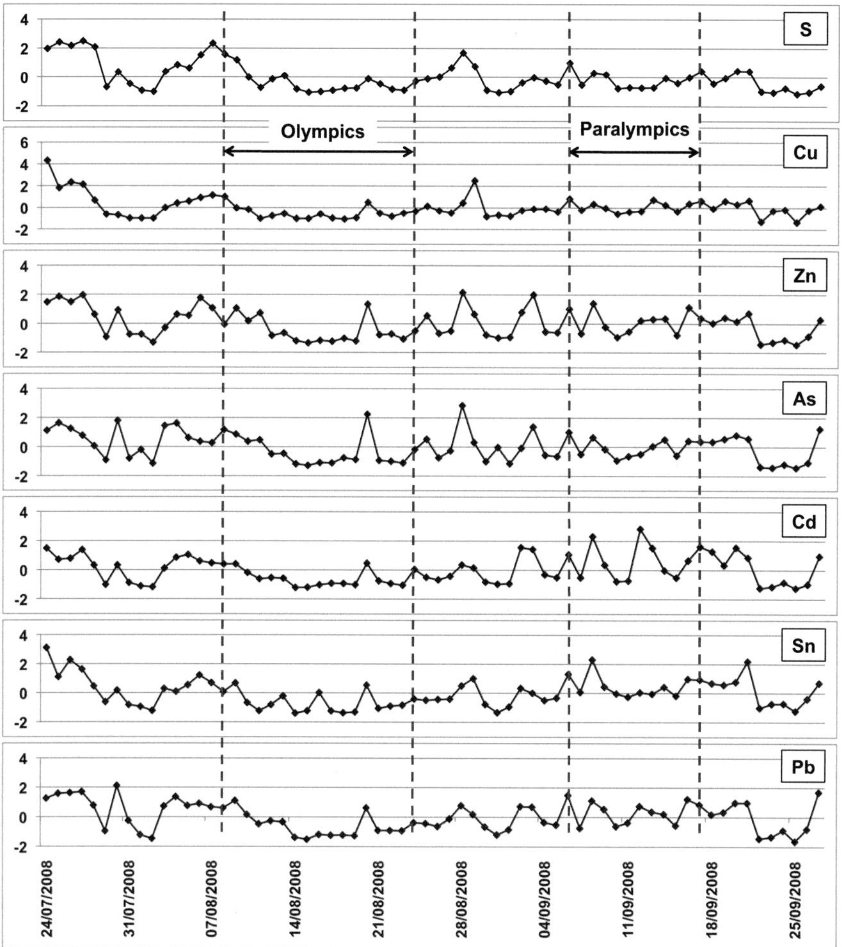

Figure 6.4: Standardized (Equation 3.4) element concentrations of seven selected elements from predominantly anthropogenic sources (S, Cu, Zn, As, Cd, Sn, Pb) of daily TSP samples during period C2 at site CRAES.

Enrichment factors (EFs)

The EFs of TSP element concentrations of the three intervals of sampling period C are summarised in Table 6.1. Elements with high EFs, and, consequently, with an important influence of anthropogenic sources, are S, Cu, Zn, As, Cd, Sn, Sb, and Pb. The EFs calculated with average crustal concentrations and those with concentrations from Chinese soils are similar for Cu, Zn, Cd, and with a little higher deviation also for Pb. For As and Sb, on the contrary, the EFs are significantly lower when calculated with average concentrations of Chinese topsoils. This means that those two elements are already highly enriched in Chinese topsoils. All elements, for which anthropogenic sources played a major role, had highest EFs during the daily sampling period C2. For weekly samples, the EFs after the Olympic Games (C3) were a little higher than before the event (C1).

Water-soluble ions

Additional to total element concentrations, also water-soluble ions were analysed for TSP samples. Nitrate and sulphate were the dominant anions, while fluoride, chloride, and phosphate concentrations were often below the detection limit of IC analysis (LODs see chapter 3.2.3). For cations, which were analysed by ICP-MS, the results of 28 elements (Li, Na, Mg, Al, P, S, K, Ca, Sc, Ti, V, Cr, Mn, Fe, Co, Cu, Zn, Ga, As, Rb, Sr, Y, Cd, Sn, Sb, Cs, Ba, Pb) were included in the study. Descriptive statistics for all water-soluble ions during the Olympic Games period C2 (daily samples) are summarized in Table C.23 as volume-related concentrations and in Table C.24 as mass-related concentrations. The course of seven selected ions (nitrate, sulphate, S, Cu, As, Cd, and Pb), which had high concentrations in the water-soluble fraction, is illustrated in Figure 6.5. For better comparison, the data were standardized using z-transformation (Equation 3.4 in chapter 3.3.2).

During the Olympic Games (8th – 24th of August 2008), sulphate concentrations of TSP samples varied from 4.1 to 57.5 $\mu g/m^3$ with an average value of 19.7 $\mu g/m^3$, while nitrate concentrations ranged from 0.6 to 6.0 $\mu g/m^3$ with an average value of 2.1 $\mu g/m^3$. The nitrate/sulphate ratio during the Olympic event ranged from 0.01 to 0.83 with an average ratio of 0.22. The average ratio for the whole Aug-08 is quite similar with 0.20. As already mentioned in chapter 4.2.3, summer is the season with the lowest nitrate/sulphate ratio in Beijing. The average ratio in Aug-06 and Aug-07 for TSP samples was 0.26 and 0.47, respectively. Consequently, the ratio in Aug-08 was lower compared to the two previous years, indicating a relatively lower importance of nitrate.

Table 6.1: Calculated enrichment factors (EF) referring to Ti for all analysed elements. EFs were calculated using the median concentrations of all TSP samples (N=42 for period C1 (weekly samples), N=66 for period C2 (daily samples), and N=21 for period C3 (weekly samples), respectively). Crustal concentrations (crust) were taken from Reimann & Caritat (1998) and soil concentrations (soil) from Chinese topsoils from Chen *et al.* (2008). For some elements (-), no soil data were available from literature.

	Period C1		Period C2		Period C3	
	EF(crust)	EF(soil)	EF(crust)	EF(soil)	EF(crust)	EF(soil)
Li	4	-	3	-	-	-
Mg	1	4	1	2	1	3
Al	1	-	1	-	1	-
P	2	-	18	-	3	-
S	90	-	274	-	83	-
K	1	-	2	-	2	-
Ca	3	-	2	-	3	-
Sc	1	-	1	-	1	-
Ti	1	1	1	1	1	1
V	1	1	1	1	1	2
Cr	1	2	2	4	1	3
Mn	2	3	3	3	2	2
Fe	1	2	1	2	1	2
Co	1	2	1	2	1	2
Ni	2	3	2	3	2	3
Cu	19	21	30	33	26	28
Zn	71	68	86	82	62	59
Ga	4	-	4	-	7	-
As	97	21	157	33	111	24
Rb	2	-	2	-	2	-
Sr	2	-	2	-	2	-
Cd	267	208	407	317	349	272
Sn	42	-	53	-	43	-
Sb	510	13	770	20	618	16
Cs	5	-	6	-	5	-
Ba	2	3	3	5	2	3
Pb	97	64	138	92	113	75

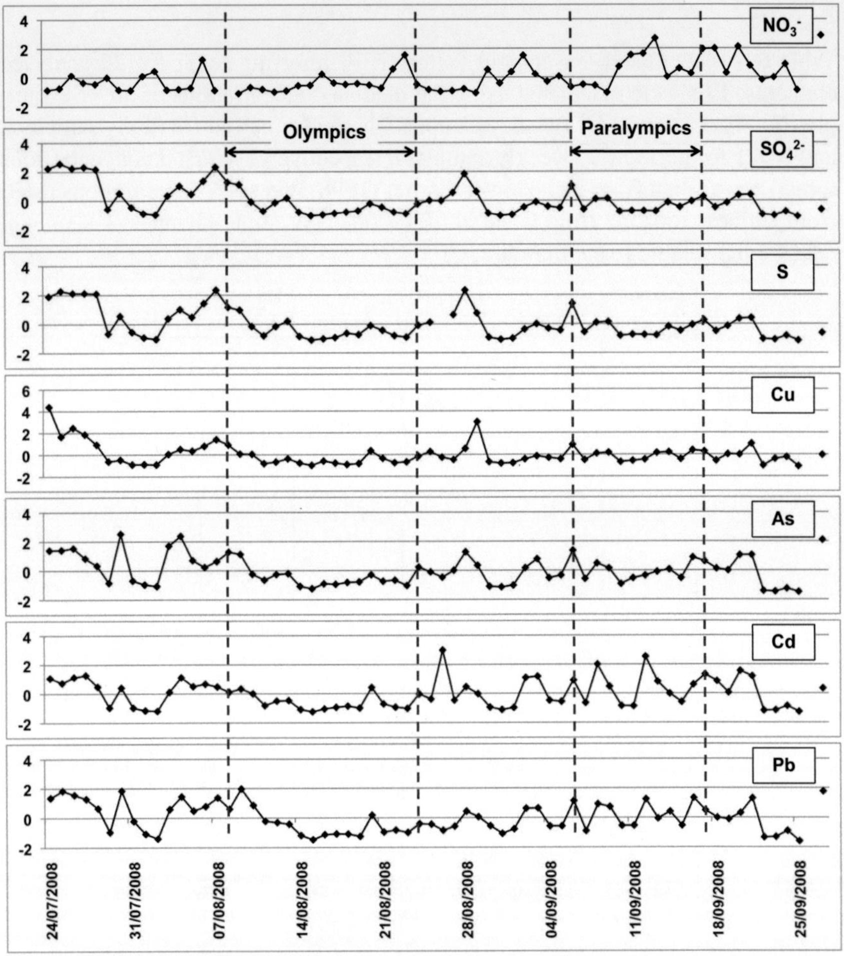

Figure 6.5: Standardized (Equation 3.4) element concentrations of seven selected water-soluble ions from predominantly geogenic sources (Sulphate, Nitrate, S, Cu, As, Cd, and Pb) of daily TSP samples of period C2 at site CRAES.

Table 6.2: Percentage of water-soluble element concentrations from total element concentrations after total digestion of all TSP samples from sampling period C2 (daily samples) at site CRAES.

	N	avg (%)	min (%)	max (%)
Al	63	10	1	35
S	62	106	90	128
K	65	76	41	130
Sc	65	13	5	49
Ti	65	3	1	6
V	65	40	15	82
Cr	60	63	9	144
Mn	65	56	38	83
Fe	65	10	5	23
Co	65	47	19	103
Cu	65	59	34	92
Zn	61	93	32	165
Ga	65	18	6	37
As	63	84	26	131
Rb	65	61	25	92
Sr	65	53	23	96
Cd	63	89	66	115
Sn	64	40	13	70
Sb	64	30	14	87
Cs	62	68	25	111
Ba	65	22	3	61
Pb	65	51	36	73

Elements, which had only low concentrations in the water-soluble fraction were Al, Sc, Ti, Fe, Ga, and Ba (on average $\leq 25\%$). Other elements, such as S, K, As, and Cd, on the contrary, had a high percentage of their total concentrations in the water-soluble share (on average $\geq 75\%$). The relative amount (in %) of the water-soluble concentration from the total concentration is summarised in Table 6.2.

6.3.2 $PM_{2.5}$ and PM_1

Mass and element concentrations

$PM_{2.5}$ samples at site CRAES were collected separately for day- and night-time before, during and after the Olympic Games (periods C1, C2, and C3, respectively). In period C1 and C3, the filters were exposed for a whole week, while in period C2 the samples were collected in 12-hour intervals. Additionally, PM_1

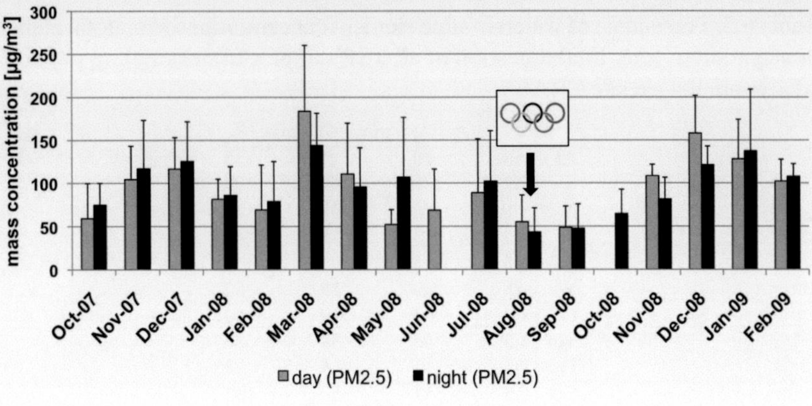

Figure 6.6: Average $PM_{2.5}$ concentration of each month of sampling period C from October 2007 to February 2009 at site CRAES. Whiskers represent the standard deviation. The Olympic Rings mark the month with the Olympic Summer Games.

sampling started for the first time in July 2008 with 24-hour intervals at site CRAES and was continued after the Olympic Games in weekly intervals (Oct-08 to Feb-09). At site CUG, $PM_{2.5}$ samples were collected in 24-hour intervals during period C2 and in weekly intervals in period C1 and C3 before and after the Olympic Games. Here, no distinction between day and night samples was made. During sampling period C2, some sampling failures occurred at both sites. In the following paragraphs, the focus will be on samples from site CRAES.

Descriptive statistics for mass and element concentrations of day-time $PM_{2.5}$ samples from site CRAES are summarized in Table C.9, Table C.10, and Table C.11, for the sampling periods C1, C2, and C3, respectively, as volume-related concentrations in ng/m^3. The corresponding night-time $PM_{2.5}$ results are listed in Table C.12, Table C.13, and Table C.14. Moreover, mass-related concentrations in μg/g are also summarized for the three sampling intervals (Table C.15, Table C.16, and Table C.17 for day-time samples, and Table C.18, Table C.19, and Table C.20 for night-time samples). The monthly average $PM_{2.5}$ concentrations of the whole period C at site CRAES are illustrated in Figure 6.6 for day- and night-time samples, respectively.

This monthly course of $PM_{2.5}$ concentrations is quite similar to the course of TSP concentrations (see Figure 6.1). However, whereas winter TSP concentrations were higher before than after the Olympic Games, this is contrary for

$PM_{2.5}$ samples. Here, the average winter concentrations were higher after than before the Olympic Games. In winter 07/08, the average $PM_{2.5}$ concentrations were 95±45 for day-time (N=13), and 100±43 $\mu g/m^3$ (N=13) for night-time samples (Dec-07, Jan-08, and Feb-08), compared to winter 08/09 (Dec-08, Jan-09, and Feb-09) with concentrations of 132±44 $\mu g/m^3$ (N=12) for day-time, and 125±47 for night-time samples (N=12).

Day-time $PM_{2.5}$ concentrations during the Olympic Games varied from 5.7 to 87.4 $\mu g/m^3$ with an average concentration of 48.0 $\mu g/m^3$ (N=15). Night-time concentrations ranged from 6.6 to 73.0 $\mu g/m^3$ with an average value of 42.5 $\mu g/m^3$ (N=12). For the 12-hour $PM_{2.5}$ samples, the course of different elements was considered and for better comparison the element concentrations were standardized using z-transformation (Equation 3.4 in chapter 3.3.2). In Figure 6.7 seven elements (Li, Al, K, Ti, Fe, Rb, and Sr), which had low EFs (Table 6.3) and are known to originate from predominantly geogenic sources, were chosen. The same course is shown in Figure 6.8 for seven elements, which had high EFs (Table 6.3) and originate mainly from anthropogenic sources (S, Cu, Zn, As, Cd, Sn, and Pb). It is noticeable, that especially for the anthropogenic elements, the course is a lot less variable during the Olympic Games. This is the case especially for S, As, and Cd, and with the exception of one peak concentration (19th of August) also for Cu. During the Paralympic Games, on the contrary, the daily variations are a lot more pronounced.

PM_1 concentrations during the Olympic Games ranged from 9.6 to 55.7 $\mu g/m^3$ with an average concentration of 27.1 $\mu g/m^3$ (N=13). The average mass concentration for the whole August was a little higher with 38.8 $\mu g/m^3$ (N=24). Descriptive statistics for mass and element concentrations of 24-hour PM_1 samples is summarized in Table C.21 as volume-related concentrations in ng/m^3 and in Table C.22 as mass-related concentration in $\mu g/g$.

Enrichment factors

The EFs of $PM_{2.5}$ and PM_1 element concentrations of sampling period C2 are summarised in Table 6.3. Elements with very high EFs, and, consequently, an important influence of anthropogenic sources, are S, Cu, Zn, As, Cd, Sn, Sb, and Pb. These are the same elements, which had also high EFs for TSP samples (Table 6.1). To a lower extend, Ni, Cs, and Ba showed EFs, which also indicate an influence of anthropogenic sources. For most elements, the EFs of $PM_{2.5}$ and PM_1 concentrations are quite similar, indicating a comparable anthropogenic influence for both size classes.

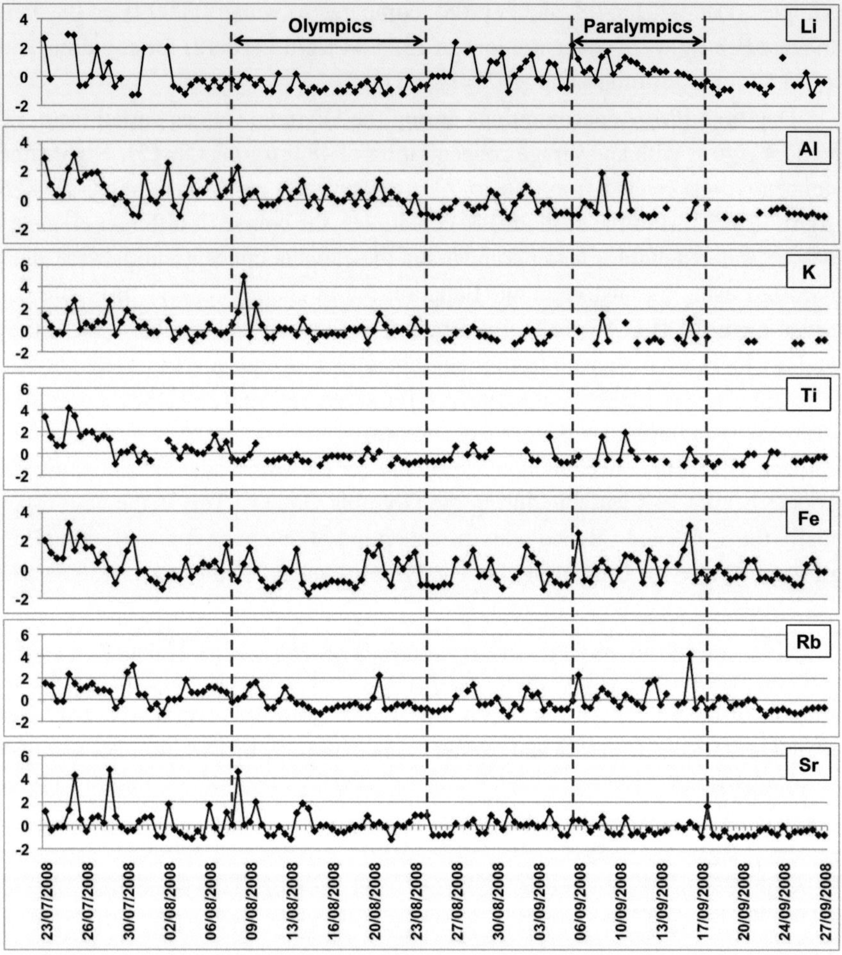

Figure 6.7: Standardized (Equation 3.4) element concentrations of seven selected elements from predominantly geogenic sources (Li, Al, K, Ca, Ti, and Fe) of 12-hour $PM_{2.5}$ samples from period C2 at site CRAES.

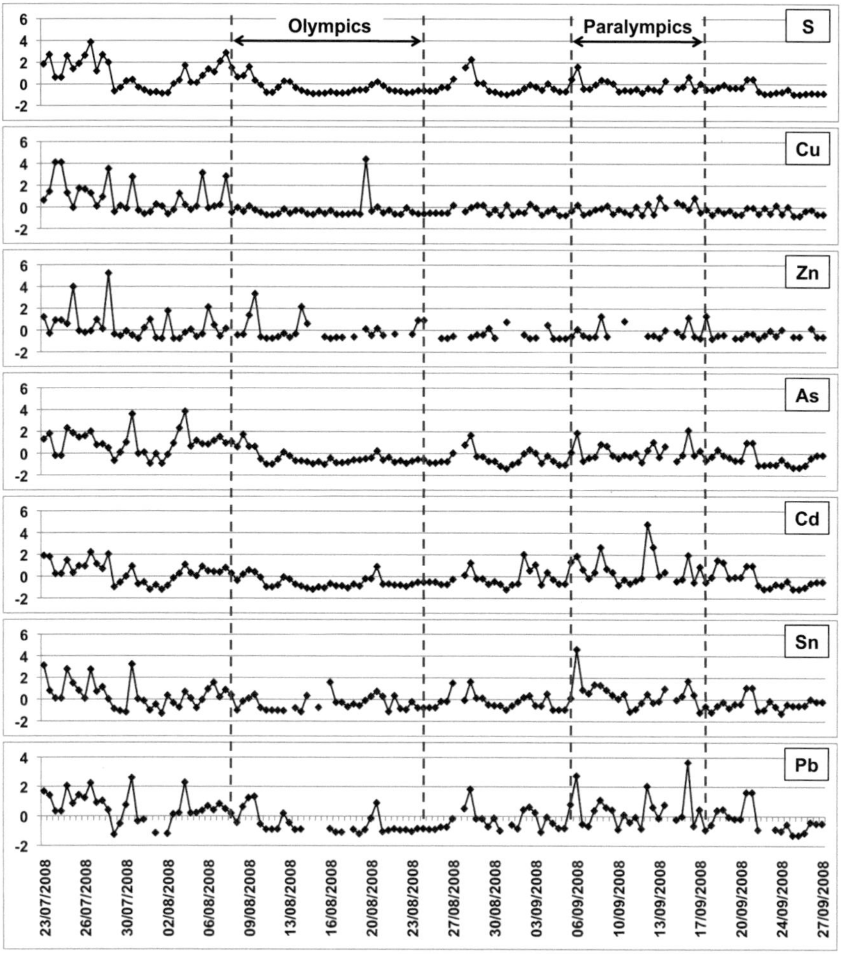

Figure 6.8: Standardized (Equation 3.4) element concentrations of seven selected elements from predominantly anthropogenic sources (S, Cu, Zn, As, Cd, Sn, Pb) of 12-hour $PM_{2.5}$ samples from period C2 at site CRAES.

Table 6.3: Calculated enrichment factors (EF) referring to Ti for all analysed elements. EFs were calculated using the median concentrations of all $PM_{2.5}$ and PM_1 samples of period C2. Crustal concentrations (crust) were taken from Reimann & Caritat (1998) and soil concentrations (soil) from Chinese topsoils from Chen *et al.* (2008). For some elements (-), no soil data were available from literature.

	$PM_{2.5}$		PM_1	
	EF(crust)	EF(soil)	EF(crust)	EF(soil)
Li	30	-	22	-
Na	73	-	91	-
Mg	14	37	7	19
Al	7	-	5	-
S	2111	-	1655	-
K	14	-	10	-
Ca	25	-	39	-
Sc	1	-	1	-
Ti	1	1	1	1
V	3	4	3	3
Cr	26	49	25	47
Mn	10	12	8	10
Fe	3	4	2	2
Co	6	10	5	8
Ni	35	66	53	99
Cu	339	370	265	289
Zn	886	847	1165	1113
Ga	35	-	37	-
As	1201	255	1250	266
Rb	14	-	11	-
Sr	13	-	11	-
Cd	3347	2606	2675	2083
Sn	398	-	367	-
Sb	2279	59	7451	194
Cs	79	-	72	-
Ba	70	105	37	55
Pb	1006	668	787	523

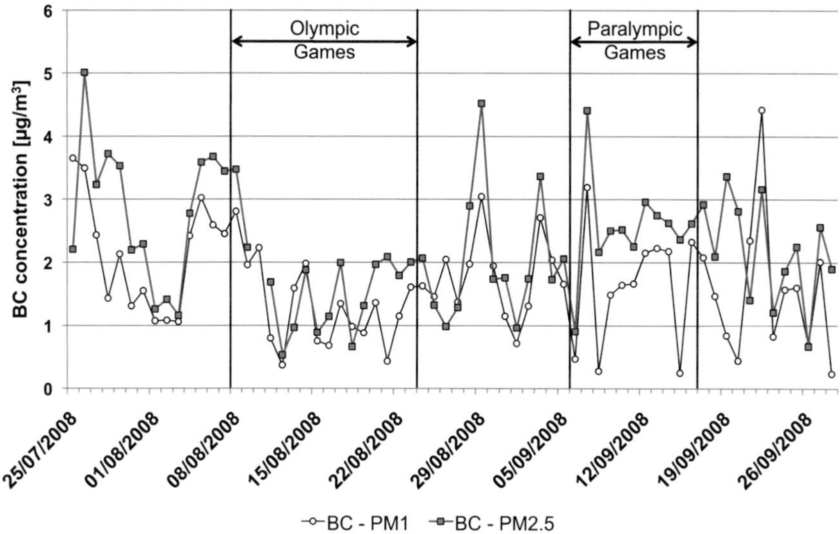

Figure 6.9: BC concentrations for 24-hour PM_1 and $PM_{2.5}$ samples from site CRAES from the 25th of July to the 28th of September 2008 (period C2).

Black carbon

Figure 6.9 shows the course of BC concentrations for 24-hour PM_1 and $PM_{2.5}$ samples from site CRAES (for $PM_{2.5}$ samples, the 24-hour-average was calculated from the 12-hour day and night samples). Black carbon concentrations decrease during the first days of the Olympic Games and remain quite low for the rest of this period. The course is quite similar for BC concentrations of both particle size classes (Pearson correlation coefficient of 0.65). A high share of BC concentration of $PM_{2.5}$ is presented in the small fraction (PM_1). In average, BC concentrations of PM_1 samples constitutes 67% of BC concentrations of $PM_{2.5}$ samples (N=53, the portion ranges from 11 to 91%).

Descriptive statistics for sampling period C2 (24th of July – 29th of September 2008) are summarised in Table C.25. Black carbon concentrations during this period ranged from 0.24 to 4.42 $\mu g/m^3$ with an average value of 1.67 $\mu g/m^3$ for PM_1 and from 0.53 to 6.21 $\mu g/m^3$ with an average value of 2.26 $\mu g/m^3$ for $PM_{2.5}$ samples. During the Olympic Games (8th – 24th of August), average BC concentrations were 1.4±0.7 for day-time $PM_{2.5}$, 1.7±0.7 for night-time $PM_{2.5}$, and 1.2±0.6 $\mu g/m^3$ for 24-hour PM_1 samples. Average BC concentrations of all 12-hour $PM_{2.5}$ samples were 2.1±1.1 for day and 2.4±1.2 $\mu g/m^3$ for night samples. Night-time BC concentrations were higher in 67% of the cases (N=54) than those of the respective day samples.

At site CUG, the BC concentration of 24-hour $PM_{2.5}$ samples range from 0.21 to 5.76 $\mu g/m^3$ with an average value of 2.33 $\mu g/m^3$ (sampling period C2). Descriptive statistics are also included in Table C.25.

6.3.3 Passively sampled atmospheric particulate matter

Descriptive statistics of passively collected $APM_{2.5-80}$ at site CUG from the 21st of July to the 26th of September 2008 are summarised in Table C.26. Transparent particles between 2.5 and 5 (size class one = SC1) and between 5 and 10 μm d_g (sice class two = SC2) showed a similar course in August and September 2008 (Figure 6.10a). Mass concentrations of coarser particles (size class three = SC3: 10 – 20 μm d_g; size class four = SC4: 20 – 40 μm d_g) are shown in Figure 6.10b.

Mass concentrations during the Olympic Games varied from 7.2 to 85.1 $\mu g/m^3$ with an average value of 21.3 $\mu g/m^3$ for SC1 and from 8.4 to 86.4 $\mu g/m^3$ with an average value of 22.6 $\mu g/m^3$ for SC2. Average concentrations for SC3 and SC4 were 17.2 $\mu g/m^3$ (range: 6.9 – 52.9 $\mu g/m^3$) and 5.2 $\mu g/m^3$ (range: 1.5 – 11.0 $\mu g/m^3$), respectively. At the 8th of August, the day of the opening ceremony of the Olympic Games, as well as at the following day, APM concentrations were particularly high (165 and 235 $\mu g/m^3$ for particles between 2.5 and 80 μm d_g). In the following days, the concentrations decreased significantly.

In Figure 6.11, the size distribution of the 9th of August, 2008, the day with highest APM concentrations in August, is compared to the size distribution of the following day, when APM concentrations were 78% lower. It is noticeable, that the transparent particle concentration of the smallest particle size class (2.5 – 5 μm d_g) decreased most strongly during this two particular days.

6.4 Discussion

6.4.1 Evaluation of the amount of particle reduction in August 2008

Comparison of the Olympic source control period in August 2008 with the pre-Olympic period

The measured $PM_{2.5}$ concentrations at site CRAES during the Olympic Games (8th–24th of August) of 48±25 $\mu g/m^3$ for day-time, and 42±22 $\mu g/m^3$ for night-

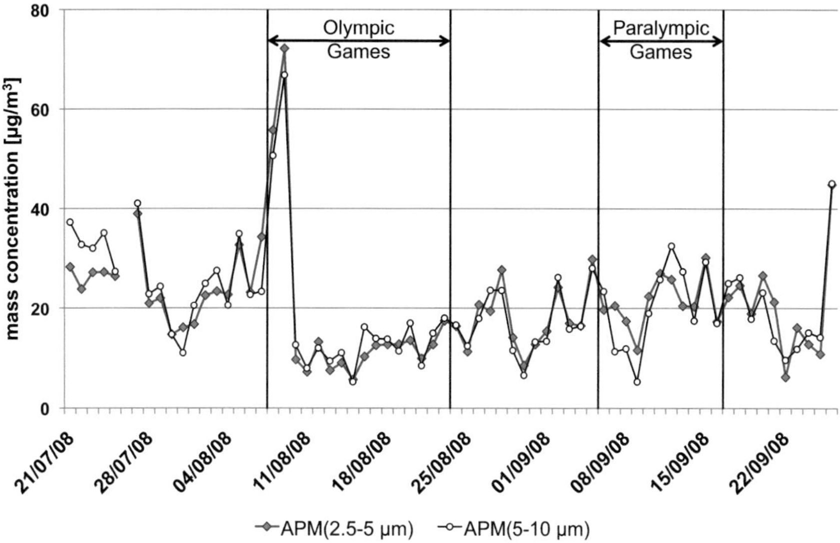

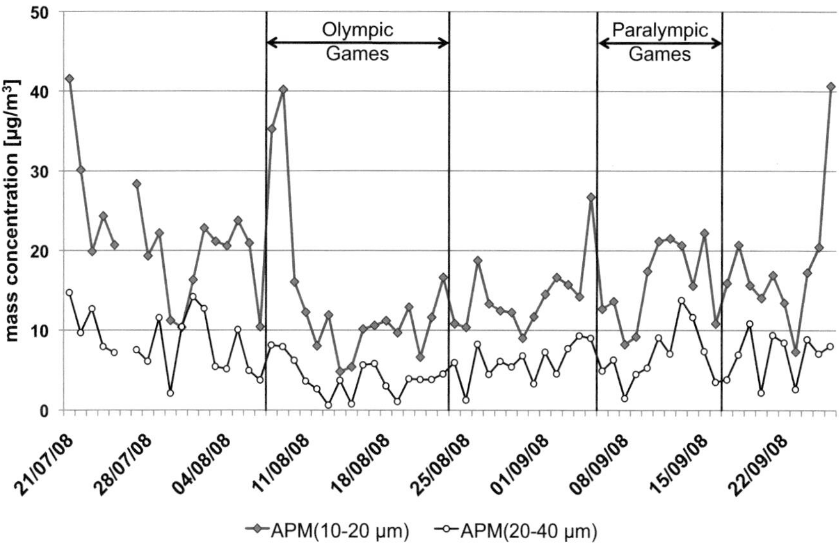

Figure 6.10: Daily concentrations of transparent APM samples from July to September 2008 (period C2) at site CUG. Top: Particle size class SC1 (2.5 – 5 μm d_g) and SC2 (5 – 10 μm d_g), bottom: particle size class SC3: 10 – 20 μm d_g) and SC4 (20 – 40 μm d_g).

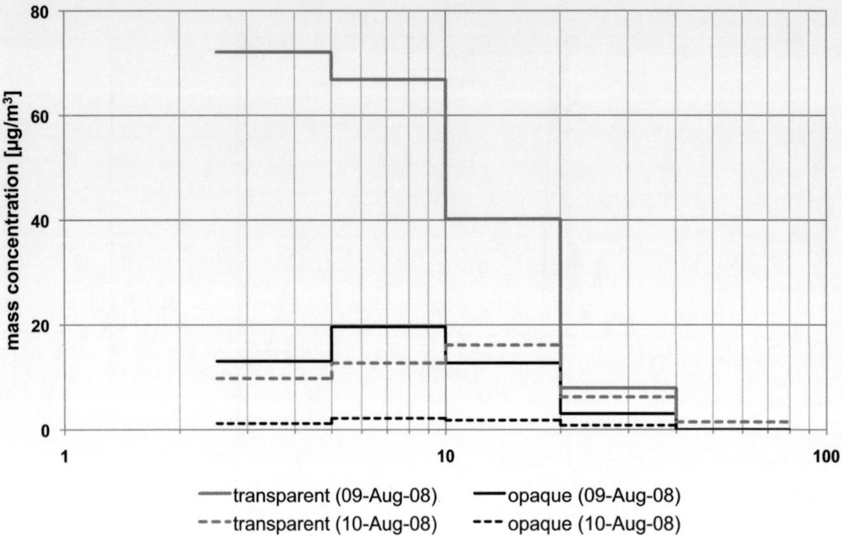

Figure 6.11: Size distribution for transparent and opaque particles at site CUG on the 9th and 10th of August, respectively.

time samples, respectively, are in line with concentrations reported by other studies. For example, Xin *et al.* (2010) measured average $PM_{2.5}$ concentrations in Beijing of 42 ± 31 $\mu g/m^3$ for the same time interval. The authors concluded, that $PM_{2.5}$ mass decreased by 52% if compared to the pre-Olympic period (1st of June–7th of August). In this study, the $PM_{2.5}$ concentrations during the Olympic Games were reduced by 48% for day-time and 46% for night-time samples at site CRAES if compared to the average concentrations of 12-hour samples before the Olympic Games (23rd of June–07th of August). For 24-hour TSP concentrations at site CRAES during the same time intervals the reduction was 49%. To what extent the mitigation measures were responsible for this reduction and what influence the weather conditions had, will be discussed later in sections 6.4.2 and 6.4.3.

Comparison of the Olympic source control period in August 2008 with the previous Augusts

For a comprehensive evaluation of the amount of air pollution reduction, the mass and element concentrations of August 2008 were compared not only to the periods directly before and after the Olympic Games, but also to those concentrations of the same time period during the previous years. For TSP samples, it is complicated to compare the values of Aug-06 and Aug-07 with those of Aug-08, because the sampling site was relocated in Oct-07 from site 4 to site CRAES (see also Figure 3.1). The same problem occurs with $PM_{2.5}$ samples from site CRAES. Therefore, the August concentrations from site CRAES can only be compared to those from the other sampling locations (e.g. site 4 during sampling period A), which operated during the previous years. PM_1 samples were not collected at all during the previous years.

At site CUG, concentrations of actively collected $PM_{2.5}$ samples and passively collected APM samples of different size classes (from 2.5 to 80 μm d_g) are available from Aug-05 till Aug-09 for comparison. Nevertheless, it should be pointed out that for both methods, the sampling intervals were different in 2008 due to a higher time-resolution during the Olympic Games period. Consequently, a certain influence of the exposure time might also contribute to differences in the measured particle concentrations. However, compared to the overall differences in concentration this effect should only play a minor role.

In Figure 6.12, the average TSP and $PM_{2.5}$ of site CRAES in Aug-08 are compared to those of site 4 in Aug-06 and Aug-07. However, it is impossible to quantify the exact effect of the relocation of the sampling site on the particle reduction. An approach for a rough estimation would be the comparison between site 4 and site 2, because the latter site is also on a roof like site CRAES and in a predominantly residential area. The difference between average $PM_{2.5}$ concentrations of site 2 and 4 in Aug-06 was only 2% for day-time samples and 1% for night-time samples in Aug-07. The differences for night-time samples in Aug-06, and for day-time samples in Aug-07 were higher (+22 and -42%, respectively), however, this higher deviation can be explained by missing data and, therefore, can be neglected. Therefore, the August values of site 4 and site CRAES should also be comparable.

The particle concentration in Aug-08 was reduced by 63% for TSP, and by 15 and 51% for day-time and night-time $PM_{2.5}$, respectively, compared to the average concentration of the previous two Augusts (average of Aug-06 and Aug-07) at site 4. The wide difference of the reduction proportion of day- and night-time $PM_{2.5}$ concentrations is noticeable. This observation shows that sources present during day-time were more difficult to control than those during night-time. One possible explanation is the higher traffic volume during day-time with all the additional "Olympic traffic" of athletes and visitors travelling to the different Olympic venues. Moreover, in 2006 and 2007, the exclusive night-time

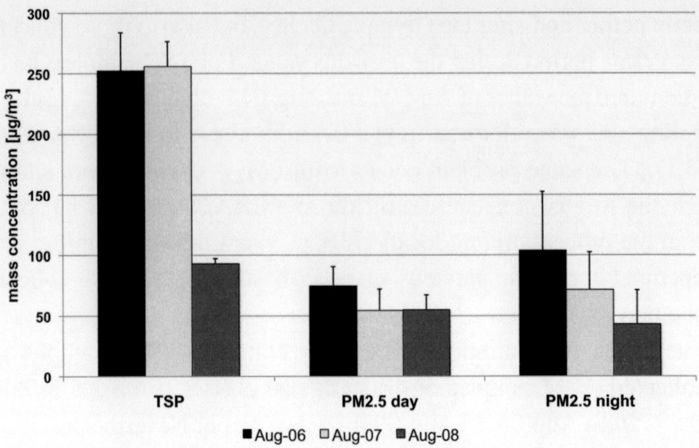

Figure 6.12: Average TSP and $PM_{2.5}$ concentrations of each August of the years 2006 to 2008. Whisker represent the standard deviation. Samples in Aug-06 and Aug-07 (N=4, weekly sampling) were from site 4, whereas samples in Aug-08 (N=31, daily sampling) were collected at site CRAES.

traffic of heavy-duty vehicles contributed to high night-time concentrations of $PM_{2.5}$ mass and also BC concentrations (see also chapter 4.4.2), which was not allowed during the Olympic Games. The predominant sources contributing to the aerosol concentrations during the Olympic Games will be discussed in more detail in section 6.4.3.

With the objective to evaluate, which particle size class was reduced most efficiently, the size distribution for the passive collected APM samples at site CUG are compared in Figure 6.13 for August 2005, 2006, 2007, and 2008.

The concentrations in Aug-08 were lower than those of the previous years for all size classes (SC). The reduction was quite similar for SC4, SC3, and SC2 with a reduction of 51, 48, and 45%, respectively, if compared to Aug-07 and a reduction of 57, 51, and 44%, respectively, if compared to the average value of Aug-05, Aug-06, and Aug-07. The finest particles in SC1 were reduced by 36% if compared to Aug-07, and 34% if compared to the average of all three previous Augusts. Consequently, the smallest size fraction between 2.5 and 5 μm d_g was reduced less than coarser particles during the source control period in August 2008. This might be due to the fact, that the long-range transport plays a more important role for this particle size fraction, because the residence time of finer particles in the atmosphere is longer (Seinfeld & Pandis, 2006) and, thus,

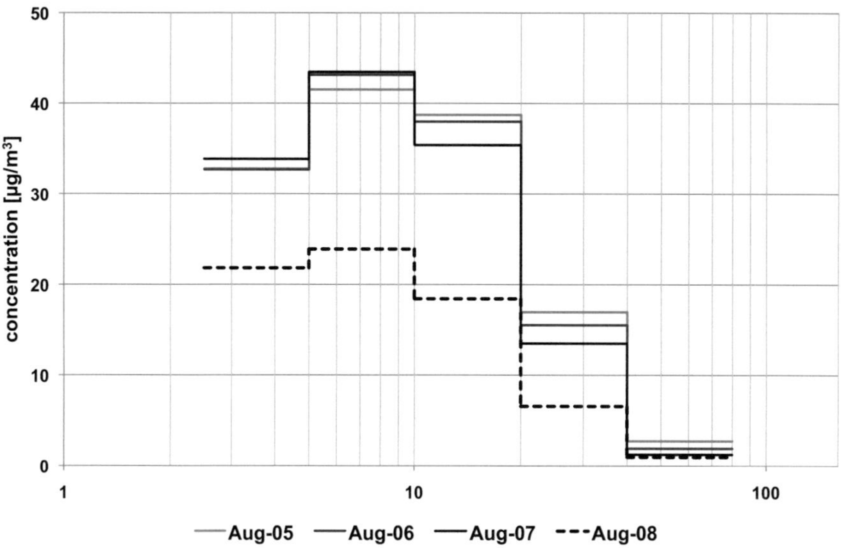

Figure 6.13: Size distribution of average APM concentrations of each August of the years 2005 to 2008.

sources out-side the abatement region around Beijing have a relatively higher contribution.

6.4.2 Evaluation of the influence of meteorological conditions on particle reduction

In order to evaluate the influence of meteorological conditions on the particle concentrations in August 2008 compared to the previous years, the meteorological parameters for each August of the years 2005–2009 are listed in Table 6.4.

The average temperature was similar for the five years and, therefore, should not have an impact on any particle mass differences between the distinct Augusts. Within these five years, maximum precipitation occurred in Aug-08. The total amount of precipitation in Aug-08 was most similar in Aug-05. Besides the total amount, also the number of days with a high amount of precipitation is of special interest, which are also included in Table 6.4. In order to illustrate the influence of weather conditions on the particle mass concentrations, the course of TSP mass concentration, together with meteorological parameters (rainfall, wind

Table 6.4: Meteorological parameters (temperature, precipitation, and wind speed) in August for the years 2005 – 2009. Weather data were obtained from the international weather station at the airport (WMO code: 54511).

		Aug-05	Aug-06	Aug-07	Aug-08	Aug-09
Temp	avg total [°C]	26.0	26.5	26.7	26.0	25.8
	avg day [°C]	26.0	26.4	26.6	25.9	25.6
	avg night [°C]	26.0	26.7	26.8	26.0	26.0
Precip	sum [mm]	127	50	99	135	61
Wind v	avg [m/s]	2.0	1.9	2.1	1.9	1.9
		Number of days in				
		Aug-05	Aug-06	Aug-07	Aug-08	Aug-09
Precip	0 [mm]	18	16	19	21	18
	<1 [mm]	5	7	8	2	8
	[1–5) [mm]	4	6	0	0	2
	[5–10) [mm]	0	0	2	3	0
	[10–20) [mm]	1	2	0	2	2
	[20–40) [mm]	2	0	1	3	1
	$\geq$ 40 [mm]	1	0	1	0	0
Wind v	< 1 [m/s]	0	0	0	0	0
	[1.0–1.5) [m/s]	4	3	1	4	6
	[1.5–2.0) [m/s]	15	18	15	11	11
	[2.0–2.5) [m/s]	9	8	9	11	8
	[2.5–3.0) [m/s]	2	1	5	5	3
	$\geq$ 3 [m/s]	1	1	1	0	2

speed, predominant wind direction), for the 24-hour samples from the 24th of July to the 28th of September 2008 is shown in Figure 6.14.

Rainfall started at the 9th of August, one day after the Opening Ceremony of the Olympic Games and continued also for the following two days. At the 10th of August almost 50 mm of precipitation was measured and, hence, this was the day with most precipitation of the whole August. A total of six days with precipitation > 1 mm was recorded during the Olympic Games. Consequently, the importance of wet deposition for particle reductions during the first days of the Olympic Games seems to be high, which is reflected by the strong decrease in TSP mass the day after a considerable amount of precipitation occurred (see Figure 6.14a).

The course of the nitrate/sulphate ratio is another indication for the influence of precipitation not only on particle concentration, but also on the composition of APM. During the sampling period C2 (24-hour), the ratio was 0.11 on average for the days with more than 5 mm precipitation (N=15) and 0.27 for those days without any precipitation (N=41).

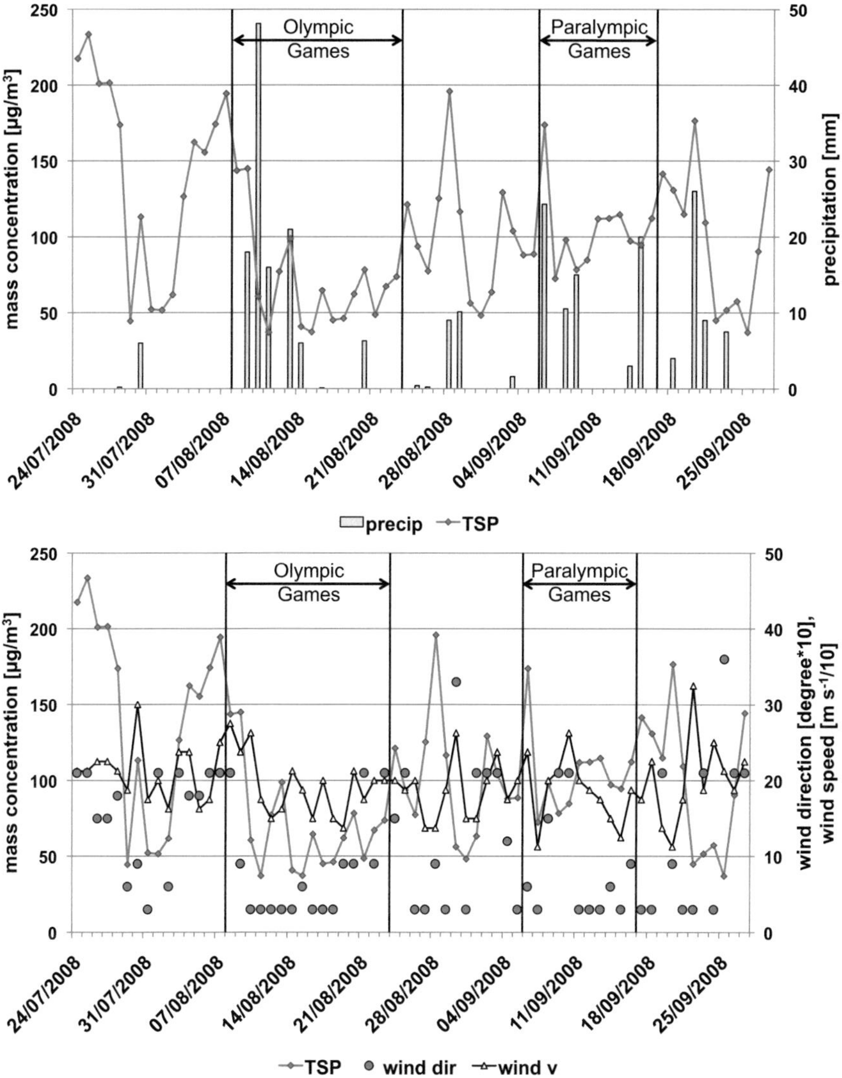

Figure 6.14: TSP concentration of the daily samples from the 24th of July to the 29th of September 2008 (period C2) at site CRAES together with (top) the amount of precipitation, as well as (bottom) wind speed and predominant wind direction.

Also the wind speed is known to influence the particle concentrations. During days with low wind speeds, APM accumulates and mass concentrations are high. An example for such a case was on the 27th and 28th of August, when the average wind speed was relative low (1.4 m/s), and the TSP mass concentration increased to 125 μg/m^3 on the 27th and even 195 μg/m^3 at the 28th of August (see Figure 6.14b). At very high wind speeds, on the contrary, especially the coarse particle fraction can show high concentrations due to resuspension processes. The concentration peak on the 30th of July might be an example for such a case. On this day, the average wind speed was with 3 m/s considerably higher than during the previous and following day (1.9 and 1.8 m/s, respectively), which might explain the more than 2-fold higher concentrations at the 30th of July (113 μg/m^3 compared to 44 μg/m^3 at the 29th and 52 μg/m^3 at the 31st of July, respectively). In Aug-08, however, no day with average wind speeds $>$ 3 m/s occurred.

6.4.3 Evaluation of the sources still contributing to particulate air pollution during the Olympic Games

With the objective to gain more detailed knowledge about the most important sources and their impact during the Olympic Games period, factor analysis (FA) was carried out. With this multivariate statistical approach it is possible to reduce the number of variables and to detect structures in the relationship between variables and, thus, to group element associations probably indicating an origin from certain sources. This was done separately for TSP and PM_1 samples during the Olympic Games period (sampling period C2) in order to gain further information about the sources responsible for particles of different size classes. Furthermore, the sources for water-soluble ion concentrations were investigated in the same way, because the mobile elements and their sources are of special interest with regard to negative health impacts due to their high bioavailability.

Source apportionment for total suspended particulates during the Olympic Games

First, FA was carried out for 24-hour TSP samples from site CRAES. Here, seven factors were extracted, which accounted for 78% of the Explained Variance (Expl. Var.) of the whole data set. The factor loadings are listed in Table 6.5 and the corresponding factor scores can be found in the appendix in Table C.27.

Factor 1 comprises numerous elements with high positive loadings (Table 6.5) and accounts for 19% of the Expl. Var. These elements are Cs, Rb, Pb, Cd,

Table 6.5: Factor (Fa) loadings of the seven extracted factors (Principal Component Analysis, substitution of missing data by average value, Varimax standardized rotation) for TSP samples (element concentrations of 24-hour samples in μg/g), together with the respective communalities (comm). Meteorological parameters (visibility, dew point, temperature, wind speed, and predominant wind direction) are included. Loadings $\geq \pm 0.55$ are marked in bold.

	Fa1	Fa2	Fa3	Fa4	Fa5	Fa6	Fa7	comm
mass	-0.06	0.31	**0.79**	0.06	-0.11	-0.17	0.15	0.90
Mg	0.12	**0.84**	-0.21	0.20	-0.03	0.02	-0.19	0.89
Al	0.10	**0.90**	-0.07	0.14	-0.01	-0.07	0.11	0.90
S	0.33	-0.35	**0.78**	0.18	0.05	-0.12	0.07	0.98
K	**0.68**	0.32	0.09	0.01	0.35	-0.04	-0.05	0.91
Ca	0.10	**0.74**	-0.26	0.04	0.11	-0.08	0.08	0.75
Sc	-0.05	**0.85**	-0.04	-0.07	0.01	0.37	0.15	0.94
Ti	0.01	**0.64**	-0.36	0.11	0.25	0.50	-0.01	0.94
V	0.48	0.51	0.18	-0.04	0.05	0.42	0.14	0.82
Cr	0.02	-0.01	-0.10	0.30	-0.09	**0.71**	0.03	0.68
Mn	**0.55**	0.45	-0.22	-0.07	-0.13	0.47	-0.02	0.86
Fe	0.30	0.54	-0.25	-0.04	0.05	**0.69**	0.03	0.95
Co	0.26	0.00	-0.23	-0.04	0.27	**0.62**	-0.28	0.80
Cu	0.12	0.00	0.21	**0.78**	0.12	0.00	-0.04	0.80
Zn	**0.68**	0.00	0.19	0.13	**0.59**	0.01	0.07	0.97
Ga	**0.64**	0.14	0.20	0.39	0.03	0.21	-0.26	0.92
As	**0.60**	-0.11	0.15	0.06	**0.68**	0.03	-0.04	0.95
Rb	**0.91**	0.08	0.05	-0.11	0.17	0.21	0.04	0.98
Sr	0.16	0.46	0.01	0.02	**0.73**	0.09	0.16	0.92
Cd	**0.70**	0.18	0.10	0.26	0.08	-0.04	-0.02	0.76
Sn	0.41	0.11	0.17	**0.66**	-0.18	0.16	0.07	0.72
Sb	-0.13	0.14	-0.10	**0.82**	0.10	0.12	-0.06	0.82
Cs	**0.92**	0.03	0.10	-0.13	0.15	0.09	0.09	0.98
Ba	0.27	-0.04	-0.12	0.02	**0.91**	0.00	-0.07	0.94
Pb	**0.90**	-0.10	0.19	0.17	0.14	0.15	0.10	0.96
vis	-0.18	0.06	**-0.81**	-0.33	0.17	0.21	0.06	0.92
dew pt	0.19	-0.25	**0.81**	-0.12	0.14	0.11	-0.27	0.96
temp	-0.13	-0.15	**0.72**	-0.50	0.17	0.17	-0.06	0.96
precip	0.53	-0.05	0.07	0.07	0.02	-0.35	-0.29	0.65
wind v	0.23	-0.05	-0.01	-0.12	-0.09	-0.06	**0.75**	0.57
wind dir	-0.23	0.21	0.08	0.08	0.15	0.01	**0.74**	0.58
Nitrate	-0.15	0.24	**-0.73**	-0.02	-0.01	0.29	-0.11	0.86
Sulphate	0.28	-0.35	**0.77**	0.16	0.03	-0.09	0.13	0.97
% of Expl. Var.	19	15	15	8	8	8	5	

K, Zn, Ga, Mn. Moreover, precipitation is also included in this factor with a lower positive loading. On the one hand, some of these elements, such as Pb, Cd, and Zn had high EFs, whereas Cs, Rb, K, Ga, Mn, on the other hand, had low EFs (Table 6.1). Therefore, this factor can be interpreted to represent mixed geogenic-anthropogenic pollution. Typical anthropogenic sources for Pb, Cd, and Zn are coal combustion and traffic (Reimann & Caritat, 1998). Rubidium, K, and Ga can also be emitted during coal combustion. Possible geogenic sources of these elements are various minerals, such as silicates. The higher factor scores (Table C.27) on days with precipitation indicate that the elements of this association were relatively stronger reduced by wet deposition.

Aluminium, Sc, Mg, Ca, Ti have high loadings in **factor 2** (Table 6.5). All these elements have low EFs (Table 6.1) and, therefore, geogenic sources seem to prevail for this element association. Consequently, factor 2 can be interpreted to represent predominantly geogenic influences. Typical host minerals for these elements are silicates, carbonates, and oxides. Possible anthropogenic sources contributing additionally to the element association of this factor might be construction activities, because of the occurrence of these elements in building materials, such as concrete, mortar, lime or bricks (Nath *et al.*, 2007).

Factor 3 has high positive loadings for TSP mass, S, and sulphate and a negative loading for nitrate (Table 6.5). Moreover, the meteorological parameters dew point and temperature are positively and visibility is negatively correlated with this factor. Sulphate and nitrate are usually formed by chemical reactions resulting in gas-to-particle conversion in the atmosphere (Seinfeld & Pandis, 2006) and are, consequently, often present as secondary aerosols. Therefore, factor 3 can be interpreted as diffuse pollution factor and secondary particles seem to dominate. The formation of secondary aerosols is favoured by high temperature and humidity (Yao *et al.*, 2002), which explains the positive correlation of this factor with temperature and dew point.

High levels of particulate air pollution are known to lead to a strong visibility impairment. This study showed that especially the sulphate concentration has a high negative correlation with visibility (Pearsons correlation coefficient of -0.74, Spearman correlation coefficient of -0.88, N=65), which is illustrated in Figure 6.15.

Antimony, Cu, and Sn have high loadings in **factor 4** (Table 6.5). All these elements are common tracers for traffic emissions. Copper, Sb, and also Sn were identified as characteristic components of brake wear (e.g. Sternbeck *et al.*, 2002; Wåhlin *et al.*, 2006). Consequently, this factor can be interpreted to represent influence from traffic. During the Olympic Games, the factor scores of this factor

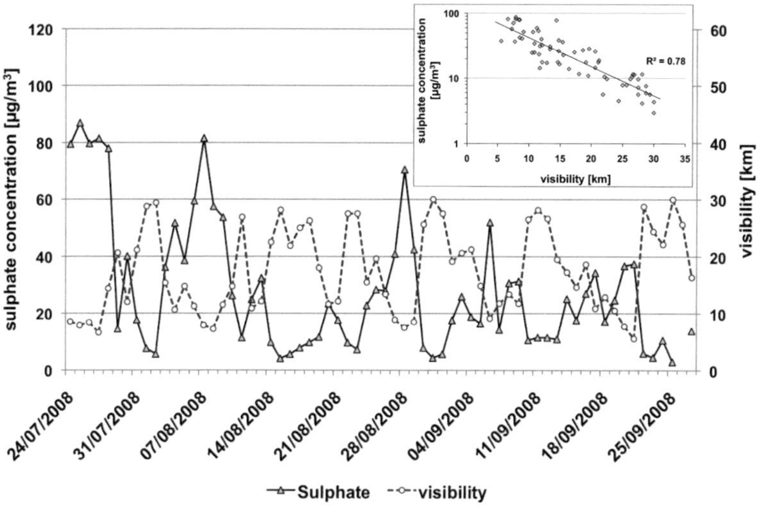

Figure 6.15: Daily course of sulphate concentrations (μg/m^3) of TSP samples at site CRAES during period C2 together with the visibility conditions (km). The scatter plot in the upper right corner of the figure further illustrates the negative correlation between sulphate concentration and visibility.

were low (Table C.27), which indicates, that the strict enforcement of traffic reduction measures was successful.

Factor 5 comprises Ba, Sr, As, and Zn, with high loadings (Table 6.5). Traffic, and especially traffic related abrasion of tyre and brake wear, but also asphalt surfaces and resuspension processes, are possible sources for all elements within this association. Zinc and Ba are often used as tracers for tyre wear (e.g. Monaci *et al.*, 2000), but were also found in brake wear (e.g. Gietl *et al.*, 2010). Construction materials were also interpreted to be a possible source for Sr and Ba (Norra *et al.*, 2008). Firework displays are other possible sources for this element association. However, this is a source, which is present only over short time spans. At the opening day of the Olympic Games, when a huge firework was displayed, factor five has a high factor score (>2), which supports the interpretation, that fireworks at least contribute to this factor temporarily. The impact of fireworks on atmospheric pollution, especially on the Opening Day of the Olympic Games, will be discussed in more detail in the next section. In general, factor 5 can be interpreted to represent diffuse anthropogenic pollution.

Chromium, Fe, and Co are the elements with high loadings in **factor 6** (Table 6.5). All three elements have low EFs (Table 6.1) and, therefore, geogenic sources seem to prevail. However, metal processing industry is a possible additional anthropogenic source contributing to the element association of this factor (Nath *et al.*, 2007).

None of the elements has high loadings in **factor 7**. Only wind speed and wind direction are positively correlated to this factor. Therefore, this factor can not be interpreted with regard to possible sources, but indicates that the wind conditions did not have any specific influence on certain TSP element concentrations.

Estimation of possible short-term impacts of firework displays

Another special, only short-term operating source for atmospheric particles are firework displays. Wang *et al.* (2007*c*) investigated the impact of fireworks on air pollution in Beijing during the lantern festival in February 2006. The authors found out that the concentration of some elements (Ba, K, Sr, Cl, Pb, Mg) as well as ions (including sulphate and nitrate) were over five times higher during the lantern days than during other days. Other studies on firework emissions were performed for example by Moreno *et al.* (2007) and Moreno *et al.* (2010) in Spain. They found the largest enrichment of Sr (86-times), K (26-times), and Ba (11-times) in $PM_{2.5}$ samples. Similar results were reported by Vecchi *et al.* (2008) from a study in Italy, where the authors found distinct enrichments of the following elements: Sr (120-times), Mg (22-times), Ba (12-times), K (11-times), and Cu (6-times).

At the opening ceremony of the Olympic Games, a huge firework was displayed at the Olympic Stadium and surroundings, which is quite close (< 6 km) to the sampling site CRAES. Moreover, the predominant wind direction at this day of 210° suggests a transport from the stadium towards the sampling site. TSP concentrations of Mg, K, and Sr had a relative high concentration at the 8th of August. For K and Sr this date is the maximum for the whole period C2, whereas other elements, such as Al, Ti or Fe have relatively lower values at this date. The concentrations of this day were 2-times higher for Mg, and 3-times higher for K, Sr, and Ba compared to the average concentration of the previous fourteen days. The water-soluble TSP concentrations were also compared to the average concentration of the previous two weeks. Here, the Ba concentration was 7-times, the Sr concentration 3.5-times, and the K concentrations 3-times higher. In PM_1 samples, Sr concentrations at the 8th of August were 4-times, and K concentrations

2-times higher compared to the average of the previous 7 days. The standardized course of K, Sr, and Ba, in the total and the water-soluble fraction of TSP samples is illustrated in Figure 6.16. Here, the concentration peak for these elements is clearly visible at Opening Day of the Olympic Games (8th of August), and, to a lower extend, also at the 6th of September, the Opening Day of the Paralympic Games.

The elevated concentrations of the mentioned indicator elements, together with the spatial proximity between the sampling site and the Olympic stadium, as well as the prevailing wind direction, lead to the conclusion that firework displays at the 8th of August significantly contributed to particulate air pollution and were evident even in the averaged 24-hour samples of this study.

Evaluation of sources contributing to the fine particle load

The chemical composition of the fine particles (e.g. PM_1) is of special interest due to their higher health-relevance (Schwartz *et al.*, 1996). Whereas the volume-related concentrations in ng/m^3 are lower for all elements if concentrations of PM_1 and TSP samples are compared, this is not the case if the mass-related concentrations in μg/g are considered. Figure 6.17 illustrates, which elements are of special importance in the fine fraction compared to total particulates. Elements with the relatively highest enrichment in the fine fraction were Na, Ni, and Cr. Generally, it can be observed, that elements with low EFs, where geogenic sources prevail, dominate in the TSP samples (e.g. Fe, Ti), whereas elements with high EFs, which originate from predominantly anthropogenic sources, had higher mass-related concentrations in the fine PM_1 fraction (e.g. Zn, As).

In order to investigate in more detail, which sources are mainly responsible for the fine particles during the Olympic Games, FA was also carried out for PM_1 samples (sampling period C2). The concentrations of 24 elements, PM_1 mass concentrations and the meteorological factors were included and eight factors were extracted. These factors accounted for 80% of the Expl. Var. The factor loadings are listed in Table 6.6.

Factor 1 comprises several elements (Cd, Ga, As, Rb, Mn, Pb, Li, Cs, Sn, and Fe) with positive loadings (Table 6.6). Some elements within this association had high EFs (Cd, As, Pb, and Sn, Table 6.3) and, consequently, anthropogenic sources seem to have a high contribution to this factor. However, also elements from predominantly geogenic sources are included within this association, such as Fe, Rb, and Mn. Noticeable is that combustion processes, and especially coal

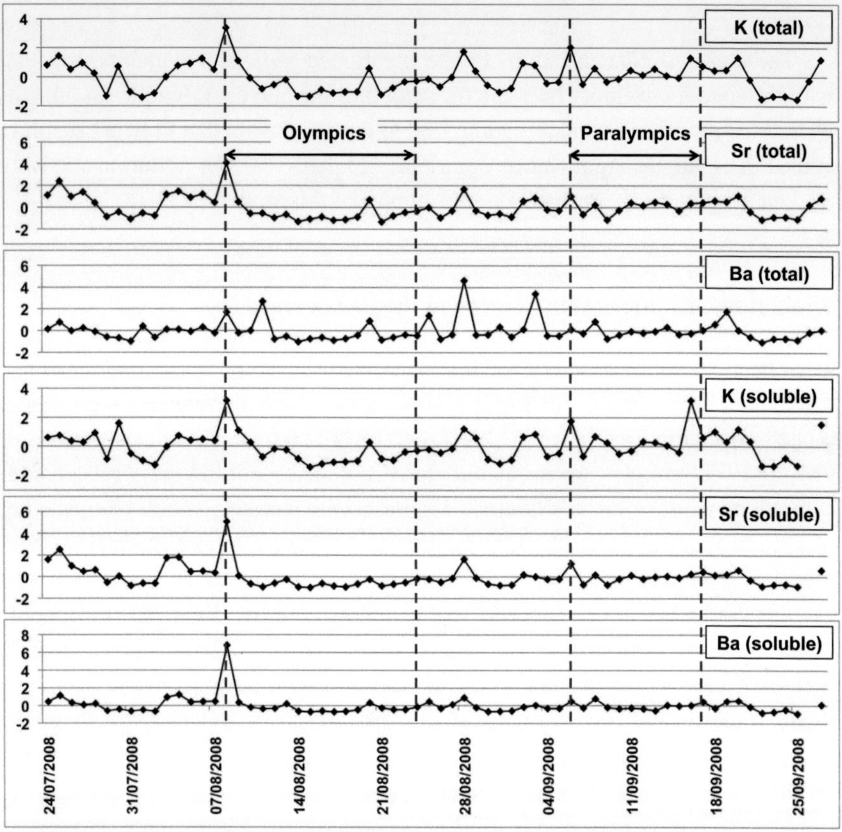

Figure 6.16: Standardized (Equation 3.4) element concentrations of three selected elements (K, Sr, and Ba) in the total and water-soluble fraction of daily TSP samples during period C2 at site CRAES.

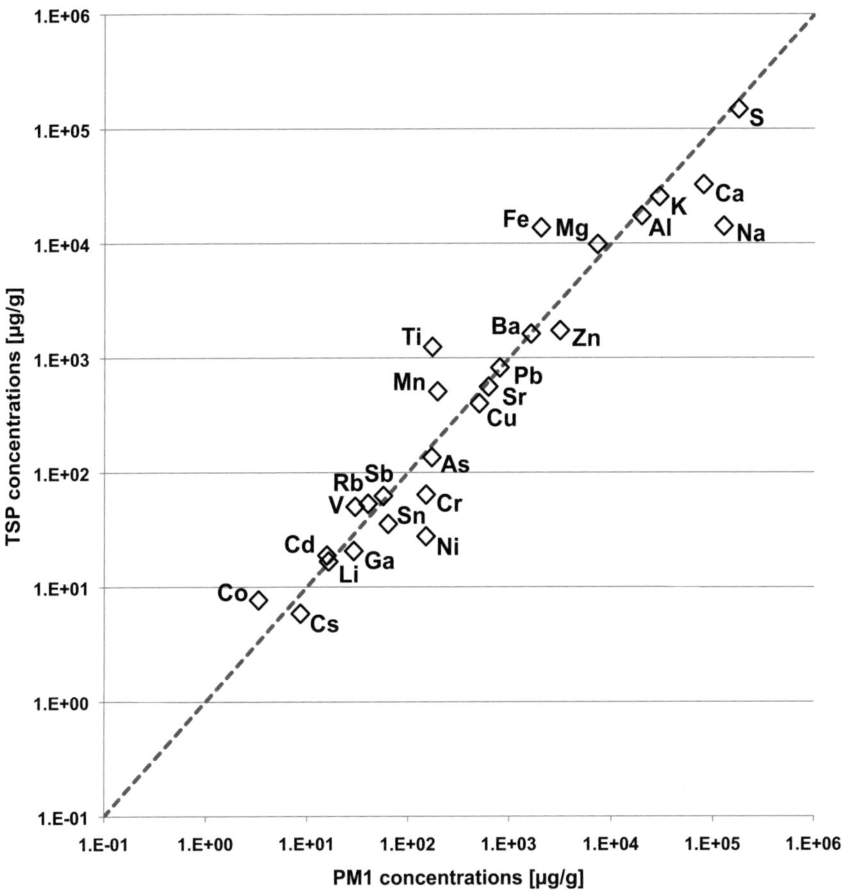

Figure 6.17: Element concentrations of PM_1 *versus* TSP samples of the 8th of August 2008 (24-hour samples) at site CRAES.

Table 6.6: Factor (Fa) loadings of the eight extracted factors (Principal Component Analysis, case by case exclusion of missing data, Varimax standardized rotation) for PM_1 samples (element concentrations of 24-hour samples in ng/m^3), together with the respective communalities (comm). Meteorological parameters (visibility, dew point, temperature, wind speed, and predominant wind direction) are included. Loadings $\geq \pm 0.55$ are marked in bold.

	Fa1	Fa2	Fa3	Fa4	Fa5	Fa6	Fa7	Fa8	comm
mass	0.27	-0.19	**0.72**	0.00	0.20	-0.01	0.01	-0.07	0.86
Li	**0.74**	0.20	-0.19	-0.06	0.00	-0.12	-0.02	0.21	0.97
Na	-0.20	**0.56**	0.08	0.41	-0.24	-0.24	0.16	0.11	0.93
Al	-0.03	**0.76**	0.05	0.40	-0.21	-0.29	-0.01	0.13	0.98
S	0.50	0.03	**0.64**	0.02	0.21	0.08	0.36	0.16	0.99
K	0.24	**0.66**	**0.59**	-0.11	0.01	0.02	0.03	0.12	0.99
Ca	-0.17	0.01	-0.11	0.07	-0.02	**-0.82**	-0.18	0.01	0.95
Ti	0.24	0.28	-0.10	0.04	**-0.74**	-0.19	-0.05	0.04	0.90
V	0.21	-0.03	0.15	0.01	-0.07	0.17	**0.77**	0.02	0.87
Cr	0.19	-0.02	-0.06	**0.74**	0.05	0.07	0.05	-0.19	0.91
Mn	**0.77**	0.13	-0.02	0.12	-0.44	0.17	0.18	-0.13	0.98
Fe	**0.58**	0.12	-0.04	0.11	**-0.70**	0.19	0.14	-0.11	0.99
Co	-0.03	-0.08	0.10	0.52	0.13	**0.62**	-0.13	-0.01	0.90
Ni	0.33	**0.62**	-0.23	0.15	-0.12	0.11	0.02	0.07	0.80
Cu	0.28	-0.02	0.22	-0.03	0.10	-0.05	**0.83**	0.09	0.97
Zn	0.07	**0.86**	0.08	-0.15	-0.13	-0.02	-0.08	-0.18	0.99
Ga	**0.83**	0.15	0.27	0.16	-0.11	0.10	0.17	-0.21	0.99
As	**0.81**	-0.03	0.45	-0.07	0.09	-0.06	0.04	0.18	0.98
Rb	**0.81**	0.16	0.22	0.09	-0.30	0.21	0.06	0.07	0.98
Sr	0.12	**0.84**	0.10	-0.13	0.07	-0.02	0.03	0.13	0.96
Cd	**0.85**	-0.02	0.24	-0.01	-0.01	0.06	0.31	-0.15	0.99
Sn	**0.68**	0.02	0.05	-0.06	0.11	-0.12	0.45	-0.31	0.93
Cs	**0.73**	0.19	-0.51	-0.01	-0.03	0.14	0.00	0.18	0.96
Ba	0.05	**0.94**	-0.07	-0.12	-0.05	0.12	-0.07	-0.11	0.99
Pb	**0.76**	-0.06	0.44	-0.08	0.08	0.02	0.27	0.01	0.98
vis	-0.34	-0.01	-0.43	-0.23	**-0.62**	-0.12	-0.33	0.15	0.96
dew pt	0.07	0.10	**0.88**	0.16	0.11	0.08	0.21	-0.09	0.98
temp	0.00	0.17	**0.86**	-0.02	-0.14	0.13	0.11	0.09	0.97
precip	0.15	-0.07	-0.09	0.19	0.28	-0.31	0.18	**-0.67**	0.91
wind v	0.07	-0.03	-0.03	-0.17	0.12	-0.21	0.18	**0.85**	0.94
wind dir	0.23	0.19	-0.26	**-0.56**	0.21	0.09	0.16	0.34	0.87
% of Expl. Var.	22	14	13	6	7	5	7	6	

combustion, is a possible source for most (Cd, Ga, As, Rb, Mn, Pb, Li and Cs) of these elements (Xu *et al.*, 2003).

Barium, Zn, Sr, Al, K, Ni, and Na have loadings in **factor 2** (Table 6.6). From this element association, Zn had by far the highest EFs (Table 6.3). The EFs of Na, Ni, and Ba also indicate a strong impact of anthropogenic sources (Table 6.3). Aluminium, K, and Sr had only low EFs (Table 6.3), and, thus, geogenic sources seem to prevails for these elements. An important source for Zn in the fine fraction is possible brake wear (Wåhlin *et al.*, 2006). Chemical industry is a possible anthropogenic source for K, Ni, and Na (Reimann & Caritat, 1998).

Factor 3 includes mass, S, and K with positive loadings (Table 6.6). Moreover, dew point and temperature are also load of this factor. Sulphur in the fine fraction had very high EFs >1500 (Table 6.3), which indicates a high contribution from anthropogenic sources. Furthermore, S concentrations were highly correlated to BC concentrations ($r=0.77$), which suggests combustion as the major source. Potassium can also originate from coal combustion (Xu *et al.*, 2003).

Only Cr has high positive loadings in **factor 4** (Table 6.6). Furthermore, with a lower loading (Table 6.6), Co is also included in factor 4. Wind direction is negatively correlated to factor 4. Possible anthropogenic sources for Cr are industrial or coal combustion emissions. However, geogenic dust is also a possible natural source.

Factor 5 comprises Ti, Fe, and visibility with negative loadings (Table 6.6). With a lower loading, Mn might also play a minor role within this association (Table 6.6). Usually, the geogenic sources, such as minerals from soils or resuspension processes, prevail for those elements, which is also reflected by their low EFs (Table 6.3). However, steel factory emissions are another possible anthropogenic source for Fe and Ti (Kuang *et al.*, 2004).

Calcium has high negative loadings in **factor 6** while Co is positively correlated to this factor (Table 6.6). Calcium had significantly higher EFs for the fine fraction (PM_1, average EF of 39) in contrast to Ca concentrations of TSP samples (EFs around 3, Table 6.1). This indicates that the anthropogenic contribution of Ca sources is higher for the finer particles. Lime and cement factories or calcareous construction waste are possible anthropogenic sources for Ca (Reimann & Caritat, 1998; Kuang *et al.*, 2004).

Copper and V are positively correlated to **factor 7** (Table 6.6). Steel production is a possible anthropogenic source for both elements (Reimann & Caritat, 1998). However, weathering and geogenic dust is a possible natural source for this element association (Reimann & Caritat, 1998). Vanadium had very low EFs, whereas the EFs of Cu indicate an important influence of anthropogenic sources (Table 6.3).

Factor 8 comprises only wind speed and precipitation and, thus, can not be interpreted with regard to source apportionment.

Evaluation of sources contributing to the mobile element concentrations

Average sulphate concentrations of TSP samples at site CRAES in Aug-08 (27.1 μg/m^3) were slightly lower than those during the same month in the previous years at site 4 (38.3 and 32.3 μg/m^3 for 2006 and 2007, respectively). For nitrate, the reduction in Aug-08 compared to Aug-06 and Aug-07 was very high (2.0 μg/m^3 in Aug-08 compared to 9.1 and 14.2 μg/m^3 in Aug-06 and Aug-07, respectively). Consequently, the nitrate/sulphate ratio in Aug-08 was also considerably lower. As mentioned in chapter 4.4.4, the nitrate/sulphate ratio is often used as an indicator for the influence of mobile traffic sources versus stationary industrial sources. The relatively higher reduction of nitrate compared to sulphate concentrations during the source control period indicates that the traffic mitigation measures were more effective than the reduction of other combustion sources, presumably coal combustion for energy generation.

Factor analysis was also carried out for the water-soluble ions of the 24-hour TSP samples (concentrations in μg/g). TSP mass concentration, water-soluble element concentrations of 28 elements, concentration of three anions (nitrate, phosphate, and sulphate), as well as some meteorological factors (visibility, dew point, temperature, precipitation, wind speed, and predominant wind direction) were taken into account. Eight factors were extracted, which accounted for 79% of the Expl. Var. of the whole data set. The factor loadings are listed in Table 6.7 and the corresponding factor scores can be found in the appendix (Table C.28).

Factor 1 comprises Rb, Cs, Pb, K, Zn, As, Mn, Al and V with positive loadings (Table 6.7). All those elements had, on average, a share of more than 50% of their total concentration in the water-soluble fraction (except for V with 40%, Table 6.2). On the one hand, some of these elements, such as Rb, Cs, K, Mn, Al, and V, had low EFs (Table 6.1). On the other hand, also elements with high EFs (Pb, Zn, and As) are included in this factor. Therefore, factor 1 can be interpreted as a mixed geogenic-anthropogenic factor that represents the diffuse atmospheric particle load.

The water-soluble cations, which have positive loadings in **factor 2** are Na, P, Ni, Li, and Cr and as an anion, phosphate is also included in this factor (Table 6.7). Moreover, visibility with a positive loading, and mass concentration with

Table 6.7: Factor (Fa) loadings of the eight extracted factors (Principal Component Analysis, case by case exclusion of missing data, Varimax standardized rotation) for water-soluble element concentrations of TSP samples (data from 24-hour samples in μg/g), together with the respective communalities (comm). Loadings > 0.55 are marked in bold.

	Fa1	Fa2	Fa3	Fa4	Fa5	Fa6	Fa7	Fa8	comm
mass	-0.08	**-0.84**	0.28	-0.04	0.01	0.05	-0.14	0.27	0.95
Li	0.19	**0.85**	0.11	0.10	0.05	0.07	-0.09	-0.11	0.93
Na	-0.02	**0.96**	-0.17	0.03	0.04	0.05	0.04	-0.13	1.00
Mg	0.35	0.27	0.25	**0.62**	-0.17	-0.22	0.27	0.06	0.92
Al	**0.61**	-0.11	**0.55**	0.32	-0.16	-0.16	0.13	0.19	0.95
P	-0.09	**0.94**	-0.21	0.06	0.04	0.06	0.01	-0.15	1.00
S	0.30	-0.20	**0.67**	-0.15	-0.10	0.25	-0.05	0.46	0.98
K	**0.81**	-0.03	0.13	-0.04	-0.39	-0.07	-0.01	-0.12	0.96
Ca	0.02	0.28	-0.33	**0.73**	-0.04	0.09	-0.06	-0.36	0.94
Sc	-0.26	0.51	-0.15	0.12	0.13	-0.04	-0.23	-0.12	0.68
Ti	0.43	0.09	0.47	**0.59**	-0.05	-0.07	0.20	0.17	0.93
V	**0.60**	0.09	0.32	0.10	0.03	0.03	-0.10	0.37	0.84
Cr	0.01	**0.61**	-0.06	0.04	0.01	-0.36	0.17	-0.18	0.70
Mn	**0.72**	0.09	-0.02	0.36	0.13	-0.02	0.06	-0.02	0.86
Fe	0.52	0.12	0.51	0.27	-0.03	-0.08	0.37	0.22	0.96
Co	0.51	0.46	0.01	0.04	0.02	0.14	0.22	0.08	0.69
Ni	0.11	**0.86**	0.14	0.13	0.05	0.00	0.01	0.09	0.93
Cu	0.09	0.01	**0.75**	-0.12	-0.04	-0.14	-0.16	-0.10	0.86
Zn	**0.78**	-0.20	0.46	0.04	0.06	-0.09	-0.14	0.03	0.97
Ga	0.18	-0.15	**0.76**	0.02	-0.08	0.03	0.07	0.15	0.85
As	**0.75**	0.08	0.17	-0.05	-0.23	-0.04	0.02	0.19	0.89
Rb	**0.96**	0.03	0.06	0.03	0.01	-0.02	0.04	0.03	0.99
Sr	-0.01	-0.17	0.00	0.07	**-0.92**	0.03	-0.24	0.07	0.95
Y	-0.19	0.06	-0.27	**0.84**	0.15	-0.11	-0.13	-0.10	0.94
Cd	0.46	-0.14	0.17	0.01	0.02	**-0.77**	0.05	-0.06	0.88
Sn	0.43	-0.29	**0.63**	-0.02	0.09	-0.20	-0.15	-0.15	0.91
Sb	-0.14	0.05	0.06	0.12	0.00	**-0.91**	0.07	0.00	0.94
Cs	**0.94**	0.04	0.16	-0.05	0.02	0.01	0.06	0.14	0.99
Ba	0.14	-0.04	0.21	-0.09	**-0.90**	-0.01	-0.03	0.16	0.95
Pb	**0.91**	0.06	0.19	-0.18	0.00	-0.01	-0.10	-0.02	0.97
vis	-0.11	**0.57**	**-0.64**	0.09	0.13	0.09	-0.13	-0.26	0.94
dew pt	0.28	-0.30	0.28	-0.08	-0.12	-0.02	0.21	**0.74**	0.97
temp	-0.02	-0.22	-0.04	-0.03	-0.09	0.10	-0.05	**0.89**	0.95
precip	0.42	0.00	0.44	-0.03	-0.02	0.20	0.36	-0.18	0.77
wind v	0.25	0.00	0.04	-0.10	-0.15	0.01	**-0.71**	-0.06	0.60
wind dir	-0.18	-0.11	0.06	0.08	-0.10	0.13	**-0.72**	0.03	0.66
NO_3^-	-0.10	0.31	-0.52	0.16	0.17	0.06	0.01	-0.53	0.84
PO_4^{3-}	0.08	**0.95**	-0.16	0.00	0.01	0.01	0.12	-0.01	0.99
SO_4^{2-}	0.30	-0.17	**0.69**	-0.18	-0.11	-0.11	-0.06	0.52	0.99
% of Expl. Var.	20	17	13	6	5	5	5	8	

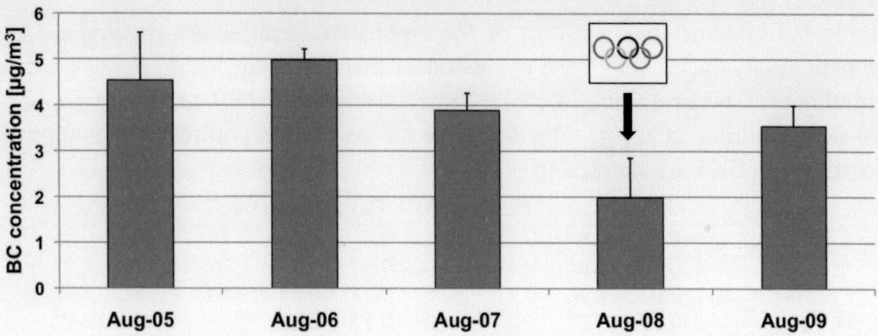

Figure 6.18: Average BC concentrations of $PM_{2.5}$ samples at site CUG in Aug-05 (N=4, weekly samples), Aug-06 (N=4, weekly samples), Aug-07 (N=5, weekly samples), Aug-08 (N=31, daily samples), Aug-09 (N=4, weekly samples), respectively. Whiskers represent the standard deviation.

a high negative loading are included in this factor. Salt particles are a possible source contributing to this factor. Chemical industry is a possible anthropogenic source for this element association (Reimann & Caritat, 1998).

Factor 3 includes Ga, Cu, sulphate, S, Sn, and Al with positive loadings and nitrate with a negative loading (Table 6.7). Combustion processes, such as coal combustion, and traffic exhaust is a possible source for this element association.

Yttrium, Ca, Mg and Ti have positive loadings in **factor 4** (Table 6.7). All these elements have low EFs (Table 6.1) and, therefore, this factor can be interpreted to represent predominantly geogenic sources. However, it is believed that anthropogenic sources, especially for the soluble shares of those elements, do have an influence. Magnesium and Ca might originate from building materials, such as mortar and concrete. Gypsum is another possible source for Ca. As white color pigment, TiO_2 is widely used in paints, paper, and plastics (Reimann & Caritat, 1998). Of all elements included in this study, Ti had the lowest share in the water-soluble fraction (average of 3%).

Factor 5 comprises Sr and Ba with high negative loadings (Table 6.7). As mentioned before, fireworks constitute a possible short-term source for both elements. The conjecture, that the firework display at the Opening Ceremony of the Olympic Games at the 8th of August released a high amount of these elements, is supported by the high factor score of factor 5 at this day (Table C.28). The daily course of the water-soluble fraction of these two elements is shown in Figure 6.16, where the concentration peak at the 8th of August is pronounced.

The water-soluble cations Sb and Cd have high negative loadings in **Factor 6** (Table 6.7). A possible anthropogenic source for the water-soluble fraction of both elements is car exhaust (Reimann & Caritat, 1998).

Only wind speed and wind direction have loadings in **factor 7** (Table 6.7). Both variables have only low communalities (0.60 and 0.66, respectively) and, therefore, no special influence of wind conditions on certain soluble element concentrations can be observed during period C2.

Factor 8 comprises dew point and temperature with high loadings (Table 6.7). With only low loadings, nitrate and sulphate, might also be included in this factor. High temperature and high humidity were reported to accelerate the secondary formation of particulate aerosols, such as nitrate and sulphate, in Beijing (Yao *et al.*, 2002). Both ions are known to dominate in the fine fraction (Seinfeld & Pandis, 2006).

Evaluation of the influence of combustion processes during the Olympic Games

Black carbon mainly originates from incomplete combustion processes and, in China, especially from the usage of coal and biofuels (Streets *et al.*, 2001). Consequently, BC can be used as a good tracer for the impact of combustion processes from industry and traffic during the source control period in Beijing. This statement is further supported by the good correlations between BC and certain indicator elements. Black carbon concentrations were highly correlated with Sb, Ga, Mn, Pb, As, and Fe ($r>0.70$ for $PM_{2.5}$ samples at site CRAES during period C2) and also showed a good correlation with Cd, Rb, and Sn ($r>0.64$). The average BC concentrations of $PM_{2.5}$ samples at site CUG for each August of the years 2005 – 2009 are displayed in Figure 6.18. The BC concentrations in Aug-08 were significantly lower if compared to the same month in the previous years (Aug-05, Aug-06, and Aug-07, respectively). This observation indicates, that the mitigation measures during the Olympic Games probably reduced combustion sources quite effectively. In Aug-09, the BC concentrations were again higher than during the Olympic Games (Figure 6.18).

6.4.4 Evaluation of the effectiveness of the applied mitigation measures

Since the days with precipitation during the first days of the Olympic Games definitively reduced atmospheric particle concentrations due to wet deposition, it is not clear what proportion of the improvements can be attributed to the mitigation measures and what to favourable weather conditions. Other studies mention the same problem (e.g. Branis & Vetvicka, 2010).

The data from the passive samples were further studied, in order to estimate, if the contribution from anthropogenic or geogenic sources decreased more in August 2008. As mentioned before, the transparent particles represent predominantly geogenic sources while the anthropogenic share for the opaque particles is high (Grobéty *et al.*, 2010) and, thus, the two superordinate source types can be distinguished. It was already shown in section 6.4.1, that the total APM concentrations for all size classes were reduced effectively in Aug-08 compared to the previous years. In order to discriminate further between geogenic and anthropogenic contributions to the total particle load in the different size fractions, also the size distributions for the average August concentrations are plotted separately for transparent and opaque particles in Figure 6.19.

As illustrated in Figure 6.19, the size distribution for the average August concentrations are very different for both particle types. For transparent particles, Aug-08 had lowest concentrations for all size classes and the amount of reduction was very high (Figure 6.19a). On the contrary, opaque particle concentrations were lower in Aug-08 compared to the previous years only for the finest size class (2.5–5 μm d_g). For opaque particles the strongest decrease already occurred between Aug-05 and Aug-06 and thereafter the August concentrations remained quite similar (Figure 6.19b). Consequently, the particles considered to originate predominantly from anthropogenic sources were reduced less than the particle group with a high share of geogenic minerals. This observation indicates that the influence of the weather conditions on particle reductions, which does not distinguish between the origin of the particles, seems to have an even stronger effect than the source reduction measures.

6.4.5 Evaluation of the increase of particulate air pollution after the Olympic Games

Many of the closed down sources operated again after the Olympic Games, e.g. the traffic volume increased, construction sites operated and many industries produced with full capacity. However, some of the mitigation measures stayed activated, e.g. some industries were relocated permanently, cleaner production technologies were applied permanently, or the public transportation system was enlarged and still widely used.

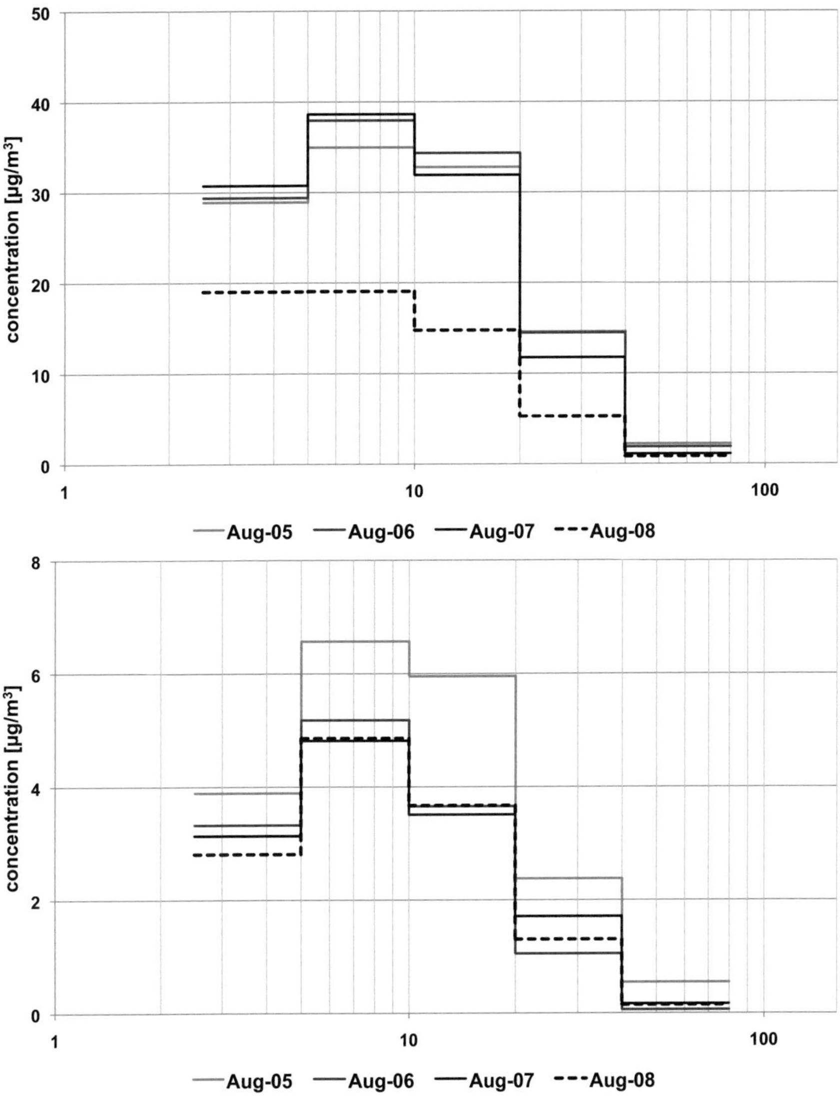

Figure 6.19: Size distribution of each August of the years 2005 to 2008 of APM samples from site CUG. (Top) average concentrations of transparent particles, (bottom) average concentrations of opaque particles.

As mentioned before, winter TSP mass concentrations were lower in winter 08/09 after the Olympic Games than in the previous winter. This might be a first indication that even without the complete set of mitigation measures, still some air quality improvements were made. Moreover, BC concentrations of $PM_{2.5}$ samples from site CUG were 80% higher again in Aug-09 compared to Aug-08 (Figure 6.18), when the strict mitigation measures were in effect. This could be expected since many of the controlled sources operated again (e.g. higher traffic volume, more industry emissions). However, it has to be noted, that BC mass was still lower than in the years 2005 – 2007 and, consequently, the BC concentrations after the Olympic Games did not reach again the levels before the first introduction of mitigation measures. This observation might be another indication, that some of the mitigation measures seem to have also long-term effects.

With the objective to investigate, which particle size group increased most after the Olympic Games, the size distribution for average concentrations of August and September 2008 and 2009 are compared in Figure 6.20. Mass concentrations were about 100% higher again for all size classes in August and September 2009. The increase was quite similar for SC1–SC4 (105, 104, 94, and 113%, respectively). Only larger particles with 40–80 μm d_g showed a different trend with an increase of only 46%. This observation indicates that the increase after the source control period was strong for atmospheric particles of different size classes.

The same comparison was done separately for transparent and opaque particles (Figure 6.21), in order to further discriminate between those two particle groups, which represent particles from predominantly geogenic and anthropogenic sources, respectively. For transparent particles, the concentrations increased again considerably for all size classes in 2009 for both months. This indicates that the overall diffuse atmospheric pollution was higher again in 2009. On the contrary, concentrations of opaque particles increased in 2009 only for the finest fraction (2.5–5 μm d_g). Within this size class, combustion residues, such as soot agglomerates and fly ashes, constitute most of the opaque particles. Consequently, this observation is an indication that the combustion sources emitted more particles in summer 2009 than during the source control period in 2008, whereas other sources for larger opaque particles, such as tyre wear, did not increase significantly. However, since the overall concentration of opaque particles is significantly lower than for transparent particles, it is important to keep in mind, that also the shorter sampling times in Aug-08 might contribute to this effect and, thus, the interpretation of the data of opaque particles is especially critical if periods with different sampling intervals are compared.

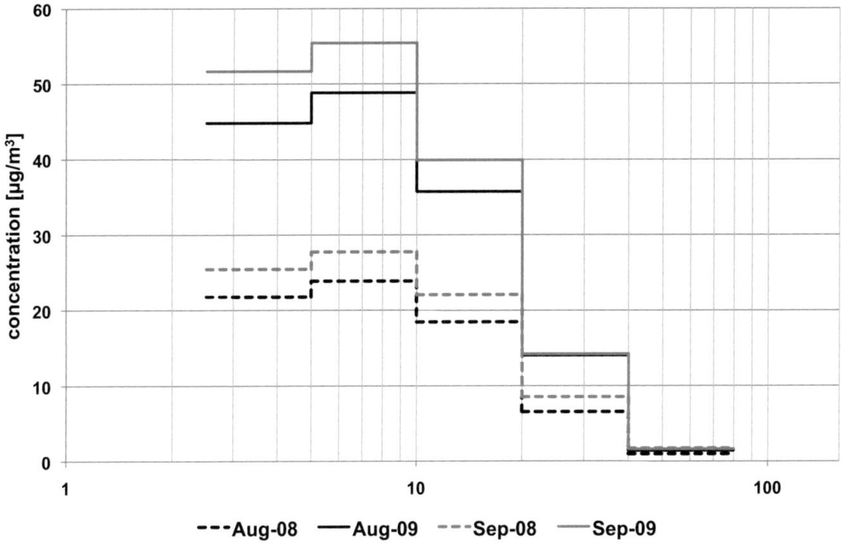

Figure 6.20: Size distribution of Aug-08, Aug-09, Sep-08, and Sep-09, respectively, of total concentrations of passive collected APM samples from site CUG.

6.5 Summary and conclusions

This study showed that particulate air pollution in Beijing decreased considerably during the Olympic Summer Games in August 2008, a period with strictly enforced mitigation measures. Mass as well as element concentrations during the Olympic source control period were lower than the respective concentrations during the time directly before and after the Olympic Games, and also lower than total August concentrations during the previous years. However, the discrimination between the influence of weather conditions and the effects of mitigation measures on the particle reductions turned out to be difficult in the case of Beijing with very heavy precipitation in the first week of the Olympic Games. Therefore, it has to be stated that the meteorological conditions certainly contributed to and probably had a high stake in the successful reduction of particulate air pollution. The large variety of sources still contributing to the particle load during the source control period makes the evaluation of the reduction success of the individual mitigation measures difficult.

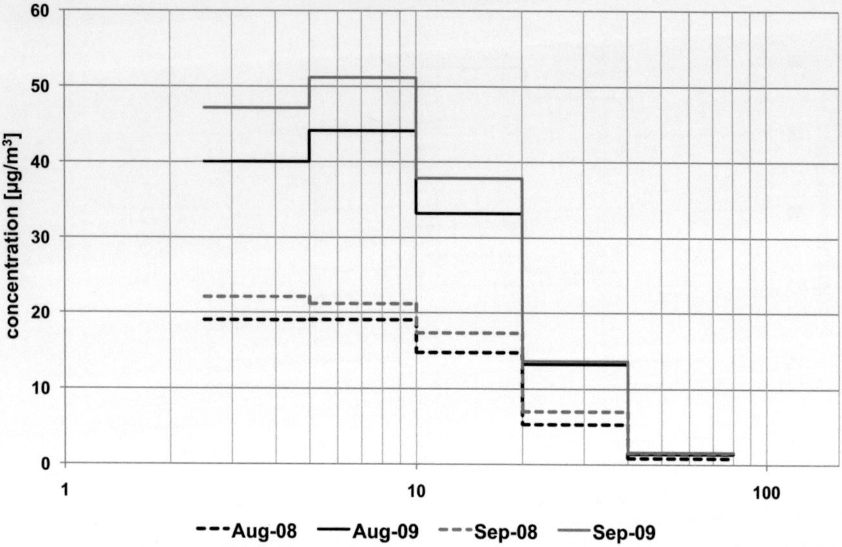

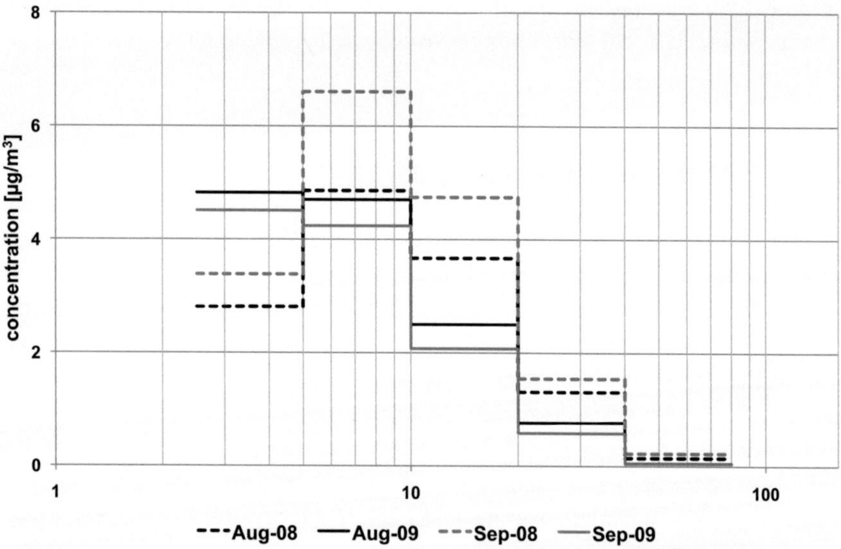

Figure 6.21: Size distribution of Aug-08, Aug-09, Sep-08, and Sep-09, respectively, of passive collected APM samples from site CUG. (Top) Average concentrations of transparent particles, (bottom) average concentrations of opaque particles.

Fine particles which are typically transported over long distances from outside of Beijing were less efficiently reduced than coarser particles. This indicates that long-range transport of atmospheric particles is difficult to control and that presumably the established mitigation area was not large enough to also reduce fine particles more efficiently. The study further showed that elements from predominantly anthropogenic sources, such as S, Cu, As, or Cd, had a less variable course during the Olympic Games than elements for which geogenic sources are more important, such as Al, Fe or Rb. Consequently, it can be assumed that the mitigation measures, as intended, were mainly successful in reducing more toxic, anthropogenic sources. Firework displays, especially at the Opening Ceremony, could be identified as a special short-time source for atmospheric particles during the Olympic Games.

After the Olympic Games, when many of the sources operated again, mass and element concentrations increased considerably. However, concentrations in 2009 remained lower than during the years prior to the source control period. Moreover, BC concentrations, which represent combustion sources, were also higher again in August 2009 than during the source control period in August 2008, but remained clearly lower than during the previous years (August 2005–2007). These observations might be a first indication, that some of the mitigation measures seem to also have long-term effects.

Conclusively, it can be said that the implementation of mitigation measures is one possible approach to improve air quality in urban areas. Those measures must be strictly enforced, should include all major pollution sources, and moreover, need to embrace a large area also outside the city. Consequently, only a comprehensive set of various measures can provide success. In Beijing, the next years will show if the indicated long-term success of the Olympic mitigation measures will go on and/or if further measures are implemented to control and further reduce the serious particulate air pollution.

Chapter 7

Towards a better understanding of inhomogeneities within single atmospheric particles and particle agglomerates

7.1 Introduction

Numerous studies on particulate air pollution in cities and megacities, including Beijing (e.g. He *et al.*, 2001; Okuda *et al.*, 2004; Sun *et al.*, 2004; Duan *et al.*, 2006; Norra *et al.*, 2010; Schleicher *et al.*, 2010*a*), were conducted in the last decade (see also chapter 1.4). Yet, all of these studies focus merely on total particle load or chemical analysis of bulk samples. On the contrary, only very limited knowledge is available on mineral and chemical composition of single atmospheric particles. An overview about results of individual-particle studies that use microscopy-based techniques, with emphasize on transmission electron microscopy, is given by Pósfai & Buseck (2010). Only few studies focussed on single atmospheric particles. Especially synchrotron radiation based mirco X-ray fluorescence analysis (μS-XRF) was not often applied for atmospheric particles, yet. One of the very rare studies using μS-XRF on $PM_{2.5}$ samples from urban atmosphere was carried out by Li *et al.* (2007) in Shanghai. The authors focussed on source identification and individual particles analysed by μS-XRF were additionally examined by Scanning Electron Microscopy (SEM) to gain more morphological information. In this work the authors focused on source identification of atmospheric particles by fingerprints obtained by μS-XRF spectra. Results of the same $PM_{2.5}$ samples from Shanghai were also discussed in Yue *et al.* (2006). The same method was already shortly presented by the same authors (Yue *et al.*, 2004) at the examples of PM_{10} particles from a roadside and an industrial area in Shanghai. Till now, no such approach was used for source identification of the complex atmospheric particle distribution in Beijing. To the author's knowledge, no study has investigated the small scale element distribution within single atmospheric particles, yet.

However, it is of particular importance to investigate the properties of **single particles** since, depending on their individual composition, these particles differ in their specific impact on climate change, negative environment and health effects, as well as acceleration of weathering of stone buildings in urban areas.

Generally, **health effects** of coarse particles are regarded as negligible. Therefore, most epidemiological studies focus on fine or ultrafine particles, because they can enter deep into the human respiratory tract and are, therefore, considered especially dangerous (e.g. Schwartz *et al.*, 1996). However, humans can inhale even larger particles through mouth breathing, which plays an important role during physical exercise, while speaking or if the nasal airways are obstructed (Bowes & Swift, 1989). Kennedy & Hinds (2002) applied wind tunnel experiments with full-torso mannequins and showed with this approach that mouth inhalability is relevant up to particle sizes of 116 μm. Bartra *et al.* (2007) investigated the effect of "air pollution" and especially diesel exhaust particles on the allergenic potential of pollen and assumed that they might increase the exposure to the allergen, the concentration and/or biological allergenic activity.

Agglomeration of fine particles on the surface of coarser particles is one of the central points investigated in this chapter. Two different aggregation processes are distinguished (Blum, 2006): (i) either that two colliding dust aggregates stick together and form a new dust agglomerate, which preserves the shapes of the educts, or, (ii) that an aggregate is formed due to restructuring processes during the collision, which does not resemble the morphologies of the two collision partners. These agglomerations between different particles are of importance because they

- can provide a scavenging effect for fine toxic particles in the atmosphere,
- change the optical properties and, thus, the climate impact of these particles, and
- produce a new particle type consisting of coarse geogenic and fine anthropogenic particles, which has not yet been investigated, thoroughly.

In order to investigate these different aspects, single particle analysis of passively and actively collected samples from Beijing was performed and the results are discussed in this chapter. On the one hand, scanning electron microscopy (**SEM**) coupled to energy-dispersive X-ray analysis (**EDX**) was applied in order to identify the most abundant geogenic and anthropogenic particles from Beijing, as well as to investigate agglomerated particles. On the other hand, the small scale element concentrations within single particles were investigated by synchrotron radiation based micro X-ray fluorescence analysis (μ**S-XRF**).

7.2 Results

7.2.1 Single particles analysed by optical microscopy

By means of automated optical microscopy, single particles between 2.5 and 80 μm d_g ($APM_{2.5-80}$), which were collected passively on transparent plates, were analysed. The methodology is described in section 3.2.6. Hereby, three particle groups were distinguished: "transparent particles", "opaque particles", as well as "pollen".

The optical microscopy was primarily used for the calculation of mass concentrations and size distributions of total, transparent, and opaque particles, respectively. Therefore, these results were described and discussed in chapter 4 with special regard to seasonal variations and in chapter 6 with special attention to applied mitigation measures. The single particles were not further characterized individually by optical microscopy.

7.2.2 Single particles analysed by SEM-EDX

Scanning Electron Microscopy was used to identify typical particles visually and to additionally analyse their elemental composition by EDX measurements. This method was applied to particles from TSP and $PM_{2.5}$ filter samples as well as to $APM_{2.5-80}$ from the adhesive collection plates (Sigma-2 sampling). In the following sections, the particles found most commonly in Beijing are presented.

Characteristic particles in Beijing's atmosphere

The most common **geogenic** minerals identified by SEM-EDX in APM from Beijing were quartz, feldspars, clay minerals, dolomite, calcite, and gypsum (Figure 7.1). The detected minerals are consistent with findings of Xu (2010). In the framework of the master thesis of Xu (2010), TSP samples from site 4 in 2006 were analysed by means of XRD analysis and the major mineral components identified in this study were calcite, chlorite, dolomite, gypsum, kaolinite, mica, K-feldspar, plagioclase and quartz. By means of SEM-EDX these minerals were found in most samples, but were more abundant during dust storm periods in spring. The study by Xu (2010) also reported a high consistency over the whole year 2006, but higher abundances in spring (March, April and May) compared to other seasons.

Anthropogenic particles (Figure 7.2) were numerous and variable in APM samples from Beijing. They comprised soot, fly ash, and different abrasion products as well as some not identifiable particles. Fly ashes of various chemical composition were detected, some consisted of Fe-oxide others of Al-Si-oxides. Soot was also very abundant in all samples from Beijing, however, more soot particles were found in winter compared to summer samples from Beijing.

Biogenic particles, such as pollen, spores, or plant debris, only were a minor fraction of the total particulate aerosols in Beijing (Figure 7.3), but spring was the season with most pollen.

7.2.3 Single particles analysed by μS-XRF

The particles analysed by μS-XRF were inhomogeneous with regard to their chemical composition. This is shown using the example of two selected particles, labelled "particle A" (Figure 7.4) and "particle B" (Figure 7.5). Both particles were collected passively in January and April 2006, respectively. The maximum concentration for each element is listed for both particles in Table 7.1.

The inhomogeneity is more pronounced in particle A (Figure 7.4). Here, the agglomerated particle can be divided into three to five different parts. Particle B (Figure 7.5) seems to be a more homogeneous particle surrounded by some kind of crust. Factor analysis helped to distinguish between these different parts of the particles and to identify the origin of the parts based on typical fingerprints. While the factor loadings show the importance of each element in the corresponding factor and the element associations, the factor scores help to illustrate the importance of each factor in a certain area of the particle. Thus, different areas of the inhomogeneous particle can be discriminated. For each of the particles three factors were extracted.

The factor loadings and factor scores for particle A are shown in Figure 7.6. For particle A, factor 1, which represents a siliceous particle, dominates in the

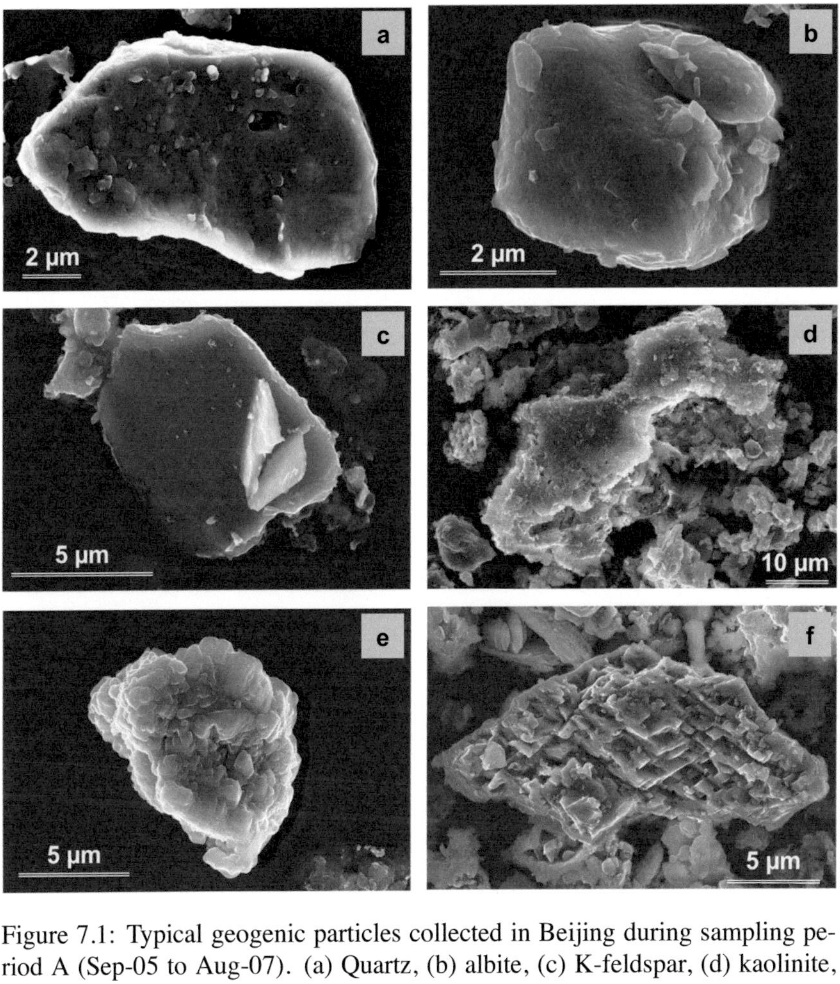

Figure 7.1: Typical geogenic particles collected in Beijing during sampling period A (Sep-05 to Aug-07). (a) Quartz, (b) albite, (c) K-feldspar, (d) kaolinite, (e) calcite, (d) dolomite.

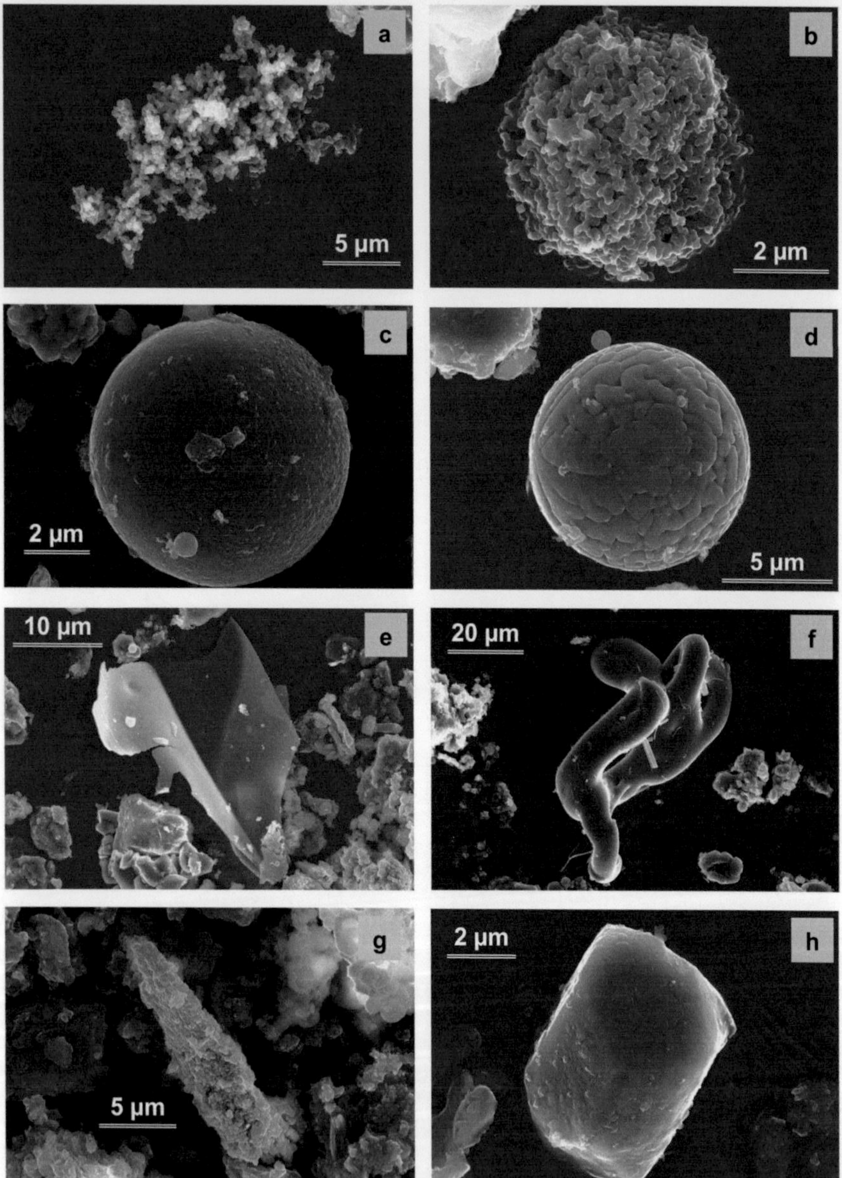

Figure 7.2: Typical anthropogenic particles collected in Beijing during sampling period A (Sep-05 to Aug-07). (a) Soot chains, (b) soot sphere, (c) fly ash (Al-Si-Oxide), (d) fly ash (Fe-Oxide), (e) thin (few hundred nm) carbon foil, (d) molten particles, (e) particle with a high amount of rare earth elements (REE, mostly Ce), (f) particle with a high concentration of Al, As, and REE.

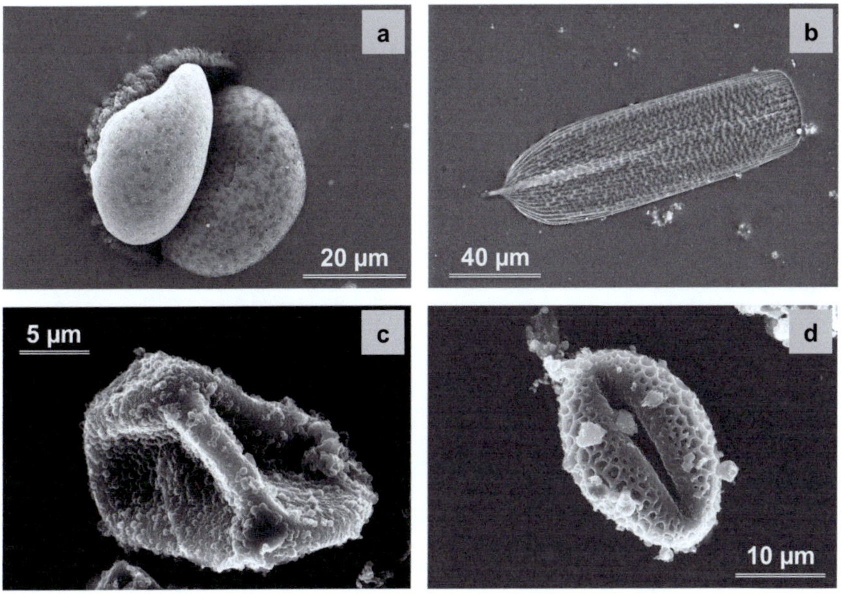

Figure 7.3: Typical biogenic particles collected in Beijing during sampling period A (Sep-05 to Aug-07). (a) Coniferous pollen, (b) insect wing, (c) not identified pollen, (d) not identified pollen.

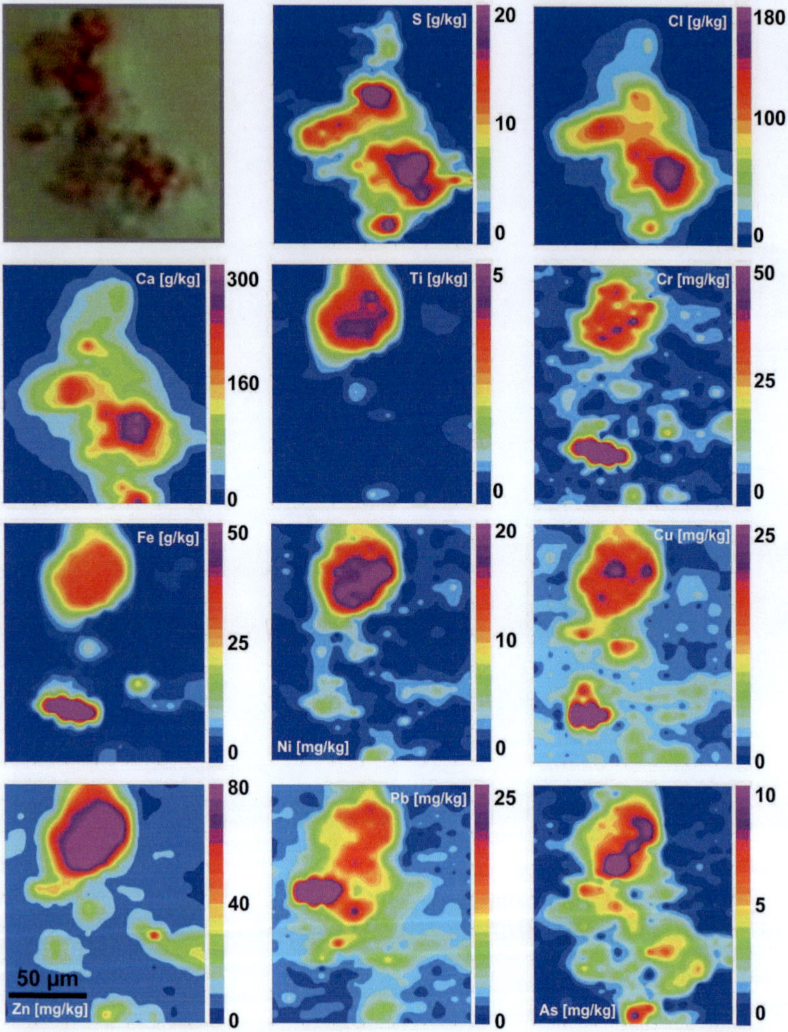

Figure 7.4: Agglomerated aerosol particle (Particle A) (approx. 150x120 μm) from Beijing (19/01/2006 to 23/01/2006). Displayed is the distribution and respective concentrations of Ca, S, Cl, Fe, Ti (in g/kg) and Cu, Ni, Zn, Cr, Pb, As (in mg/kg).

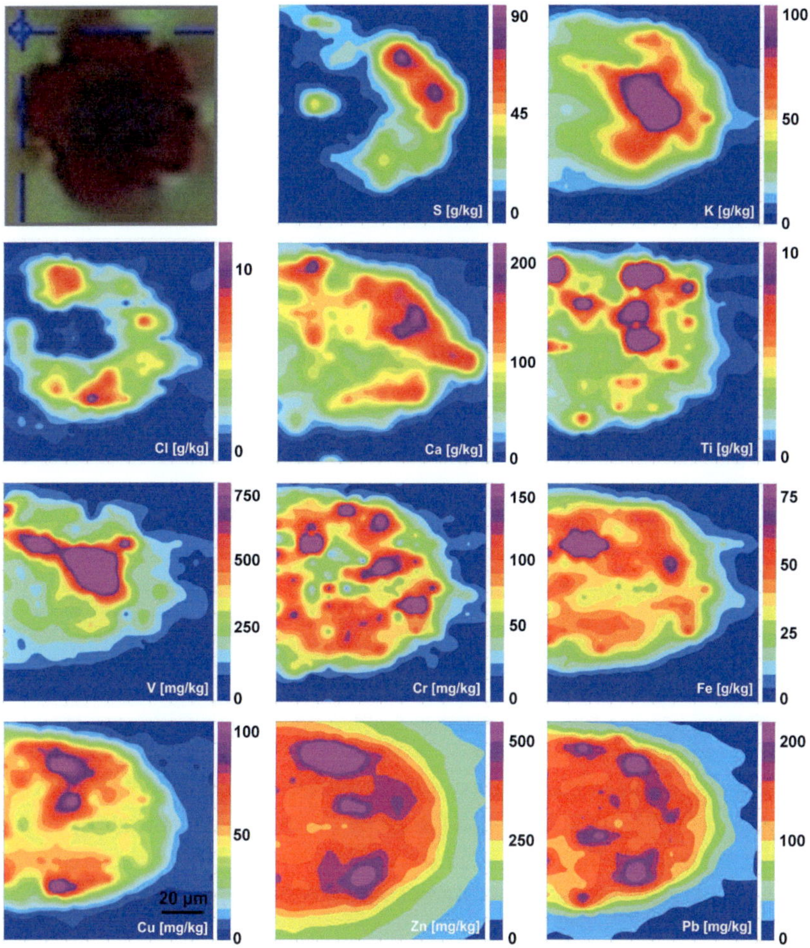

Figure 7.5: Atmospheric particle (Particle B) (approx. 100x100 μm) from Beijing (06/04/2006 to 10/04/2006). Displayed is the distribution and respective concentrations of Ca, S, Cl, Fe, K, Ti (in g/kg) and Cu, Pb, Zn, Cr, V (in mg/kg).

Table 7.1: Maximum concentrations of each element in the selected particles A and B. Concentrations in italic are in g/kg while all other values are in mg/kg.

Element	Max. conc. in particle A	Max. conc. in particle B
P	*15*	*40*
S	*40*	*100*
Cl	*200*	*10*
K	*30*	*190*
Ca	*310*	*220*
Ti	*5*	*70*
V	200	1580
Cr	140	440
Mn	590	5570
Fe	*130*	*280*
Ni	30	180
Cu	180	160
Zn	100	780
Ga	10	90
As	20	70
Rb	60	410
Sr	260	2300
Y	10	80
Zr	45	920
Nb	5	200
Mo	5	20
Pb	80	300

upper area of the agglomerate, while the middle part consists of a salt particle, characterized by factor 2. In factor 3, the metals Cr, Fe, and Mn are dominant.

The factor loadings and factor scores for particle B are shown in Figure 7.7. The main part of particle B is represented by factor 1, which comprises many different elements and no mineral phase can be clearly identified. Factor 3, on the other hand illustrates a salt crust that surrounds the particle.

7.3 Discussion

7.3.1 Coarse particles as scavenger for smaller toxic particles

Identified interactions between geogenic and anthropogenic particles were numerous, which highlights the importance to investigate these agglomerated particles more detailed. As discovered with SEM, small anthropogenic particles like soot or fly ash were often adsorbed to surfaces of larger minerals (Figure 7.8).

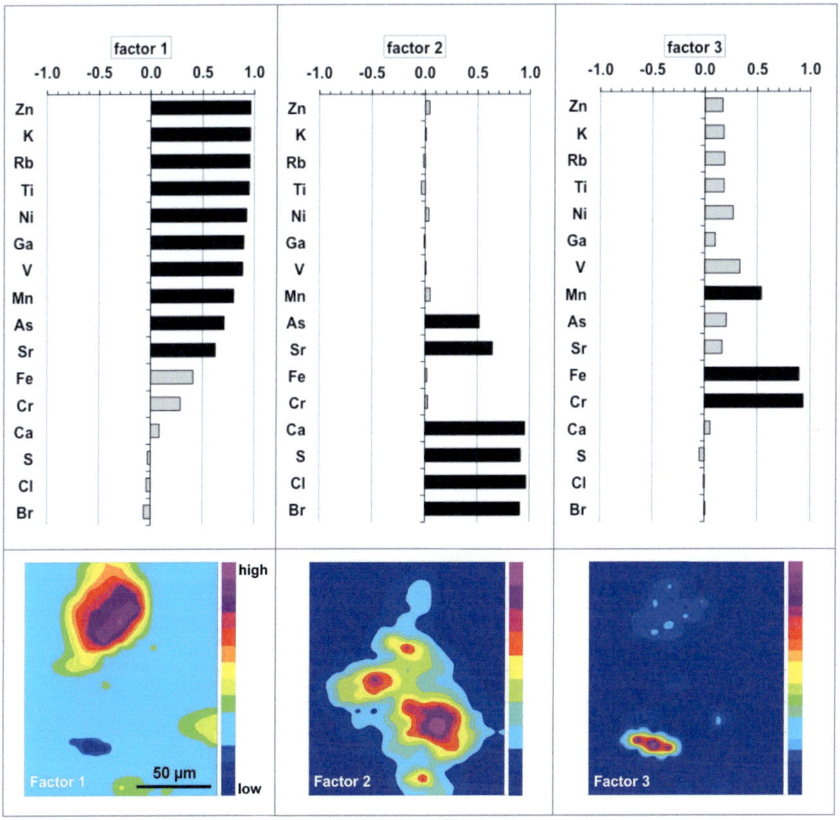

Figure 7.6: Agglomerated aerosol particle (Particle A) (approx. 150x120 μm) from Beijing (19/01/2006 to 23/01/2006). Displayed are the factor loadings and the factor scores.

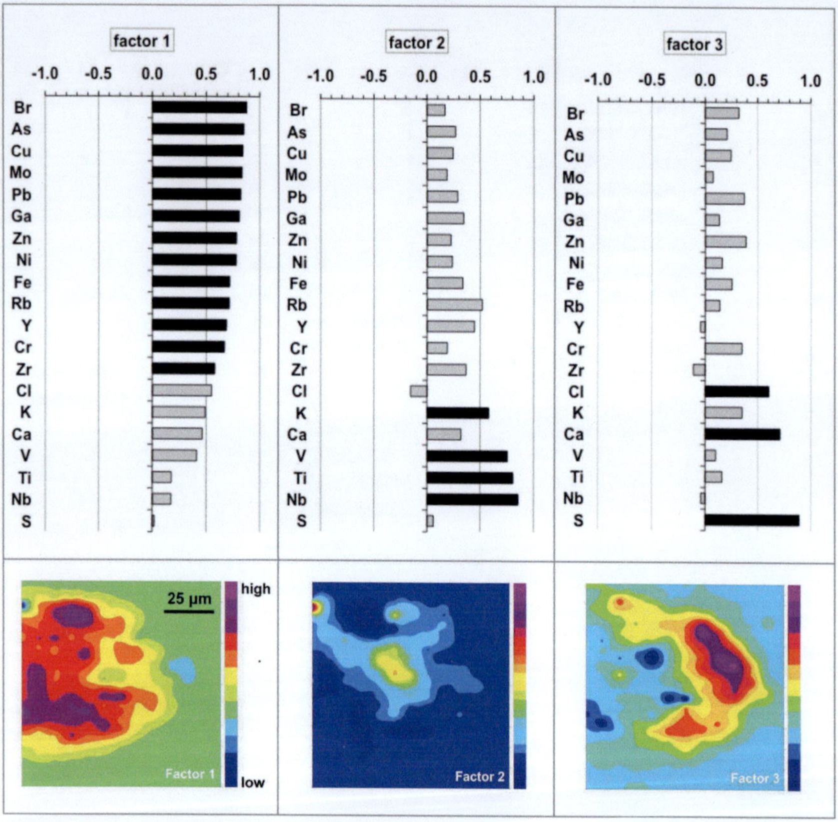

Figure 7.7: Atmospheric particle (Particle B) (approx. 100x100 μm) from Beijing (06/04/2006 to 10/04/2006). Displayed are the factor loadings and the factor scores.

These coarse particles were mainly of geogenic origin and can be seen as less toxic due to their composition and because they are less capable to infiltrate deep into the lungs. Such absorption can occur early during transport processes by impaction. However, generally it has to be mentioned that fine particles also can be deposited on coarse particles during sampling. The distinction between both processes forming aerosol agglomerations in the environment or on the filter is fairly difficult. Nevertheless, in cases of complex nested agglomerates (Figure 7.8) it is highly probable that these aerosol agglomerates were formed in the atmosphere and not on the filter, where just plain overlays are most likely to occur.

The most important effect of the agglomeration of fine particles on the surface of larger ones, is the **scavenging effect** of these fine and with regard to adverse health effects more dangerous particles, by the coarse ones. Since the frequency with which such agglomerations occur was not yet investigated in detail, also the removal efficiency of fine particles by coarser ones is also not known. Air pollution control normally starts with reducing the coarse particle concentration. This was also observed in Beijing during the Olympic Games periods with the applied mitigation measures (chapter 6). Thus, in this context an open question is how the fine particle mass concentration will respond when the coarse particle concentration is lowered. Since it was observed that many toxic elements, such as Cd or As, were especially enriched in the fine particle fraction (see chapter 4 or 5) the reduction of coarse particles and, consequently, a lower scavenging effect, could lead to an even more critical air pollution situation with regard to adverse health effects. As seen from these examples, a better knowledge about the amount of this scavenging effect would help to predict the changes in aerosol composition expected by the introduction of mitigation measures in the future.

The mapping of particle agglomerates by means of μS-XRF also highlighted the importance of the mentioned scavenging effect. Particle A is a good example of an agglomeration, where a small particle with a high concentration of potential harmful metals (Factor 3 in Figure 7.6) is agglomerated to the surface of a larger particle of different composition.

7.3.2 Accelerated stone decay in urban areas

Stone decay of monuments is favoured by wet and dry deposition of APM. Often the formation of salt efflorescence, e.g. sulphates, plays a decisive role for the soiling of buildings. Particles with sulphate and nitrate coatings together with sufficient moisture increase metal solubility and possibly catalyse further surface reactions on stone facades of buildings. For example, McAlister *et al.* (2008)

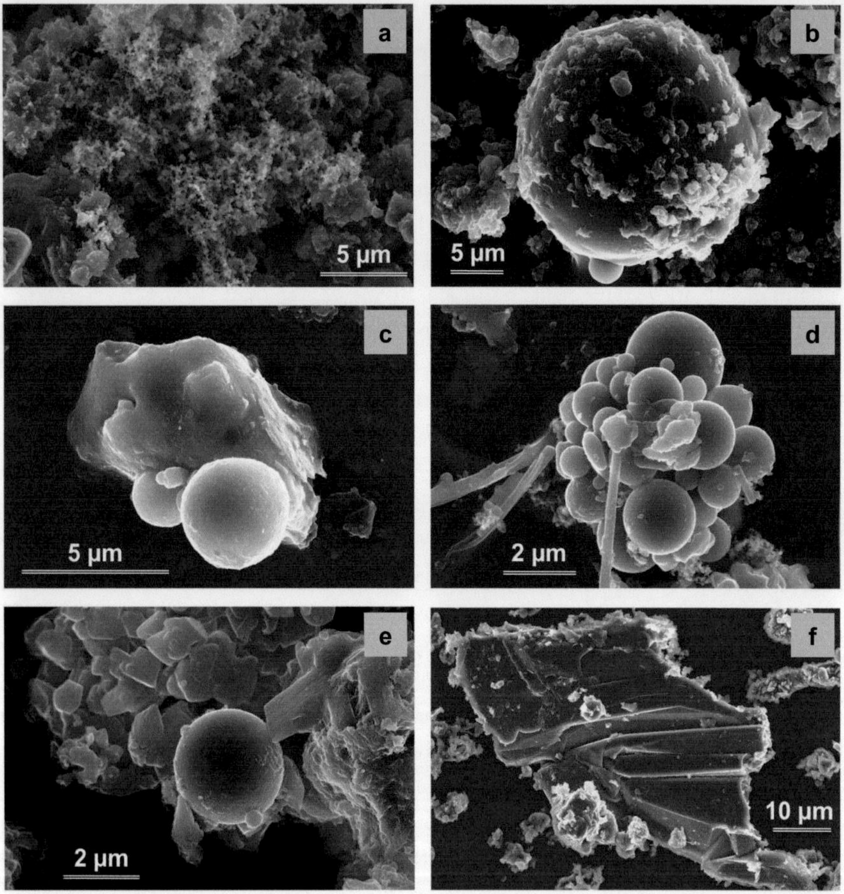

Figure 7.8: Agglomerated particles collected in Beijing during sampling period A (Sep-05 to Aug-07). (a) Soot on particle surface, (b) fly ash with many different particles on its surface (c) fly ash on a feldspar, (d) cluster of many fly ashes, (e) fly ash on gibbs particle, (d) carbon particles with soot on the rim and fly ashes and other particles on its surface.

investigated transition metals and water-soluble ions in deposits on a building and their potential catalysis of stone decay in Budapest, Hungary. Schleicher & Recio Hernández (2010) focused on sulphate formation on sandstones from monuments in Salamanca, Spain, and concluded that atmospheric pollution was one of the main sulphur sources for salt formation in their case study.

Abundant gypsum and salt particles were found by single particle analysis within this study. Consequently, a contribution of these particles to the weathering of buildings in Beijing can be expected. Particle B (Figure 7.5) is another example of a particle with a salt coating and high sulphur concentrations (Factor 3 in Figure 7.7), which can be regard as critical for stone decay.

7.4 Summary and conclusions

Single particle analysis highlighted the large variety of different geogenic and anthropogenic particles contributing to the overall air pollution in Beijing. On the contrary to bulk chemical analysis, it was possible to identify the most important individual particles. Furthermore, μS-XRF analysis showed varying concentrations of trace and main elements within the particles. Conclusively, this study demonstrates that the μS-XRF is an excellent tool for the detection of spatial distribution of main and trace elements within individual atmospheric particles.

The high abundance of minerals, such as quartz, feldspars, or carbonates, emphasizes the high importance of geogenic particles in Beijing. The strong anthropogenic input from a variety of sources, such as traffic, industry, coal burning and intensive construction activities, was reflected by the various anthropogenic particles. Especially the high amount of fly ashes and soot in most samples from Beijing, is considered to be dangerous with regard to adverse health effects for the population.

Abundant interactions between geogenic and anthropogenic particles were observed. This observation is of special interest in the context of an expected scavenging effect of fine particles by larger ones. Since a higher reduction of coarse particles can be expected by air pollution control, the open question is how the fine particle mass concentration will respond and if even more hazardous small particles can be expected in the atmosphere due to missing scavengers.

Generally, this study was only a first attempt to apply a set of combined tools of single particle analysis to atmospheric particles. More analysis have to be carried out in order to classify specific particles according to their chemical composition and to use this approach for source identification and the characterization of pollution levels of atmospheric particles. Consequently, μS-XRF analysis of more atmospheric particles from Beijing should be applied in future in order to gain more knowledge about the complex urban mixture of airborne particles and their respective toxicities. A quantitative statement about elemental distribution within individual particles in Beijing should then be realistic. Another aim for further studies could be to illustrate, (i) which particles are responsible for the spatial and temporal variations in chemical compositions of atmospheric aerosols at ground level, (ii) which particles are responsible for health impairment, and (iii) which particles accelerate the stone decay on buildings in urban areas.

Executive summary and outlook

The scope of this work was to develop a comprehensive understanding of the anthropogenic and geogenic aerosol pollution in the megacity Beijing. The results of this study emphasize the high complexity of the urban aerosol pollution in Beijing and the large spatio-temporal variations in mass concentration but also elemental composition.

It was shown that for a comprehensive assessment of environmental effects of atmospheric particles it is important not to solely monitor and reduce the total burden but to focus on toxic metal concentrations. Furthermore, the study highlighted the importance to discriminate further between mobile and immobile metal concentration in order address the bioavailable and, thus, health-relevant fraction of metal pollution. Based on the results of this study it is highly recommended that future health assessment studies should focus on the bioavailable metal concentrations instead of applying only total concentrations. This approach would enhance the quality of predictions with regard to the health harming potential of aerosol pollution.

The study shows that even the long-term detailed measurements carried out here are not sufficient to completely assess all factors influencing the spatio-temporal variations and the exact shares of the different sources. For a better understanding of the numerous aerosol sources, better source profiles and detailed information about industrial emissions are needed. Therefore, more source apportionment studies should be carried out in future in order to assess the origin of major pollutants more precisely. In this context, especially the anthropogenic sources are of interest. As seen in this study, industry, traffic and combustion processes contributed most to the toxic particle concentrations. This became especially apparent by the high burden of air pollutants in the inner city related to traffic, in the southeastern parts of Beijing related to industry and, additionally, by the dominance of night-time pollution due to heavy duty vehicles. However, the meteorological conditions, such as precipitation, wind speed, wind direction and mixing height, which can not be addressed by mitigation measures, also strongly influence the spatio-seasonal pattern of air pollution. Through a better understanding of the extent of influence of these parameters and possible interactions, the worst pollution episodes could be predicted and, thus, the urban population could at least be forewarned.

Over the annual course, the inhabitants of Beijing are exposed to varying air pollution levels. High mass concentration episodes in spring can be considered less harmful due to lower concentrations of toxic particles from anthropogenic sources and especially a lower percentage of bioavailable metal fractions. Winter pollution episodes, on the contrary, constitute a great threat for human health because concentrations of bioavailable toxic elements, such as Cd, As, or Pb, are high.

It was shown that coal combustion is responsible for high particulate air pollution in Beijing, especially in winter, and is the dominant source for bioavailable metal fractions. About ten years after phasing out leaded gasoline, Beijing still has a higher Pb pollution than most cities worldwide, which can largely be attributed to coal combustion. The switching to natural gas for domestic heating purposes, which was started in preparation of the Olympic Games, will probably not be sufficient because of a high industrial usage of coal. Since China's natural resources are rich in coal but poor in natural gas deposits, it is likely that coal combustion will remain on high levels in China for the next decades.

The detailed investigation of the atmospheric particulate matter during a period with strictly enforced mitigation measures during the Olympic Games offered a unique possibility to assess the effects and potential of abatement measures. The stronger reduction of coarse particle concentrations compared to the finer particle fraction highlights the importance of large abatement regions in order to improve air quality, since long-range transport of fine particles is difficult to reduce. It is well known that fine particles constitute a high health risk because they can transport their toxic constituents deep into the human body. By means of sequential extractions it could be shown that health effects of fine atmospheric particles may even be more adverse than previously known due to a high portion of bioavailable toxic element concentrations from anthropogenic sources. Naturally, some of these fine particles are mitigated by coarse particles as shown within this study. This prevents a deep inhalation of the fine particles and enhances their removal from the atmosphere due to sedimentation. It remains to investigate, however, to what extend the concentrations of these hazardous small particles might increase due to missing scavengers after reducing mainly coarse particles by air pollution control. In order to clarify this point, more detailed single particle studies are necessary in the future. An investigation of a limited number of particles showed that a good tool for these studies is μS-XRF analysis, which could be performed in addition to common SEM-EDX analysis.

Conclusively, it can be stated that urban air pollution remains a complex challenge for scientists and policy makers. Very detailed information is needed in order to improve air quality and, thus, to provide liveable conditions for the growing urban population worldwide. Since the meteorological conditions, the geographical setting, and the aerosol sources are different in each urban area, the monitoring of air pollution should be individually adjusted to each city in order to assess the most harmful sources and to apply appropriate mitigation measures.

Appendix A

Tables corresponding to chapter 4

Table A.1: Descriptive statistics (valid number, average, standard deviation, lower and upper quartiles, median, 10th and 90th percentile) for TSP mass and element concentrations at site 4 during period A from Sep-05 to Aug-07. Mass concentrations in $\mu g/m^3$, all element concentrations in ng/m^3.

SITE 4 TSP	N	avg	stdev	lower QT	upper QT	median	10th PCT	90th PCT
mass	103	369	154	250	451	359	204	567
Na	101	3080	1350	1970	3890	2960	1470	4940
Mg	101	5640	2820	3670	7130	5080	2740	8780
Al	101	13100	6810	8480	15900	12000	6390	21700
K	101	6040	2970	3710	7250	5700	2810	9570
Ca	101	23000	12200	14400	28600	20600	10900	34400
Sc	101	2.76	1.49	1.65	3.41	2.42	1.22	4.97
Ti	101	664	328	403	826	599	326	1100
V	101	23.1	10.6	14.5	28.7	20.8	12.5	39.3
Cr	101	33.4	14.5	24.0	41.7	31.0	17.0	51.6
Mn	101	294	124	198	375	278	151	451
Fe	101	10200	4330	7030	12400	9280	4850	15600
Co	101	5.31	2.39	3.29	7.13	4.95	2.78	8.35
Ni	100	19.8	9.09	12.3	25.9	17.1	10.5	32.7
Cu	101	130	62.9	90.4	156	121	63.7	196
Zn	101	880	415	588	1110	806	424	1550
Ga	101	11.9	7.73	6.30	14.2	9.57	5.27	25.9
As	101	58.5	46.1	31.0	70.6	49.4	20.6	101
Rb	101	32.8	14.7	22.3	41.2	30.7	16.4	50.8
Sr	101	137	64.4	90.9	181	129	65.0	227
Y	100	4.83	2.51	2.94	6.42	4.39	2.06	8.19
Zr	101	26.3	11.4	17.1	32.4	23.9	13.7	38.0
Nb	100	2.87	1.41	1.84	3.75	2.58	1.39	4.67
Mo	51	3.42	3.48	1.30	4.31	2.29	0.77	6.71
Ag	100	1.21	0.55	0.83	1.42	1.06	0.66	2.08
Cd	101	9.90	7.08	4.32	12.5	8.34	3.41	18.1
Sn	101	29.9	16.8	18.8	38.3	25.9	12.1	53.3
Sb	101	29.3	16.7	16.9	37.6	27.0	12.4	45.5
Cs	101	3.15	1.40	2.18	3.93	2.84	1.68	5.24
Ba	99	263	131	168	344	230	121	425
Pb	101	353	181	237	438	309	157	628

Table A.2: Descriptive statistics (valid number, average, standard deviation, lower and upper quartiles, median, 10th and 90th percentile) for TSP mass and element concentrations at site 4 during period A from Sep-05 to Aug-07. Mass concentrations in $\mu g/m^3$, all element concentrations in $\mu g/g$.

SITE 4 TSP	N	avg	stdev	lower QT	upper QT	median	10th PCT	90th PCT
mass	103	369	154	250	451	359	204	567
Na	102	8650	3310	7100	9660	7930	6430	11300

Mg	102	15300	3940	12600	17300	15400	9590	20500
Al	102	35700	11500	28800	39700	33100	24500	51600
K	102	16900	7250	13500	18900	15200	11700	22600
Ca	102	63000	16900	53200	74000	64400	42200	85500
Sc	102	7.34	2.38	5.98	8.41	6.87	5.12	10.1
Ti	102	1790	527	1430	2020	1770	1230	2410
V	102	62.8	15.4	53.8	69.9	59.0	45.5	87.6
Cr	102	95.2	48.2	74.4	99.7	88.7	69.7	114
Mn	102	807	148	714	887	781	658	994
Fe	102	27900	6920	23100	31300	27100	19900	37400
Co	102	14.4	2.75	12.6	16.0	14.1	11.4	17.5
Ni	101	58.4	39.6	43.4	63.3	48.7	39.3	71.2
Cu	102	373	183	281	406	330	221	605
Zn	102	2500	962	1870	3050	2450	1390	3510
Ga	102	31.6	13.4	23.1	41.3	26.4	20.7	48.6
As	102	159	99.5	102	188	132	73.3	273
Rb	102	90.5	24.9	73.9	103	86.1	65.0	120
Sr	102	370	92.6	308	426	363	260	471
Y	101	12.9	3.44	10.6	14.7	12.4	9.16	18.8
Zr	101	73.1	17.8	61.0	81.9	68.7	55.4	99.7
Nb	100	7.73	1.82	6.48	8.88	7.34	5.64	9.94
Mo	52	15.2	37.5	4.45	15.4	7.93	1.93	26.8
Ag	100	3.39	1.03	2.64	3.94	3.34	2.23	4.71
Cd	102	27.0	15.9	15.7	32.7	23.1	12.5	47.1
Sn	102	83.9	41.6	52.7	109	79.9	42.9	142
Sb	102	80.3	38.3	56.4	94.6	74.8	47.5	121
Cs	102	8.79	2.81	6.99	9.94	8.51	6.19	12.8
Ba	100	733	320	586	768	700	526	880
Pb	102	969	321	796	1120	969	559	1430

Table A.3: Descriptive statistics (valid number, average, standard deviation, lower and upper quartiles, median, 10th and 90th percentile) for day-time $PM_{2.5}$ mass, BC and element concentrations at site 1 during period A from Sep-05 to Aug-07. Mass and BC concentrations in $\mu g/m^3$, all element concentrations in ng/m^3.

SITE 1 DAY	N	avg	stdev	lower QT	upper QT	median	10th PCT	90th PCT
mass	93	67.9	33.6	46.7	88.2	64.8	26.9	106
Na	94	543	311	302	755	491	199	994
Mg	94	447	340	169	623	325	120	1010
Al	95	1270	1190	365	1770	877	213	2780
K	94	2060	1270	1350	2520	1850	845	3150
Ca	90	1950	1520	873	2740	1490	324	4500
Sc	93	0.25	0.23	0.07	0.35	0.17	0.05	0.56
Ti	95	69.4	60.8	24.3	98.8	45.5	16.2	135
V	95	3.75	2.43	1.82	5.25	3.09	1.24	7.21
Cr	68	5.83	9.02	2.42	6.27	3.78	1.32	10.5
Mn	94	72.3	38.8	42.7	101	68.8	29.2	125
Fe	94	1160	795	556	1590	902	409	2160
Co	94	0.80	0.58	0.29	1.27	0.63	0.15	1.74
Ni	83	10.6	15.4	3.28	11.5	6.28	1.72	20.5
Cu	95	38.7	32.2	19.2	43.1	33.2	12.7	67.4
Zn	95	268	136	163	353	256	98.4	426
Ga	95	4.08	3.25	1.62	6.10	3.01	1.21	8.69
As	95	15.6	10.3	7.69	21.7	13.4	4.54	29.1
Rb	95	9.83	4.73	6.90	12.3	9.18	4.59	16.4
Sr	94	12.0	12.1	3.59	15.6	8.94	2.74	24.1
Y	66	1.35	1.05	0.52	1.94	1.10	0.18	3.16
Zr	79	18.1	11.6	8.42	25.8	17.1	4.85	31.7
Nb	53	0.51	0.36	0.21	0.64	0.51	0.10	1.15
Mo	35	14.3	15.6	3.31	17.3	8.66	1.47	34.1
Ag	82	0.88	0.57	0.46	1.08	0.78	0.32	1.82
Cd	95	3.29	2.08	1.80	4.51	2.92	0.87	5.79
Sn	95	10.3	7.09	4.82	13.6	8.74	3.37	18.6
Sb	94	8.01	5.98	4.36	9.89	7.10	2.40	14.0
Cs	94	1.47	1.50	0.87	1.69	1.25	0.58	2.25
Ba	73	46.4	78.5	11.6	52.8	23.6	5.60	80.9
Pb	94	154	82.3	95.0	180	145	53.4	280
BC	93	3.77	2.16	2.43	5.38	4.24	1.57	6.22

Table A.4: Descriptive statistics (valid number, average, standard deviation, lower and upper quartiles, median, 10th and 90th percentile) for night-time $PM_{2.5}$ mass, BC and element concentrations at site 1 during period A from Sep-05 to Aug-07. Mass and BC concentrations in $\mu g/m^3$, all element concentrations in ng/m^3.

SITE 1 NIGHT	N	avg	stdev	lower QT	upper QT	median	10th PCT	90th PCT
mass	95	57.1	32.2	36.5	76.9	50.5	21.6	108
Na	93	450	322	236	596	350	137	878
Mg	93	307	255	122	400	223	88	698
Al	93	901	830	360	1180	549	192	2230
K	93	1810	1660	906	2110	1540	586	2540
Ca	87	1650	1520	473	2340	1210	151	3540
Sc	90	0.17	0.17	0.05	0.23	0.10	0.04	0.48
Ti	92	43.8	39.9	14.6	55.1	29.7	11.6	106
V	92	2.57	1.74	1.21	3.47	1.96	0.92	5.39
Cr	59	4.25	3.87	1.42	5.60	3.54	0.67	9.20
Mn	92	55.9	31.0	34.0	68.4	50.6	20.1	109
Fe	93	798	580	469	961	595	226	1750
Co	92	0.69	0.94	0.24	0.90	0.45	0.16	1.40
Ni	76	11.2	22.0	2.79	12.2	4.60	1.56	24.6
Cu	93	31.6	18.2	18.0	37.1	30.8	11.2	57.8
Zn	93	238	122	134	313	238	78.7	396
Ga	93	3.43	2.63	1.73	4.13	2.22	1.15	8.04
As	93	13.2	8.73	7.58	17.1	11.4	3.99	26.8
Rb	93	7.86	4.16	5.19	10.1	7.31	3.03	13.0
Sr	92	9.63	14.4	2.97	9.92	5.36	1.80	22.8
Y	62	0.67	0.55	0.23	0.84	0.60	0.15	1.31
Zr	75	15.8	9.24	9.18	22.1	13.6	5.40	27.1
Nb	44	0.42	0.31	0.17	0.57	0.34	0.11	0.86
Mo	34	11.3	13.9	1.72	12.5	6.50	1.05	39.6
Ag	79	0.79	0.69	0.35	0.95	0.66	0.23	1.37
Cd	93	2.68	1.55	1.61	3.56	2.46	0.74	4.74
Sn	93	9.67	6.35	4.97	12.7	9.11	2.66	17.4
Sb	92	6.19	3.77	3.54	7.90	5.61	2.03	10.8
Cs	92	1.05	0.50	0.68	1.35	1.00	0.42	1.69
Ba	61	42.3	85.9	9.08	39.9	19.2	7.13	69.0
Pb	93	132	81.0	85.9	172	111	51.8	235
BC	92	3.39	1.77	1.88	4.75	3.11	1.32	5.69

Table A.5: Descriptive statistics (valid number, average, standard deviation, lower and upper quartiles, median, 10th and 90th percentile) for day-time $PM_{2.5}$ mass, BC and element concentrations at site 2 during period A from Sep-05 to Aug-07. Mass and BC concentrations in $\mu g/m^3$, all element concentrations in ng/m^3.

SITE 2 DAY	N	avg	stdev	lower QT	upper QT	median	10th PCT	90th PCT
mass	82	94.3	47.2	67.3	115	85.7	47.7	150
Na	83	779	483	425	1030	641	327	1430
Mg	83	820	729	327	1060	538	219	2040
Al	83	2260	2340	839	2500	1420	524	5970
K	83	2200	1160	1530	2600	2030	1030	3530
Ca	83	3530	3260	1220	4310	2650	690	7720
Sc	83	0.44	0.44	0.15	0.57	0.30	0.09	1.14
Ti	83	116	114	40.9	144	69.8	23.0	269
V	83	4.65	3.42	2.42	5.78	3.64	1.71	10.0
Cr	67	6.60	4.78	3.05	8.39	5.47	1.65	12.3
Mn	83	87.9	42.1	55.1	119	74.2	45.9	151
Fe	83	1860	1390	929	2290	1410	677	3840
Co	82	1.33	1.03	0.58	1.83	1.05	0.31	2.64
Ni	70	7.78	9.38	3.61	8.11	5.80	1.88	13.1
Cu	83	49.9	22.3	33.6	67.1	44.2	24.7	80.9
Zn	82	379	147	257	491	357	207	573
Ga	82	6.11	6.12	2.42	8.41	3.88	1.69	12.5
As	83	19.6	12.7	9.88	24.9	16.5	7.96	39.4
Rb	83	11.6	4.83	7.77	14.4	11.7	6.25	17.1
Sr	83	21.1	20.1	8.69	29.0	14.3	4.81	41.0
Y	67	1.09	1.01	0.42	1.43	0.74	0.11	2.67
Zr	69	17.9	9.10	11.8	24.5	16.5	5.84	31.2
Nb	54	0.62	0.53	0.25	0.86	0.46	0.17	1.28
Mo	42	9.25	8.94	3.45	13.7	5.95	0.80	22.4
Ag	73	0.85	0.59	0.49	0.95	0.74	0.31	1.76
Cd	83	3.74	1.91	2.42	4.76	3.49	1.74	5.90
Sn	83	38.1	230	8.74	16.5	12.8	5.11	20.3

Sb	83	13.6	6.98	8.33	18.7	12.0	5.60	21.8
Cs	83	1.41	0.59	0.97	1.75	1.41	0.73	2.12
Ba	64	51.5	38.6	23.3	71.1	38.7	14.6	107
Pb	83	181	88.8	118	225	169	79.1	293
BC (2007)	31	4.32	2.00	2.85	5.72	3.74	2.39	6.41

Table A.6: Descriptive statistics (valid number, average, standard deviation, lower and upper quartiles, median, 10th and 90th percentile) for night-time $PM_{2.5}$ mass, BC and element concentrations at site 2 during period A from Sep-05 to Aug-07. Mass and BC concentrations in $\mu g/m^3$, all element concentrations in ng/m^3.

SITE 2 NIGHT	N	avg	stdev	lower QT	upper QT	median	10th PCT	90th PCT
mass	87	105	56.8	64.0	140	91.3	49.0	179
Na	88	878	530	457	1120	837	336	1450
Mg	88	753	788	289	787	447	188	2090
Al	87	2080	2470	629	2030	1150	380	5970
K	88	2530	2080	1420	2910	2020	879	3750
Ca	88	3140	2980	1330	4200	2240	738	6540
Sc	88	0.41	0.49	0.13	0.43	0.22	0.08	1.10
Ti	88	108	117	40.4	111	61.4	24.0	280
V	88	4.98	4.07	2.20	5.56	3.56	1.65	11.7
Cr	72	6.94	5.55	2.35	10.5	5.25	0.98	14.6
Mn	88	87.3	46.1	57.5	110	76.2	41.0	145
Fe	88	1720	1540	834	1870	1180	604	4010
Co	88	1.41	1.27	0.43	2.01	0.96	0.26	3.00
Ni	75	8.91	10.7	3.29	10.4	5.13	1.82	16.4
Cu	88	75.8	37.1	52.2	93.4	71.1	35.5	118
Zn	88	427	185	285	549	402	186	630
Ga	87	7.33	6.40	2.93	11.6	4.86	2.18	15.1
As	88	24.1	16.7	11.5	31.3	18.2	8.03	48.2
Rb	88	11.0	4.81	7.36	14.3	10.1	5.53	18.0
Sr	88	25.9	33.9	7.60	28.1	11.7	4.61	97.2
Y	67	1.21	1.25	0.43	1.36	0.82	0.25	3.26
Zr	73	19.4	11.3	10.4	29.4	18.8	5.58	32.7
Nb	58	0.63	0.61	0.18	0.86	0.45	0.06	1.61
Mo	42	9.89	8.55	4.48	12.2	7.74	2.45	18.6
Ag	76	1.06	0.75	0.53	1.34	0.82	0.35	1.99
Cd	88	4.70	3.04	2.72	5.22	4.27	1.81	8.50
Sn	88	18.5	10.4	12.0	24.0	15.7	6.57	32.1
Sb	88	24.1	12.9	15.6	31.7	21.1	8.08	42.3
Cs	88	1.33	0.57	0.91	1.68	1.29	0.59	2.06
Ba	67	101	184	19.7	74.8	49.4	12.7	180
Pb	88	237	121	146	315	209	94.4	412
BC (2007)	35	6.69	3.30	4.49	8.30	5.24	3.44	11.83

Table A.7: Descriptive statistics (valid number, average, standard deviation, lower and upper quartiles, median, 10th and 90th percentile) for day-time $PM_{2.5}$ mass, BC and element concentrations at site 3 during period A from Sep-05 to Aug-07. Mass and BC concentrations in $\mu g/m^3$, all element concentrations in ng/m^3.

SITE 3 DAY	N	avg	stdev	lower QT	upper QT	median	10th PCT	90th PCT
mass	94	83.2	37.2	52.3	106	77.4	42.6	133
Na	94	756	468	407	1030	589	305	1330
Mg	94	679	659	294	799	451	210	1410
Al	94	1920	2360	737	1940	1040	510	4790
K	94	2230	1210	1450	2680	2020	1040	3530
Ca	94	3080	3170	1320	4105	2200	791	5750
Sc	94	0.36	0.44	0.13	0.37	0.19	0.08	0.84
Ti	94	96.9	115	36.8	98.8	56.4	24.0	201
V	94	3.97	3.53	2.00	4.46	2.73	1.45	8.69
Cr	66	7.05	7.50	2.92	7.56	5.14	2.20	17.2
Mn	94	78.8	39.3	52.1	103	66.1	37.8	135
Fe	94	1650	1460	823	1770	1120	628	3070
Co	92	1.02	0.86	0.39	1.36	0.72	0.27	2.03
Ni	71	11.2	36.1	2.44	7.35	3.91	1.36	16.3
Cu	94	53.2	35.2	31.6	67.6	41.6	21.1	92.7
Zn	94	418	199	262	532	394	185	697
Ga	94	6.47	14.3	2.28	6.58	3.39	1.79	11.2
As	94	21.6	13.8	13.1	27.1	17.6	10.1	37.5
Rb	94	12.1	6.31	8.23	15.5	11.0	5.12	19.2
Sr	94	17.1	14.1	7.85	22.1	11.8	6.14	34.3

Y	68	1.00	1.19	0.32	1.02	0.61	0.12	3.03
Zr	78	18.4	13.0	10.7	25.0	15.8	4.35	33.1
Nb	61	0.53	0.61	0.14	0.65	0.33	0.05	1.33
Mo	41	11.6	13.8	3.49	13.7	7.75	1.43	24.2
Ag	82	0.89	0.50	0.53	1.11	0.81	0.34	1.58
Cd	94	4.36	2.75	2.37	5.35	3.74	1.75	7.49
Sn	94	15.6	9.83	9.02	18.8	14.3	6.11	25.9
Sb	94	11.1	6.55	6.35	14.3	9.34	4.28	18.2
Cs	94	1.48	0.73	1.04	1.80	1.35	0.73	2.48
Ba	72	42.3	31.7	21.1	54.0	33.7	12.9	95.8
Pb	94	211	104	133	279	181	101	327
BC	94	4.32	2.00	2.89	5.80	4.02	2.07	6.42

Table A.8: Descriptive statistics (valid number, average, standard deviation, lower and upper quartiles, median, 10th and 90th percentile) for night-time $PM_{2.5}$ mass, BC and element concentrations at site 3 during period A from Sep-05 to Aug-07. Mass and BC concentrations in $\mu g/m^3$, all element concentrations in ng/m^3.

SITE 3				lower	upper		10th	90th
NIGHT	N	avg	stdev	QT	QT	median	PCT	PCT
mass	95	82.8	43.7	52.0	106	73.1	31.0	148
Na	96	687	440	346	962	565	205	1390
Mg	96	491	458	174	617	282	136	1330
Al	96	1640	2560	406	1540	677	284	4520
K	96	2320	2040	1190	2720	1760	718	4050
Ca	93	2250	1960	796	2980	1750	439	4670
Sc	95	0.27	0.34	0.08	0.26	0.12	0.05	0.86
Ti	96	68.7	86.5	22.5	67.5	32.3	17.5	213
V	96	3.71	3.25	1.49	4.60	2.49	1.04	9.21
Cr	70	6.62	14.2	1.87	6.47	4.00	0.76	11.7
Mn	96	70.8	40.2	41.3	95.3	58.4	31.0	139
Fe	96	1240	1150	525	1390	783	399	3120
Co	95	1.06	1.32	0.31	1.44	0.59	0.18	2.08
Ni	74	9.10	17.1	2.30	8.11	4.45	1.01	13.0
Cu	96	55.5	37.9	30.4	72.0	45.1	19.5	95.7
Zn	95	390	222	225	549	347	123	663
Ga	96	5.58	5.10	2.31	8.14	3.10	1.38	12.4
As	96	24.5	20.8	12.6	29.5	19.9	7.43	48.4
Rb	96	10.2	5.82	5.85	13.5	8.63	4.13	18.9
Sr	96	14.5	13.6	6.02	16.4	9.97	3.43	33.8
Y	64	0.79	0.67	0.33	1.09	0.57	0.15	1.81
Zr	77	17.1	10.3	9.52	22.3	15.6	6.22	28.2
Nb	50	0.48	0.46	0.12	0.62	0.30	0.05	1.25
Mo	44	9.82	9.90	3.31	12.8	7.43	0.30	27.3
Ag	82	0.91	0.80	0.45	1.12	0.72	0.27	1.73
Cd	96	5.24	4.59	2.48	6.99	3.78	1.35	9.84
Sn	96	17.2	10.9	9.64	21.4	15.3	6.40	31.6
Sb	96	12.5	8.32	7.11	15.7	10.6	4.07	20.1
Cs	96	1.24	0.68	0.74	1.62	1.07	0.56	2.22
Ba	76	48.4	68.7	13.5	58.8	28.1	6.01	75.5
Pb	96	223	155	130	263	174	71.3	418
BC	95	5.24	2.89	3.26	6.85	4.57	2.00	8.60

Table A.9: Descriptive statistics (valid number, average, standard deviation, lower and upper quartiles, median, 10th and 90th percentile) for day-time $PM_{2.5}$ mass, BC and element concentrations at site 4 during period A from Sep-05 to Aug-07. Mass and BC concentrations in $\mu g/m^3$, all element concentrations in ng/m^3.

SITE 4				lower	upper		10th	90th
DAY	N	avg	stdev	QT	QT	median	PCT	PCT
mass	102	102	44.9	68.8	121	95.6	54.0	160
Na	103	826	433	528	1100	716	375	1320
Mg	102	757	526	365	1020	590	231	1450
Al	102	1980	1830	830	2372	1471	563	3820
K	102	2500	1450	1480	3270	2190	935	4420
Ca	100	3330	2240	1590	4490	2990	878	6570
Sc	102	0.37	0.34	0.15	0.49	0.29	0.10	0.77
Ti	102	102	92.5	39.1	124	76.1	21.2	208
V	103	4.28	3.25	2.18	5.24	3.48	1.48	7.77
Cr	84	8.88	5.53	5.20	12.5	7.78	2.62	17.0
Mn	101	87.9	44.3	59.8	107	79.0	41.1	142
Fe	101	1770	1270	955	2210	1510	636	3090
Co	102	1.14	0.88	0.49	1.57	0.97	0.33	1.92

Ni	86	15.5	17.2	6.86	17.4	10.3	4.38	28.8
Cu	103	65.6	34.7	43.3	85.8	58.7	27.4	112
Zn	103	452	228	296	565	410	182	705
Ga	102	5.47	4.75	2.63	6.84	3.62	1.83	10.3
As	102	28.5	20.7	15.2	34.0	25.1	9.98	49.5
Rb	102	12.9	7.10	7.73	17.4	11.5	5.99	22.2
Sr	101	19.3	13.1	9.67	26.4	17.0	5.38	37.1
Y	80	0.92	0.72	0.43	1.20	0.75	0.23	1.88
Zr	85	19.6	11.7	10.5	27.4	18.4	5.45	34.2
Nb	71	0.53	0.51	0.17	0.65	0.40	0.07	1.33
Mo	37	11.8	11.7	3.57	15.3	6.83	0.67	33.9
Ag	88	0.93	0.58	0.52	1.17	0.82	0.35	1.76
Cd	103	4.91	3.37	2.54	6.41	4.07	1.55	9.05
Sn	102	17.4	11.3	9.78	22.1	16.6	5.99	26.8
Sb	103	14.1	8.86	8.12	18.4	13.4	4.90	24.5
Cs	102	1.57	0.82	1.00	2.08	1.42	0.76	2.67
Ba	79	71.0	131	28.5	71.6	47.2	21.6	115
Pb	102	228	138	142	307	190	89.2	399
BC	101	7.86	3.83	4.86	9.73	7.44	3.86	13.20

Table A.10: Descriptive statistics (valid number, average, standard deviation, lower and upper quartiles, median, 10th and 90th percentile) for night-time $PM_{2.5}$ mass, BC and element concentrations at site 4 during period A from Sep-05 to Aug-07. Mass and BC concentrations in $\mu g/m^3$, all element concentrations in ng/m^3.

SITE 4 NIGHT	N	avg	stdev	lower QT	upper QT	median	10th PCT	90th PCT
mass	98	96.7	59.6	51.8	117	87.3	32.4	190
Na	101	738	532	373	850	589	288	1540
Mg	100	521	442	228	745	401	140	1080
Al	100	1280	1300	497	1670	843	304	2560
K	100	2250	2040	1230	2620	1690	719	4220
Ca	99	2340	1920	959	3200	1770	506	4760
Sc	99	0.23	0.24	0.09	0.32	0.16	0.06	0.49
Ti	100	61.9	63.4	25.0	80.8	43.7	14.4	118.8
V	101	2.94	2.28	1.41	4.00	2.32	0.89	5.20
Cr	77	6.74	6.50	2.86	9.75	4.74	1.67	13.2
Mn	99	71.6	45.9	45.1	86.2	61.4	24.4	120
Fe	99	1170	893	620	1550	954	353	2040
Co	100	1.05	1.55	0.39	1.16	0.64	0.21	2.14
Ni	84	15.2	15.5	6.25	17.3	11.7	2.64	32.2
Cu	101	72.6	42.1	43.6	95.4	64.6	25.0	128
Zn	101	398	228	244	507	380	143	680
Ga	101	5.19	4.66	2.37	6.60	3.17	1.76	12.7
As	100	28.7	25.6	14.2	35.2	22.9	7.17	53.3
Rb	100	9.40	5.58	5.70	11.5	8.42	2.97	16.7
Sr	101	16.0	17.4	6.59	18.3	10.7	3.21	31.4
Y	70	0.67	0.62	0.25	0.97	0.49	0.10	1.50
Zr	83	19.7	11.0	11.4	27.2	17.5	6.74	36.3
Nb	52	0.49	0.37	0.23	0.65	0.38	0.12	0.98
Mo	39	11.4	12.6	3.22	15.0	6.57	1.45	33.3
Ag	81	0.98	0.83	0.49	1.14	0.74	0.30	1.95
Cd	101	5.29	4.30	2.51	6.26	4.15	1.25	10.1
Sn	100	19.4	13.5	9.46	25.4	16.7	6.14	35.4
Sb	101	18.0	15.2	8.86	22.0	14.9	5.18	31.6
Cs	100	1.13	0.62	0.72	1.36	1.11	0.38	1.97
Ba	79	53.1	96.1	14.5	52.5	31.0	5.72	95.7
Pb	100	217.8	143	113	269	182	68.5	393
BC	100	10.21	6.28	4.94	13.0	9.36	3.38	19.32

Table A.11: Descriptive statistics (valid number, average, standard deviation, lower and upper quartiles, median, 10th and 90th percentile) for day-time $PM_{2.5}$ mass, BC and element concentrations at site 5 during period A from Sep-05 to Aug-07. Mass and BC concentrations in $\mu g/m^3$, all element concentrations in ng/m^3.

SITE 5 DAY	N	avg	stdev	lower QT	upper QT	median	10th PCT	90th PCT
mass	94	94.8	54.8	60.2	119	80.0	37.7	160
Na	94	863	1400	405	1040	654	279	1380
Mg	94	666	695	227	845	419	156	1470
Al	94	1670	1695	519	2240	1060	329	4230
K	94	2820	1840	1590	3550	2540	884	4590
Ca	91	2630	2190	987	3380	2000	563	5010

Sc	93	0.32	0.33	0.09	0.42	0.20	0.06	0.84
Ti	94	81.1	81.0	24.0	109	49.6	15.9	189
V	94	3.73	2.82	1.86	4.72	2.98	1.01	7.77
Cr	69	6.11	6.61	2.39	7.46	4.87	0.95	11.4
Mn	94	74.9	40.4	43.2	98.9	64.4	31.2	125
Fe	94	1500	1140	681	1900	1200	441	2960
Co	92	0.98	0.70	0.39	1.36	0.84	0.22	2.03
Ni	79	10.0	9.81	4.26	13.3	7.19	2.15	19.8
Cu	94	86.7	60.9	45.6	106	70.1	36.7	137
Zn	94	401	191	246	535	385	185	673
Ga	94	4.72	3.58	2.05	7.02	3.24	1.65	9.29
As	94	23.7	17.5	12.9	27.4	19.8	7.38	42.5
Rb	94	13.1	7.33	8.19	18.0	11.4	5.04	23.1
Sr	94	14.6	12.4	5.57	18.0	11.4	3.46	31.3
Y	67	0.90	0.92	0.33	1.13	0.64	0.22	1.78
Zr	78	17.9	10.7	10.3	23.1	15.8	5.43	31.0
Nb	56	0.54	0.43	0.18	0.74	0.44	0.11	1.10
Mo	46	12.9	22.1	2.84	14.5	5.71	1.47	28.5
Ag	82	1.07	1.06	0.58	1.28	0.86	0.42	1.80
Cd	94	4.74	2.83	2.38	6.29	4.55	1.63	8.55
Sn	94	16.3	9.74	9.64	20.0	14.4	5.98	28.3
Sb	94	12.1	7.45	6.84	15.3	11.4	4.14	21.3
Cs	94	1.57	0.88	0.92	2.12	1.41	0.58	2.61
Ba	71	62.1	78.1	22.2	69.5	34.3	11.7	113
Pb	94	217	123	128	287	194	80.1	376
BC	94	6.00	3.78	3.15	7.55	5.17	2.16	10.8

Table A.12: Descriptive statistics (valid number, average, standard deviation, lower and upper quartiles, median, 10th and 90th percentile) for night-time $PM_{2.5}$ mass, BC and element concentrations at site 5 during period A from Sep-05 to Aug-07. Mass and BC concentrations in $\mu g/m^3$, all element concentrations in ng/m^3.

SITE 5 NIGHT	N	avg	stdev	lower QT	upper QT	median	10th PCT	90th PCT
mass	95	80.0	57.9	40.7	97.6	70.7	17.1	149
Na	93	567	339	311	752	507	200	1030
Mg	95	350	326	116	522	227	65	883
Al	94	999	1070	282	1400	536	168	2410
K	96	2280	2300	1010	2940	1830	476	3720
Ca	85	1410	1470	539	1820	1010	146	2930
Sc	90	0.19	0.20	0.05	0.25	0.12	0.03	0.49
Ti	96	44.7	43.9	14.1	55.1	28.4	6.35	114
V	96	2.59	2.00	1.07	3.65	2.05	0.57	5.81
Cr	69	7.10	18.2	2.09	6.06	3.60	0.53	10.7
Mn	96	60.3	35.7	27.6	87.1	61.6	13.2	108
Fe	96	915	711	384	1350	726	201	1750
Co	96	0.80	0.62	0.36	1.09	0.66	0.19	1.67
Ni	76	10.1	13.2	3.27	9.25	5.33	1.89	29.1
Cu	96	71.6	50.3	39.9	87.0	59.2	22.5	124
Zn	96	328	208	149	449	328	70.5	659
Ga	96	3.92	2.87	2.14	5.36	2.94	1.28	7.51
As	96	22.5	24.8	11.5	25.7	17.4	7.10	35.1
Rb	96	9.21	6.08	4.29	13.6	7.89	2.22	18.2
Sr	95	9.24	11.0	3.08	12.3	6.12	1.56	17.2
Y	58	0.61	0.42	0.36	0.78	0.49	0.24	1.13
Zr	72	17.9	9.70	10.5	24.6	15.7	7.39	34.1
Nb	49	0.37	0.31	0.13	0.56	0.25	0.08	0.75
Mo	40	12.0	17.6	1.75	12.3	4.71	0.81	30.9
Ag	81	0.84	0.57	0.43	1.15	0.78	0.16	1.41
Cd	95	4.69	4.12	1.91	6.20	3.76	1.12	9.38
Sn	96	15.5	11.6	7.61	20.7	13.0	4.07	28.3
Sb	96	10.7	7.67	4.90	13.4	8.72	2.56	23.0
Cs	96	1.10	0.69	0.56	1.62	0.93	0.37	1.98
Ba	65	53.6	121	12.6	51.5	26.1	6.29	73.7
Pb	96	181	121	84.2	257	164	50.9	301
BC	87	6.49	4.11	3.79	9.01	5.41	2.14	10.9

Table A.13: Descriptive statistics (valid number, average, standard deviation, lower and upper quartiles, median, 10th and 90th percentile) for day-time $PM_{2.5}$ mass and element concentrations at site 1 during period A from Sep-05 to Aug-07. Mass concentrations in $\mu g/m^3$, all element concentrations in $\mu g/g$.

SITE 1 DAY	N	avg	stdev	lower QT	upper QT	median	10th PCT	90th PCT
mass	93	68.4	33.4	46.7	88.2	65.1	26.9	106
Na	92	8490	5490	5140	9940	7120	3880	14400
Mg	92	7310	7090	3320	9230	5360	1990	12800
Al	92	20200	20500	6940	26100	14400	5350	44000
K	92	33300	20800	22900	37400	28000	19000	46100
Ca	88	32000	32900	13600	38800	25000	5290	62200
Sc	91	3.79	3.76	1.21	4.87	2.70	0.84	7.19
Ti	92	1120	1140	447	1330	843	292	2280
V	92	61.2	49.6	36.8	71.6	51.6	25.2	92.9
Cr	67	101	159	37.0	105	55.4	27.7	184
Mn	92	1210	887	756	1350	1070	603	1590
Fe	92	18000	13700	10200	20200	15400	8270	28800
Co	92	12.9	11.4	6.45	14.5	10.6	4.28	22.1
Ni	82	174	227	56.2	166	94.6	29.9	435
Cu	92	670	502	353	787	508	300	1138
Zn	92	4630	2930	2980	5150	4070	2080	6760
Ga	92	66.2	57.6	34.8	82.2	51.1	27.5	107
As	92	261	194	163	298	216	127	354
Rb	92	166	88.1	116	179	152	99.3	242
Sr	91	179	172	74.1	240	120	48.5	312
Y	66	21.2	18.8	7.84	27.2	17.1	4.70	41.3
Zr	78	344	347	164	372	212	79.3	756
Nb	51	8.69	10.8	3.37	10.0	5.37	1.69	16.6
Mo	32	376	786	74.3	303	231	34.9	647
Ag	81	16.5	17.5	7.14	19.2	11.3	5.30	31.5
Cd	92	55.4	42.4	38.1	58.8	47.9	27.2	86.5
Sn	92	179	155	107	201	145	76.6	304
Sb	92	136	113	80.4	137	106	59.3	214
Cs	92	23.4	15.0	15.2	26.2	20.1	13.1	33.9
Ba	70	636	798	153	698	381	103	1490
Pb	92	2540	1260	1900	2950	2410	1490	3490

Table A.14: Descriptive statistics (valid number, average, standard deviation, lower and upper quartiles, median, 10th and 90th percentile) for night-time $PM_{2.5}$ mass and element concentrations at site 1 during period A from Sep-05 to Aug-07. Mass concentrations in $\mu g/m^3$, all element concentrations in $\mu g/g$.

SITE 1 NIGHT	N	avg	stdev	lower QT	upper QT	median	10th PCT	90th PCT
mass	95	58.1	32.1	37.9	76.9	50.7	22.7	108
Na	91	8750	5260	4990	10900	7780	3300	14500
Mg	91	6450	5640	2810	9180	5190	1860	11300
Al	91	18100	18100	6950	24300	11100	4000	38900
K	91	35000	23000	21500	39300	29100	17300	61100
Ca	85	35400	42000	10400	41500	23200	4270	83000
Sc	91	3.25	3.32	1.12	5.04	1.98	0.66	6.89
Ti	91	883	857	318	1300	624	207	1811
V	91	50.0	31.2	30.8	64.1	45.1	23.3	84.0
Cr	56	96.8	129	31.1	104	62.6	16.4	203
Mn	91	1120	559	729	1390	1040	534	1710
Fe	91	16000	11900	9010	21200	12900	7150	26600
Co	91	13.4	13.6	6.13	16.7	11.1	3.44	21.5
Ni	75	288	957	49.9	236	77.8	36.1	523
Cu	91	657	405	397	759	574	312	1100
Zn	91	4970	2690	3100	6260	4340	2520	8190
Ga	91	69.3	49.0	34.5	86.3	59.3	27.1	119
As	91	263	139	184	323	241	130	398
Rb	91	159	69.5	106	199	146	87.9	252
Sr	91	164	135	62	229	135	38.0	318
Y	61	13.6	12.9	6.58	19.3	10.1	2.27	24.7
Zr	73	384	394	162	497	293	96.2	775
Nb	44	10.8	17.0	3.49	12.3	5.91	1.66	24.6
Mo	33	262	411	31.7	196	107	18.8	769
Ag	77	17.1	14.8	8.39	22.0	13.0	5.75	36.0
Cd	91	54.0	25.8	37.5	65.5	51.5	25.3	82.3
Sn	91	203	172	123	228	167	68.5	354
Sb	91	123	61.9	79.4	155	114	63.2	189

Cs	90	21.2	8.56	14.4	25.6	19.5	12.6	31.1
Ba	60	646	794	187	778	346	153	1450
Pb	91	2720	1690	1820	3220	2210	1450	4140

Table A.15: Descriptive statistics (valid number, average, standard deviation, lower and upper quartiles, median, 10th and 90th percentile) for day-time $PM_{2.5}$ mass and element concentrations at site 2 during period A from Sep-05 to Aug-07. Mass concentrations in $\mu g/m^3$, all element concentrations in $\mu g/g$.

SITE 2 DAY	N	avg	stdev	lower QT	upper QT	median	10th PCT	90th PCT
mass	88	89.9	49.1	58.5	113	82.1	29.2	150
Na	84	8560	4530	5450	9920	7810	4370	13200
Mg	86	9050	8990	4080	10900	6940	2500	14700
Al	86	24800	28600	11000	28400	17900	5110	43700
K	86	24700	12900	17700	28900	23800	13300	34700
Ca	85	40800	56100	13700	46400	29600	7140	64800
Sc	86	4.75	5.02	2.02	5.74	3.65	1.06	7.73
Ti	86	1250	1270	550	1390	973	256	2080
V	86	52.3	41.1	30.9	59.3	42.8	20.4	86.1
Cr	71	94.9	140	35.1	91.0	61.7	20.4	152
Mn	86	954	448	743	1100	904	517	1340
Fe	86	20700	16200	12100	23900	17600	7970	33300
Co	86	14.8	12.7	7.70	19.0	11.4	4.21	22.2
Ni	74	101	124	40.7	109	68.9	21.6	148
Cu	86	606	372	344	803	541	261	1020
Zn	83	4480	2330	2900	5630	4190	1870	6910
Ga	85	65.8	64.2	28.4	81.8	42.5	21.7	135
As	85	214	137	126	272	195	70.4	351
Rb	86	131	54.9	92.0	161	132	62.7	202
Sr	86	223	207	104	284	194	58.7	395
Y	69	11.4	11.8	4.88	14.3	8.07	1.34	23.1
Zr	67	199	145	117	235	156	82.7	378
Nb	54	6.55	6.25	3.02	8.56	4.99	1.41	12.7
Mo	42	104	117	26.5	159	54.4	7.73	214
Ag	76	10.1	8.06	5.89	11.2	8.12	3.04	15.9
Cd	86	41.8	24.0	25.9	54.5	39.7	14.2	62.5
Sn	84	150	88.4	89.4	194	137	44.2	270
Sb	84	155	98.8	92.9	195	144	61.6	265
Cs	86	15.7	6.96	11.1	20.5	16.3	5.39	24.9
Ba	65	547	406	292	624	438	240	921
Pb	86	1990	904	1570	2550	2080	705	3180

Table A.16: Descriptive statistics (valid number, average, standard deviation, lower and upper quartiles, median, 10th and 90th percentile) for night-time $PM_{2.5}$ mass and element concentrations at site 2 during period A from Sep-05 to Aug-07. Mass concentrations in $\mu g/m^3$, all element concentrations in $\mu g/g$.

SITE 2 NIGHT	N	avg	stdev	lower QT	upper QT	median	10th PCT	90th PCT
mass	91	103	56.9	64.0	133	90.8	41.3	172
Na	88	8910	4260	6410	9850	8380	4750	12700
Mg	88	7580	7590	3680	9810	6310	1980	14100
Al	87	21100	25300	8350	24500	15400	5000	42000
K	88	24900	15700	18000	28000	21700	12600	39100
Ca	87	34300	40900	16400	41300	25500	9190	62700
Sc	88	4.00	4.52	1.51	4.45	3.22	0.98	7.98
Ti	88	1090	1170	563	1240	864	317	1970
V	88	50.5	36.0	31.0	60.8	41.5	22.0	84.2
Cr	72	72.5	71.8	32.6	85.5	59.7	10.7	127
Mn	88	914	376	709	1060	867	537	1370
Fe	88	18000	14900	10200	21000	14700	7190	28700
Co	88	14.3	12.7	5.97	17.2	11.8	3.67	30.4
Ni	75	102	137	36.5	87.0	61.6	22	278
Cu	88	813	368	565	1020	773	403	1240
Zn	87	4830	2755	2770	6090	4360	2260	7200
Ga	87	74.9	62.3	37.7	94.1	52.7	28.1	135
As	88	239	120	159	306	219	108	395
Rb	88	118	46.9	86.4	143	106	68.8	175
Sr	88	224	218	98.1	269	171	59.6	417
Y	68	11.3	10.9	6.07	13.9	9.91	1.43	20.4
Zr	72	213	152	125	282	157	68.9	388
Nb	56	6.35	6.49	2.21	8.42	4.23	0.89	14.1

Mo	40	127	129	39.8	171	85.0	21.3	302
Ag	75	12.1	11.3	5.33	13.0	9.68	3.69	22.5
Cd	88	48.8	25.6	33.2	56.2	46.8	21.8	75.6
Sn	87	207	123	137	243	186	60.7	334
Sb	88	259	130	171	330	256	88.4	436
Cs	88	14.1	5.79	9.70	17.6	13.0	7.96	22.0
Ba	66	718	894	259	675	475	170	1210
Pb	88	2450	996	1820	2980	2530	901	3500

Table A.17: Descriptive statistics (valid number, average, standard deviation, lower and upper quartiles, median, 10th and 90th percentile) for day-time $PM_{2.5}$ mass and element concentrations at site 3 during period A from Sep-05 to Aug-07. Mass concentrations in $\mu g/m^3$, all element concentrations in $\mu g/g$.

SITE 3 DAY	N	avg	stdev	lower QT	upper QT	median	10th PCT	90th PCT
mass	94	83.8	37.2	52.5	106	78.4	42.6	133
Na	92	9700	7420	6000	10600	7740	4620	15500
Mg	92	8020	6060	4000	9800	6370	2610	14700
Al	92	21700	20500	9620	24200	15200	7170	44100
K	92	27800	15100	20600	31000	24900	16100	40900
Ca	92	37500	35900	19600	44900	27800	12000	62500
Sc	92	4.08	3.82	1.80	4.91	3.12	1.02	8.16
Ti	92	1100	972	527	1360	834	328	2010
V	92	47.2	33.1	27.9	53.4	40.5	19.1	79.0
Cr	64	77.5	75.5	35.6	86.5	60.2	26.7	125
Mn	92	1000	501	699	1120	916	579	1560
Fe	92	19600	13000	12000	22500	16500	8840	32100
Co	90	13.5	15.1	5.78	14.5	9.7	4.42	23.7
Ni	69	153	476	26.5	89.1	53.1	11.8	266
Cu	92	682	409	402	844	590	321	1130
Zn	92	5540	3100	3680	6320	4700	2930	8550
Ga	92	81.6	137	31.0	84.0	40.3	26.9	147
As	92	287	211	177	318	251	112	475
Rb	92	150	65.2	107	171	138	90.5	222
Sr	92	207	152	126	249	172	80.1	321
Y	66	11.5	12.3	4.30	12.7	8.56	1.76	21.7
Zr	76	263	258	121	295	191	66.7	525
Nb	61	5.25	5.18	2.03	6.01	3.86	0.40	12.1
Mo	41	175	241	40.3	186	80.5	25.5	501
Ag	80	11.6	7.36	7.11	13.2	10.4	5.05	17.4
Cd	92	55.9	40.6	36.3	62.9	45.4	25.6	81.8
Sn	92	206	139	117	244	181	100	317
Sb	92	143	85.0	89.0	160	130	64.3	251
Cs	92	19.0	10.0	12.9	21.7	16.4	10.5	26.5
Ba	77	554	676	228	594	443	150	1000
Pb	92	2820	1790	1800	3260	2330	1490	4230

Table A.18: Descriptive statistics (valid number, average, standard deviation, lower and upper quartiles, median, 10th and 90th percentile) for night-time $PM_{2.5}$ mass and element concentrations at site 3 during period A from Sep-05 to Aug-07. Mass concentrations in $\mu g/m^3$, all element concentrations in $\mu g/g$.

SITE 3 NIGHT	N	avg	stdev	lower QT	upper QT	median	10th PCT	90th PCT
mass	96	82.6	44.1	52.0	106	75.3	30.2	148
Na	96	8750	5550	5280	10300	7490	3770	16000
Mg	96	6380	5020	2770	8370	5010	2000	11700
Al	96	18300	20200	6660	20900	11700	4550	48400
K	96	29400	18400	20400	31900	24600	14900	46500
Ca	93	30600	28800	11700	43800	22900	7110	59500
Sc	95	3.19	3.18	1.29	3.76	2.03	0.94	7.13
Ti	96	841	831	354	1000	565	240	2240
V	96	46.3	33.7	24.9	50.2	37.7	19.0	90.2
Cr	70	77.3	102	26.1	90.8	43.2	12.6	156
Mn	96	948	522	657	1060	864	492	1350
Fe	96	16100	12800	8250	18600	12400	6900	31500
Co	95	15.2	24.6	5.28	15.3	9.95	3.09	26.7
Ni	74	104	131	28.4	112	52.3	16.7	325
Cu	96	795	671	465	864	625	343	1360
Zn	95	5230	2860	3380	6320	5060	2270	7520
Ga	96	78.6	74.0	33.0	104	43.7	26.2	173
As	96	325	198	183	414	284	127	582

Rb	96	135	66.0	95.3	160	124	78.8	208
Sr	96	182	142	100	210	144	67.7	327
Y	64	9.6	7.17	4.49	11.4	7.78	2.70	20.0
Zr	77	230	163	130	288	192	84.5	417
Nb	52	5.60	5.67	1.66	9.23	3.77	0.39	10.8
Mo	44	236	460	29.0	196	81.2	7.63	571
Ag	82	12.9	12.1	6.09	15.7	9.94	4.24	20.1
Cd	96	70.2	55.3	36.3	77.1	57.5	26.6	149
Sn	96	242	163	150	293	215	104	380
Sb	96	175	120	112	206	153	76.0	271
Cs	96	16.8	8.27	11.8	20.3	15.4	8.91	23.5
Ba	83	576	791	208	657	388	113	858
Pb	96	2990	1680	1950	3890	2570	1450	4680

Table A.19: Descriptive statistics (valid number, average, standard deviation, lower and upper quartiles, median, 10th and 90th percentile) for day-time $PM_{2.5}$ mass and element concentrations at site 4 during period A from Sep-05 to Aug-07. Mass concentrations in $\mu g/m^3$, all element concentrations in $\mu g/g$.

SITE 4 DAY	N	avg	stdev	lower QT	upper QT	median	10th PCT	90th PCT
mass	102	102	44.6	68.8	121	95.6	54.0	160
Na	102	8650	3690	6220	10300	7900	4610	13300
Mg	101	7890	4690	4230	10100	7660	2840	15300
Al	101	20200	15600	10500	23500	17900	6210	39700
K	100	26200	13500	19700	30600	23500	13300	38800
Ca	99	35100	24600	19800	46600	31900	8440	59900
Sc	101	3.82	2.81	1.73	4.73	3.44	1.15	6.99
Ti	101	1020	760	462	1380	927	247	1820
V	102	44.4	34.8	28.4	50.4	35.8	18.6	69.5
Cr	83	89.6	49.6	50.5	120	76.6	29.9	158
Mn	101	905	397	690	1040	876	568	1270
Fe	101	17900	9910	10200	22300	16900	8590	27900
Co	101	11.4	7.24	7.20	13.8	11.5	4.24	17.3
Ni	85	180	264	75.9	173	106	52.6	408
Cu	102	712	403	429	851	636	347	1190
Zn	102	4890	2880	3460	5410	4200	2550	7300
Ga	101	54.7	47.9	29.2	68.1	40.5	24.3	95.0
As	101	294	184	179	362	270	134	513
Rb	101	134	68.1	98.4	160	121	74.6	191
Sr	100	198	124	118	260	174	69.4	328
Y	78	9.9	7.53	4.84	12.4	8.37	1.79	19.9
Zr	84	207	148	101	270	199	67.2	355
Nb	71	5.33	4.61	1.62	7.26	4.44	0.85	11.2
Mo	34	145	158	33.4	220	76.8	8.15	430
Ag	87	10.2	7.83	6.13	11.8	7.78	4.56	17.5
Cd	102	51.1	37.0	32.8	63.1	43.6	23.1	73.8
Sn	101	185	128	110	221	168	85.8	276
Sb	102	149	86.5	98.8	181	126	59.9	251
Cs	101	16.5	8.44	12.3	18.1	15.7	9.08	23.2
Ba	78	903	2060	291	770	482	206	1440
Pb	102	2300	1270	1670	2720	2140	1180	3290

Table A.20: Descriptive statistics (valid number, average, standard deviation, lower and upper quartiles, median, 10th and 90th percentile) for night-time $PM_{2.5}$ mass and element concentrations at site 4 during period A from Sep-05 to Aug-07. Mass concentrations in $\mu g/m^3$, all element concentrations in $\mu g/g$.

SITE 4 NIGHT	N	avg	stdev	lower QT	upper QT	median	10th PCT	90th PCT
mass	98	97.6	59.1	55.5	117	87.9	32.4	190
Na	98	8530	4760	5370	11200	6940	4170	15400
Mg	98	6380	5470	2860	8790	4540	1700	13800
Al	98	16000	16300	6390	19300	10800	4010	38800
K	97	25300	16100	16700	28600	20100	13800	45100
Ca	96	33900	38300	12200	39400	18900	6130	95800
Sc	98	2.78	3.02	1.01	3.39	1.79	0.55	7.02
Ti	97	769	781	294	869	503	194	1900
V	98	35.7	28.3	18.5	38.9	26.3	14.3	66.8
Cr	75	88.3	142	33.3	87.1	56.8	20.3	148
Mn	97	791	376	599	962	719	455	1210
Fe	97	14100	11000	7470	16900	10300	5764	26000
Co	97	10.7	8.58	5.45	12.7	9.15	3.34	17.1

Ni	81	252	395	64.1	235	112	37.4	387
Cu	98	896	694	522	1090	715	434	1350
Zn	98	4740	2690	3090	5730	4070	2510	7620
Ga	98	56.5	32.7	33.7	80.7	42.9	26.4	104
As	97	303	145	221	348	290	156	463
Rb	97	110	53.9	78.7	132	100	57.4	183
Sr	98	176	127	84.7	226	152	50.1	322
Y	70	8.35	8.06	3.47	11.5	5.51	1.29	17.9
Zr	80	271	243	117	304	186	80.8	725
Nb	51	7.19	7.71	1.53	10.7	4.44	0.56	18.9
Mo	40	180	277	37.0	220	77.6	16.6	478
Ag	78	12.0	13.6	5.41	12.3	7.40	3.96	21.6
Cd	98	57.9	33.9	37.3	67.8	49.0	23.7	104
Sn	97	221	111	146	279	197	111	388
Sb	98	204	150	128	227	176	80.6	304
Cs	97	13.5	6.60	9.6	16.3	12.6	7.54	25.1
Ba	77	607	743	195	691	369	101	1220
Pb	98	2450	1240	1810	2770	2290	1290	3510

Table A.21: Descriptive statistics (valid number, average, standard deviation, lower and upper quartiles, median, 10th and 90th percentile) for day-time $PM_{2.5}$ mass and element concentrations at site 5 during period A from Sep-05 to Aug-07. Mass concentrations in $\mu g/m^3$, all element concentrations in $\mu g/g$.

SITE 5 DAY	N	avg	stdev	lower QT	upper QT	median	10th PCT	90th PCT
mass	93	96.2	54.3	61.7	119	81.0	43.4	160
Na	92	11000	28600	5960	9270	7500	4690	14100
Mg	92	7690	10500	2900	9370	5550	1910	13200
Al	91	18300	14600	7580	25200	15000	4160	34600
K	92	31900	17400	22600	37700	28800	14700	48400
Ca	89	30000	29600	14400	34000	22700	7450	55700
Sc	92	3.33	2.62	1.37	4.91	2.56	0.66	6.57
Ti	92	853	685	328	1120	644	197	1720
V	92	41.9	28.3	24.6	52.8	33.4	14.4	84.3
Cr	67	69.7	81.1	32.1	77.8	53.3	14.9	110.5
Mn	92	827	340	631	952	763	478	1208
Fe	92	16500	9500	9560	21600	13670	6020	30600
Co	91	10.6	7.16	5.90	12.2	9.9	4.06	16.8
Ni	77	143	216	46.1	152	68.7	35.4	303
Cu	92	1190	1200	541	1240	888	349	2510
Zn	92	4760	2530	3150	5540	4230	2150	7860
Ga	92	50.6	29.7	30.2	67.0	40.9	24.2	93.5
As	92	263	145	166	342	234	105	496
Rb	92	153	80.0	104	179	139	79.9	255
Sr	92	160	117	77.4	207	150	50.9	266
Y	67	9.60	10.0	3.51	10.6	6.55	2.40	19.8
Zr	76	238	283	109	274	184	68.2	382
Nb	57	5.40	4.44	2.89	7.62	4.22	1.25	11.1
Mo	45	136	199	29.2	147	56.1	19.8	289
Ag	80	12.5	10.0	6.21	15.3	9.17	4.61	23.8
Cd	92	53.6	28.2	30.8	67.3	50.8	24.0	81.1
Sn	92	192	114	110	242	157	75.9	345
Sb	92	139	81.6	81.8	176	124	55.8	252
Cs	92	18.5	10.2	13.3	22.2	16.6	8.01	28.7
Ba	70	787	1280	235	582	419	157	1690
Pb	92	2480	1110	1820	3090	2380	1190	3880

Table A.22: Descriptive statistics (valid number, average, standard deviation, lower and upper quartiles, median, 10th and 90th percentile) for night-time $PM_{2.5}$ mass and element concentrations at site 5 during period A from Sep-05 to Aug-07. Mass concentrations in $\mu g/m^3$, all element concentrations in $\mu g/g$.

SITE 5 NIGHT	N	avg	stdev	lower QT	upper QT	median	10th PCT	90th PCT
mass	96	80.0	57.9	41.8	96.3	72.5	13.4	149
Na	89	8160	4050	5450	10000	7830	3900	12500
Mg	91	5360	5920	2090	6510	3430	1160	10200
Al	90	14600	17300	5280	16600	8920	2770	36400
K	92	31800	19000	20400	39500	28400	12700	52100
Ca	81	22400	28300	6910	27700	11900	2300	48500
Sc	91	2.55	3.27	0.85	3.03	1.32	0.44	6.63
Ti	92	655	669	253	722	389	163	1610

V	92	37.8	34.5	18.2	47.9	26.1	12.7	78.0
Cr	66	107	244	30.9	75.3	47.1	15.9	162
Mn	92	880	412	637	1060	808	419	1430
Fe	92	13800	10900	6860	17400	9870	4790	29200
Co	92	12.1	9.90	6.42	13.6	9.21	4.00	20.3
Ni	72	175	276	44.6	145	62.9	33.6	510
Cu	92	1300	1310	607	1378	807	349	3150
Zn	92	4840	2730	2850	6040	4460	2230	8050
Ga	92	57.8	32.4	35.1	71.8	52.7	28.0	108
As	92	324	185	202	377	286	151	516
Rb	92	137	75.4	80.3	179	120	64.8	234
Sr	91	129	117	51.4	163	98.2	28.7	269
Y	54	10.0	9.76	4.42	11.0	7.42	2.37	25.1
Zr	70	286	283	121	324	210	83.5	456
Nb	46	6.36	9.30	2.13	7.67	4.28	0.86	10.8
Mo	39	260	488	21.9	216	65.8	8.63	1110
Ag	79	14.4	20.2	5.66	15.6	9.81	2.66	24.4
Cd	91	68.6	43.5	40.3	86.3	56.6	27.3	112
Sn	92	248	193	132	305	200	92.2	456
Sb	92	157	89.0	104	190	139	60.0	263
Cs	92	17.3	11.7	10.6	20.6	15.0	7.91	27.7
Ba	61	651	956	165	712	315	103	1580
Pb	92	2660	1170	1860	3350	2490	1580	4060

Table A.23: Descriptive statistics (valid number, average, standard deviation, lower and upper quartiles, median, 10th and 90th percentile) for APM mass concentrations at site CUG from the 25th of April 2005 to the 30th of August 2007 (sampling in intervals of 3 and 4 days). Mass concentrations in $\mu g/m^3$ for the different size classes for total, transparent, and opaque particles, respectively.

Size class	N	avg	stdev	lower QT	upper QT	median	10th PCT	90th PCT
Total								
2.5 – 5	243	50.0	23.4	34.8	59.8	44.1	27.2	77.2
5 – 10	243	62.6	25.6	44.7	75.5	58.9	36.2	91.9
10 – 20	243	61.7	27.8	42.6	76.9	55.0	31.9	94.4
20 – 40	243	35.5	26.2	18.6	43.6	26.6	14.0	68.1
40 – 80	243	8.66	12.8	2.14	8.15	4.53	1.23	23.2
2.5 – 80	243	218	104	148	267	193	117	329
2.5 – 10	243	113	48.3	79.1	135	104	64.0	169
Transparent								
2.5 – 5	243	44.5	22.0	30.6	54.3	39.4	23.8	68.6
5 – 10	243	55.0	24.6	37.8	66.8	50.9	30.4	82.4
10 – 20	243	55.0	26.9	36.2	68.4	48.9	28.4	85.6
20 – 40	243	32.1	25.2	16.2	39.0	23.3	11.9	60.8
40 – 80	243	7.84	12.2	1.71	7.53	3.90	1.07	19.3
2.5 – 80	243	194	100	128	236	169	99.7	291
2.5 – 10	243	99.6	45.8	68.4	119	92.7	54.2	151
Opaque								
2.5 – 5	243	5.44	2.53	3.70	6.56	4.82	2.82	8.86
5 – 10	243	7.56	2.92	5.39	9.03	6.98	4.33	11.6
10 – 20	243	6.74	2.86	4.53	8.64	6.41	3.33	10.5
20 – 40	243	3.50	2.35	1.75	4.54	3.13	1.09	6.12
40 – 80	243	0.82	1.09	0.00	1.11	0.45	0.00	1.99
2.5 – 80	243	24.1	9.57	16.5	29.8	22.1	12.7	36.7
2.5 – 10	243	13.0	5.33	9.3	16.0	11.9	7.11	20.6

Table A.24: Descriptive statistics (valid number, average, standard deviation, lower and upper quartiles, median, 10th and 90th percentile) for $PM_{2.5}$ mass and element concentrations at site CUG during period B in Spring 2007. Mass concentrations in $\mu g/m^3$, all element concentrations in ng/m^3.

	valid N	avg	stdev	lower QT	upper QT	median	10th PCT	90th PCT
mass	36	149	35	124	168	138	113	201
Li	33	1.7	1.4	0.5	2.7	1.5	0.3	3.9
Na	32	688	542	255	987	581	144	1490
Mg	36	342	286	139	404	276	98	695
Al	33	897	572	416	1340	849	252	1750
P	31	31	21	14	47	28	2.5	56
K	36	1550	1330	502	2180	1080	351	3790
Ca	33	2360	2250	754	3040	1490	371	6540
Sc	34	0.2	0.1	0.1	0.2	0.1	0.02	0.3
Ti	36	49	52	21	59	38	10	87

V	36	2.6	1.9	1.0	3.3	2.5	0.6	4.4
Cr	31	9.0	9.7	2.5	9.3	6.5	1.4	20
Mn	36	61	39	30	77	54	18	124
Fe	36	1040	662	524	1360	894	370	2090
Co	36	0.5	0.3	0.3	0.8	0.5	0.1	1.0
Ni	17	11	8.7	5.6	15	8.1	1.6	24
Cu	34	36	68	11	34	22	5.4	47
Zn	34	366	234	206	486	344	71	727
Ga	36	2.0	1.3	1.0	2.7	1.9	0.5	3.8
As	36	9.4	11	2.8	12	8.2	0.7	19
Rb	36	8.2	8.3	2.0	11	4.6	1.2	22
Sr	34	8.0	5.7	3.9	11	6.5	2.1	14
Y	20	0.5	0.4	0.2	0.8	0.4	0.1	1.1
Zr	30	13	8.4	5.5	17	11	3.0	26
Mo	30	21	13	12	28	20	2.1	38
Ag	35	0.6	0.6	0.1	0.8	0.5	0.1	1.6
Cd	33	2.6	3.4	0.3	2.7	1.9	0.1	4.6
Sn	34	11	11	5.2	14	8.3	2.4	19
Sb	36	8.9	14	3.7	7.6	6.1	1.2	14
Cs	30	1.2	1.1	0.3	1.9	0.6	0.1	3.0
Ba	28	72	82	21	101	33	8.6	202
Pb	36	104	98	27	151	86	10	243

Table A.25: Descriptive statistics (valid number, average, standard deviation, lower and upper quartiles, median, 10th and 90th percentile) for PM_{10} mass and element concentrations at site CUG during period B in Spring 2007. Mass concentrations in $\mu g/m^3$, all element concentrations in ng/m^3.

	N	avg	stdev	lower QT	upper QT	median	10th PCT	90th PCT
mass	35	222	80	155	267	211	140	312
Li	36	5.2	4.1	3.0	6.7	4.3	1.4	8.3
Na	34	946	666	411	1345	766	279	1890
Mg	36	1710	1270	879	2140	1560	473	2690
Al	36	4420	3040	2100	6830	3640	1320	8280
P	36	104	66	57	140	98	20	196
K	36	2830	1830	1290	3940	2640	692	5340
Ca	36	9100	5820	5550	10400	7410	3960	16700
Sc	36	0.8	0.6	0.4	1.3	0.7	0.2	1.6
Ti	36	196	133	99	297	165	57	365
V	36	9.1	5.9	4.9	12	7.7	2.7	17
Cr	35	18	21	8.8	17	13	5.4	26
Mn	36	131	83	84	160	117	39	248
Fe	36	4340	2590	2540	5980	3430	1630	6850
Co	36	2.2	1.9	1.1	2.5	1.7	0.8	3.7
Ni	33	14	19	5.5	15	9.3	1.8	18
Cu	36	51	77	13	52	39	7.9	66
Zn	35	422	278	221	628	378	38	860
Ga	36	3.3	2.0	2.0	4.2	3.3	0.9	5.9
As	36	13	13	5.6	16	9.8	2.2	25
Rb	36	14	11	5.1	20	12	2.8	28
Sr	36	32	22	20	38	29	9.1	62
Y	27	1.3	1.3	0.4	2.0	1.0	0.1	2.2
Zr	35	19	12	12	23	18	3.2	33
Mo	30	19	15	11	25	16	3.1	35
Ag	35	0.5	0.3	0.2	0.7	0.5	0.1	0.9
Cd	32	3.2	3.9	0.5	3.8	2.4	0.2	5.8
Sn	36	12	12	3.7	16	9.2	2.1	25
Sb	36	12	16	5.6	13	9.5	2.2	16
Cs	36	1.6	1.4	0.4	2.3	1.3	0.3	3.2
Ba	28	83	71	33	105	71	7.1	209
Pb	36	128	105	36	188	112	16	268

Table A.26: Descriptive statistics (valid number, average, standard deviation, lower and upper quartiles, median, 10th and 90th percentile) for $PM_{2.5}$ mass and element concentrations at site CUG during period B in Spring 2007. Mass concentrations in $\mu g/m^3$, all element concentrations in $\mu g/g$.

	N	avg	stdev	lower QT	upper QT	median	10th PCT	90th PCT
mass	36	149	35	124	168	138	113	201
Li	33	11	7.2	3.6	16	10	1.9	19
Na	32	4560	3700	1830	6390	3110	1080	9050

Mg	36	2130	1330	1100	2820	1810	697	4300
Al	33	5840	3410	2930	8840	5140	1890	9650
P	31	191	116	108	281	193	15	335
K	36	9490	6470	3950	12800	8020	2540	20000
Ca	33	16200	16800	5280	19200	10700	2900	47300
Sc	34	1.0	0.7	0.4	1.5	0.8	0.2	2.2
Ti	36	311	286	165	394	238	82	573
V	36	16	9.5	7.2	21	16	4.6	29
Cr	31	58	64	19	62	37	12	121
Mn	36	381	182	241	498	402	142	577
Fe	36	6600	3160	3860	9010	6290	2900	11300
Co	36	3.5	1.8	2.2	4.5	3.3	1.0	6.4
Ni	17	68	52	31	114	49	8.8	128
Cu	34	208	317	86	222	141	40	299
Zn	34	2380	1350	1470	3010	2660	691	3750
Ga	36	13	6.1	8.9	17	12	4.0	20
As	36	57	52	23	80	47	4.9	95
Rb	36	49	41	15	68	36	8.8	120
Sr	34	51	30	30	71	44	15	97
Y	20	3.4	3.2	1.3	3.9	2.9	0.5	9.1
Zr	30	84	59	40	111	78	20	149
Mo	30	145	111	83	165	137	19	258
Ag	35	3.5	3.3	0.8	5.0	2.8	0.5	9.5
Cd	33	16	18	2.2	17	13	0.6	40
Sn	34	68	57	38	80	57	18	138
Sb	36	52	62	29	55	41	9.2	72
Cs	30	6.8	5.6	2.5	11	4.4	1.1	15
Ba	28	520	642	141	633	238	47	1490
Pb	36	624	472	254	924	579	98	1180

Table A.27: Descriptive statistics (valid number, average, standard deviation, lower and upper quartiles, median, 10th and 90th percentile) for PM_{10} mass and element concentrations at site CUG during period B in Spring 2007. Mass concentrations in μg/m^3, all element concentrations in μg/g.

	N	avg	stdev	lower QT	upper QT	median	10th PCT	90th PCT
mass	35	222	80	155	267	211	140	312
Li	35	21	8.7	16	26	20	12	28
Na	33	3880	1960	2310	4770	3740	1320	6140
Mg	35	6910	2640	4310	8610	7000	3530	9470
Al	35	18000	7830	10700	23000	16900	9430	28300
P	35	436	208	298	546	434	142	690
K	35	11700	4580	7190	15300	12100	5380	17700
Ca	35	39200	13200	31100	44100	38200	21300	59400
Sc	35	3.4	1.6	2.0	4.4	3.3	1.5	5.3
Ti	35	800	345	485	1030	774	400	1250
V	35	38	16	24	44	38	19	53
Cr	34	76	71	45	74	52	38	167
Mn	35	560	303	424	656	566	251	714
Fe	35	18200	6170	13900	21500	17000	11400	26300
Co	35	9.6	7.6	5.7	10	7.3	5.2	16
Ni	32	59	66	24	74	38	12	111
Cu	35	200	233	102	213	158	56	286
Zn	34	1830	1040	1140	2710	1800	196	3070
Ga	35	14	4.6	11	17	15	6.7	19
As	35	54	43	30	66	44	16	80
Rb	35	57	29	26	82	56	20	99
Sr	35	133	46	106	157	136	64	187
Y	26	4.9	3.7	2.1	6.7	3.9	0.6	10
Zr	34	80	38	64	105	81	21	122
Mo	29	98	71	42	140	87	7.4	224
Ag	34	2.1	1.3	1.2	2.6	2.1	0.4	3.5
Cd	31	12	13	3.2	16	9.6	1.3	24
Sn	35	50	43	24	65	35	11	95
Sb	35	49	42	34	52	41	15	63
Cs	35	6.4	4.4	2.3	9.0	5.8	1.9	13
Ba	27	348	241	123	555	315	25	695
Pb	35	522	334	230	787	526	102	932

Appendix B

Tables corresponding to chapter 5

Table B.1: Descriptive statistics (valid number, average, standard deviation, lower and upper quartiles, median, 10th and 90th percentile) for TSP element concentrations in ng/m^3 at site TSP and TSP-2 for all 4 fractions obtained by sequential extraction.

	N	avg	stdev	lower QT	upper QT	median	10th PCT	90th PCT
Fraction 1								
Mg	35	1360	436	1040	1680	1330	752	1810
Al	35	270	302	82	295	138	60	756
K	35	3780	2980	1790	4410	3110	1300	7290
Ca	35	14500	4790	11100	19200	14000	8780	21400
Ti	35	6.4	3.6	3.5	8.2	5.8	1.8	11
V	35	3.0	1.4	1.9	4.4	2.4	1.6	4.8
Cr	35	2.2	1.2	1.2	3.1	1.9	1.1	3.9
Mn	35	135	52	96	182	129	71	203
Fe	35	144	67	104	178	137	57	250
Co	35	1.2	0.5	0.7	1.5	1.0	0.6	1.9
Ni	35	3.4	2.4	1.9	4.3	2.7	1.2	5.9
Cu	35	34	15	23	42	30	18	48
Zn	35	665	463	283	935	625	102	1290
As	35	21	12	10	28	19	8.1	40
Rb	35	14	8.9	7.4	18	12	5.1	23
Sr	35	65	29	42	77	61	34	112
Cd	35	6.9	6.4	2.1	10	5.3	1.0	16
Pb	35	37	33	13	50	24	9.8	73
Fraction 2								
Mg	35	2990	1600	1600	3900	2740	1330	4530
Al	35	1400	531	943	1710	1400	679	2290
K	35	1690	1040	1150	1820	1380	904	2930
Ca	35	12400	7360	7170	15700	11500	5320	18700
Ti	35	15	6.3	9.2	18	16	7.2	22
V	35	4.9	1.8	3.6	5.7	4.8	2.6	7.7
Cr	34	2.4	0.8	1.8	3.0	2.3	1.3	3.7
Mn	35	110	51	74	135	103	55	159
Fe	35	1040	392	702	1270	1060	478	1510
Co	35	1.2	0.5	0.8	1.5	1.0	0.6	1.9
Ni	33	2.9	1.8	1.4	3.8	2.8	0.8	5.6
Cu	35	45	20	31	54	39	26	83
Zn	35	498	210	319	652	533	243	780
As	35	31	26	15	39	25	10	52
Rb	35	7.6	3.2	5.5	9.6	6.9	4.5	12
Sr	35	50	28	34	64	43	21	91
Cd	35	5.7	4.1	2.9	7.7	4.9	1.9	9.7
Pb	35	262	123	161	339	265	124	421
Fraction 3								
Mg	35	2070	1050	1340	2530	2110	983	2930
Al	35	2150	1080	1460	2600	2040	937	4080
K	35	1230	472	887	1480	1150	720	2020

Ca	34	4470	2610	2360	5500	4170	1920	6130
Ti	35	110	55	76	133	95	53	188
V	35	6.7	2.9	4.7	7.6	5.8	3.7	12
Cr	34	15	10	11	16	13	8.1	20
Mn	35	108	46	72	118	105	59	175
Fe	35	4360	1940	3080	4930	4380	1960	6810
Co	35	2.4	1.0	1.6	3.1	2.3	1.3	3.6
Ni	34	9.9	5.5	6.3	12	8.4	5.3	15
Cu	35	62	28	41	86	60	28	103
Zn	35	285	126	190	334	287	136	463
As	35	11	7.9	5.3	15	9.8	3.0	24
Rb	35	6.5	2.3	4.9	7.6	6.5	4.1	9.5
Sr	34	29	16	19	37	25	11	48
Cd	34	3.1	2.8	1.0	4.1	2.2	0.8	6.7
Pb	35	104	77	38	141	90	31	246
Fraction 4								
Mg	35	951	458	567	1260	871	506	1430
Al	35	1840	836	1200	2360	1750	956	2870
K	6	712	675	21	973	726	9.3	1810
Ca	34	2380	1290	1230	3080	2230	1030	3910
Ti	35	96	42	58	121	93	53	168
V	34	3.0	1.8	1.5	4.3	2.7	1.1	5.2
Cr	32	5.4	5.9	3.4	5.7	4.2	2.7	6.9
Mn	34	45	18	31	54	44	24	65
Fe	35	3430	1320	2360	3980	3490	2190	4780
Co	34	1.0	0.4	0.7	1.2	1.0	0.6	1.5
Ni	30	2.6	3.0	0.5	3.3	1.8	0.2	6.3
Cu	34	30	13	18	36	28	16	48
Zn	35	136	243	67	127	96	42	154
As	34	11	8.6	4.8	17	8.6	3.2	20
Rb	0	–	–	–	–	–	–	–
Sr	35	15	10	7.5	20	10	6.1	32
Cd	35	1.5	1.3	0.5	2.2	1.2	0.4	3.4
Pb	35	72	39	39	93	64	33	144
Sum 1–4								
Mg	35	7370	3180	4740	8600	7140	3400	11100
Al	35	5670	2250	4220	7020	5710	2790	8350
K	35	6830	4340	3950	7920	5900	3180	11800
Ca	35	33600	14600	20100	40000	32900	18500	48500
Ti	35	227	73	175	250	207	156	334
V	35	18	6.3	13	21	17	11	25
Cr	35	24	17	18	25	21	13	32
Mn	35	396	146	304	498	389	207	598
Fe	35	8970	3330	6610	10300	8960	4280	14300
Co	35	5.7	2.2	3.9	7.5	5.5	3.1	8.3
Ni	35	18	10	10	22	16	8.4	31
Cu	35	169	64	112	208	165	87	263
Zn	35	1580	688	1080	2050	1350	754	2630
As	35	74	50	39	99	61	26	127
Rb	35	28	13	19	34	23	16	43
Sr	35	157	78	101	191	148	68	293
Cd	35	17	13	7.6	23	14	5.8	33
Pb	35	475	248	260	649	417	210	861

Table B.2: Descriptive statistics (valid number, average, standard deviation, lower and upper quartiles, median, 10th and 90th percentile) for TSP element concentrations in μg/g at site TSP and TSP-2 for all 4 fractions obtained by sequential extraction.

	N	avg	stdev	lower QT	upper QT	median	10th PCT	90th PCT
Fraction 1								
Mg	35	3800	1170	2960	4440	3550	2720	4880
Al	35	718	736	267	1050	372	187	1780
K	35	10400	8650	6130	10400	7960	4040	16700
Ca	35	39900	8280	33600	45000	38800	30000	52100
Ti	35	18.1	9.3	10	23	17	7.0	31
V	35	8.4	3.5	5.7	10	7.0	5.1	15
Cr	35	6.6	3.9	4.0	8.6	5.8	2.5	11
Mn	35	369	87	323	422	360	308	476
Fe	35	412	188	260	563	408	144	691
Co	35	3.1	1.1	2.4	3.8	2.8	2.1	4.5

Ni	35	9.4	5.4	4.8	12	8.0	3.9	18
Cu	35	99	58	60	120	80	55	153
Zn	35	1830	1060	973	2630	1840	363	3130
As	35	59	33	31	76	53	26	93
Rb	35	37	20	24	47	33	14	63
Sr	35	174	45	136	196	174	119	217
Cd	35	19	14	4.8	28	16	3.5	42
Pb	35	104	83	38	157	79	25	239
Fraction 2								
Mg	35	7910	2690	5880	9580	7880	5180	11200
Al	35	3820	957	3150	4320	3760	2700	4760
K	35	4850	3390	3010	5280	3600	2830	7730
Ca	35	32800	12700	24900	36900	30700	18400	50500
Ti	35	44	24	26	55	39	18	74
V	35	13	2.1	12	14	13	11	17
Cr	34	6.6	2.2	5.2	7.8	6.0	4.3	10
Mn	35	291	79	225	322	282	203	408
Fe	35	2820	688	2420	3100	2720	2020	3700
Co	35	3.1	0.9	2.4	3.8	3.1	2.0	4.1
Ni	33	7.7	4.5	4.5	10	6.8	3.1	15
Cu	35	129	63	95	150	109	66	236
Zn	35	1330	448	1030	1550	1350	913	1760
As	35	82	56	44	100	65	30	126
Rb	35	21	8.6	15	27	20	12	33
Sr	35	131	55	97	164	114	85	190
Cd	35	15	9.3	8.8	20	13	6.3	25
Pb	35	703	230	548	839	641	454	1020
Fraction 3								
Mg	35	5460	1560	4500	6870	5200	3420	7550
Al	35	5700	1860	4220	6700	5160	3970	8250
K	35	3450	1280	2770	3840	3050	2440	4880
Ca	34	11600	3980	9540	12800	11500	6530	16500
Ti	35	319	163	204	431	322	112	505
V	35	19	6.0	14	21	18	11	30
Cr	34	38	16	31	40	35	29	45
Mn	35	289	58	253	335	276	219	371
Fe	35	11400	2920	10300	12500	11400	8840	15500
Co	35	6.3	1.3	5.4	7.1	6.2	4.6	8.2
Ni	34	25	7.3	21	26	24	20	32
Cu	35	164	52	134	189	158	106	246
Zn	35	766	254	626	853	769	473	1090
As	35	29	17	16	43	24	9.6	54
Rb	35	18	4.9	14	22	17	12	24
Sr	34	75	34	52	87	65	49	107
Cd	34	7.7	6.1	3.7	9.9	6.0	2.4	14
Pb	35	260	141	158	389	203	99	464
Fraction 4								
Mg	35	2550	840	1920	3090	2610	1530	3490
Al	35	5020	1740	3770	5730	4470	3310	7860
K	6	1480	1210	69	2350	1730	18	3000
Ca	34	6360	2760	4200	8540	6070	3230	9680
Ti	35	264	95	206	304	255	157	367
V	34	7.9	3.5	4.7	10	7.5	3.3	13
Cr	32	14	11	9.5	16	12	6.9	21
Mn	34	119	27	105	133	119	84	153
Fe	35	9420	2400	7670	11000	9000	6100	12700
Co	34	2.7	0.7	2.2	3.0	2.7	1.8	3.8
Ni	30	5.9	5.4	1.3	7.5	5.1	0.8	14
Cu	34	82	29	62	100	79	49	116
Zn	35	492	1346	186	299	254	139	445
As	34	28	17	15	40	23	11	48
Rb	0	–	–	–	–	–	–	–
Sr	35	39	20	24	51	28	20	68
Cd	35	4.1	3.0	1.6	5.6	3.6	1.0	7.6
Pb	35	195	79	128	261	198	97	304
Sum 1–4								
Mg	35	19700	4160	16400	23200	20100	15000	24500
Al	35	15300	3520	12800	17400	13800	11600	19200
K	35	18900	12800	12500	21400	14800	11100	25200
Ca	35	90000	21400	75600	109000	87500	65100	115000
Ti	35	645	202	507	753	587	434	877
V	35	48	8.8	41	53	46	37	58
Cr	35	63	27	51	69	57	45	80

Mn	35	1060	156	956	1170	1060	852	1260
Fe	35	24100	4040	21500	26000	23500	20300	29600
Co	35	15	2.9	13	18	15	12	19
Ni	35	46	15	37	57	42	29	66
Cu	35	472	165	373	490	445	327	766
Zn	35	4420	1750	3220	5360	4160	2770	5740
As	35	197	111	109	257	182	89	291
Rb	35	77	29	55	87	69	47	125
Sr	35	416	137	337	478	382	266	560
Cd	35	45	27	27	55	39	17	83
Pb	35	1260	435	880	1510	1280	660	1910

Table B.3: Descriptive statistics (valid number, average, standard deviation, lower and upper quartiles, median, 10th and 90th percentile) for $PM_{2.5}$ element concentrations in ng/m^3 at site CUG for all 4 fractions obtained by sequential extraction.

	N	avg	stdev	lower QT	upper QT	median	10th PCT	90th PCT
Fraction 1								
mass	32	68	24	54	80	63	43	93
Mg	32	190	74	128	225	182	112	299
Al	32	37	25	20	49	25	17	83
K	32	1720	922	1190	2240	1420	774	3070
Ca	32	1210	731	672	1540	894	634	2190
Ti	32	1.1	1.0	0.4	1.4	0.8	0.3	2.4
V	32	2.0	2.5	0.9	2.0	1.3	0.8	3.2
Cr	32	2.1	0.8	1.5	2.7	1.9	1.0	3.2
Mn	32	36	12	27	43	33	26	56
Fe	32	65	44	35	84	55	26	104
Co	32	0.4	0.3	0.2	0.5	0.4	0.2	0.7
Ni	32	2.0	1.4	1.2	2.2	1.6	0.6	4.5
Cu	32	18	8.0	12	23	16	9.9	28
Zn	32	360	130	266	432	336	219	524
As	32	13	5.8	7.6	16	13	6.3	23
Rb	32	7.5	3.6	5.2	8.6	7.0	4.2	12
Sr	32	6.5	5.9	2.8	6.8	5.5	2.7	9.7
Cd	32	3.0	1.5	1.9	4.4	2.5	1.6	4.9
Pb	32	47	46	17	66	32	9.9	110
Fraction 2								
mass	32	68	24	54	80	63	43	93
Mg	32	136	89	79	158	112	53	211
Al	32	131	62	80	171	113	68	204
K	32	983	638	535	1200	808	441	1690
Ca	31	838	529	471	978	730	303	1394
Ti	32	2.0	1.2	1.3	2.4	1.8	1.0	3.0
V	32	1.3	1.5	0.5	1.5	0.9	0.5	2.1
Cr	30	1.8	0.8	1.2	2.2	1.5	1.0	3.0
Mn	32	21	8.5	15	24	19	12	31
Fe	32	145	45	110	162	145	99	207
Co	32	0.2	0.2	0.1	0.3	0.1	0.1	0.4
Ni	22	42	33	12	83	27	6.5	86
Cu	32	26	27	13	29	20	8.4	36
Zn	32	402	271	92	644	515	56	668
As	32	7.1	3.8	4.0	9.8	6.5	3.4	12
Rb	32	3.9	2.1	2.4	5.1	3.2	2.1	6.5
Sr	32	4.3	3.1	2.5	5.2	3.5	0.9	7.6
Cd	31	1.3	0.9	0.6	2.1	1.0	0.5	2.5
Pb	32	69	48	28	90	64	17	145
Fraction 3								
mass	32	68	24	54	80	63	43	93
Mg	32	168	145	94	212	125	45	299
Al	32	111	133	41	110	77	34	235
K	31	624	417	298	889	425	205	1170
Ca	26	1700	3670	457	1030	666	110	1870
Ti	32	18	12	9.3	22	13	8.0	32
V	32	1.1	0.9	0.6	1.4	0.9	0.4	2.1
Cr	23	3.2	1.9	1.5	4.3	3.5	1.0	4.9
Mn	32	22	9.0	16	25	19	15	35
Fe	31	141	112	47	195	154	13	223
Co	32	0.3	0.2	0.2	0.4	0.3	0.1	0.6
Ni	27	16	14	6.8	31	11	0.5	36
Cu	32	12	8.6	6.7	17	11	4.2	18
Zn	32	384	288	67	643	466	19	692

As	28	8.1	9.2	1.5	10	4.0	1.0	28
Rb	32	2.7	0.9	2.2	3.2	2.6	1.7	4.0
Sr	32	12	11	2.9	16	7.4	1.2	34
Cd	28	6.0	5.9	2.1	8.6	4.5	0.2	12
Pb	32	52	43	17	95	35	12	117
Fraction 4								
mass	32	68	24	54	80	63	43	93
Mg	32	78	61	35	104	60	20	169
Al	32	158	140	77	163	118	54	338
K	0							
Ca	25	968	527	679	1410	994	125	1650
Ti	32	21	11	13	25	22	4	33
V	8	2.7	1.0	2.2	3.3	3.2	0.7	3.5
Cr	0							
Mn	11	6.5	5.8	2.2	9.1	5.3	0.2	12
Fe	32	480	279	253	672	394	205	843
Co	14	0.2	0.2	0.1	0.4	0.2	0.1	0.5
Ni	5	34	68	2.1	7.8	2.6	1.7	156
Cu	32	42	26	16	61	48	8.9	67
Zn	23	350	142	341	440	390	25	457
As	22	2.0	1.5	0.8	3.0	1.9	0.3	4.2
Rb	0							
Sr	32	13	9.0	6.3	19	11	3.7	26
Cd	32	1.1	0.8	0.5	1.5	0.9	0.5	2.1
Pb	32	32	25	18	38	24	9.0	62
Sum 1–4								
mass	32	68	24	54	80	63	43	93
Mg	32	572	316	365	750	436	269	948
Al	32	438	309	256	586	335	195	713
K	32	4100	5220	2230	3790	2710	1800	6040
Ca	32	4160	3420	2510	4210	3270	1350	7110
Ti	32	42	20	32	52	36	25	65
V	32	5.1	4.4	2.8	6.0	4.6	2.4	7.2
Cr	32	6.0	2.7	4.1	7.2	5.7	3.1	9.1
Mn	32	82	29	63	95	73	53	125
Fe	32	826	338	602	1010	692	499	1200
Co	32	1.1	0.6	0.7	1.4	0.8	0.5	1.8
Ni	32	50	62	6	80	24	1.8	123
Cu	32	98	48	60	123	99	45	141
Zn	32	1400	723	640	1990	1630	440	2140
As	32	29	12	17	38	31	12	42
Rb	32	14	5	10	16	13	8.5	20
Sr	32	36	19	23	44	37	15	51
Cd	32	11	6	6	13	9.5	4.5	17
Pb	32	200	83	133	254	179	103	317

Table B.4: Descriptive statistics (valid number, average, standard deviation, lower and upper quartiles, median, 10th and 90th percentile) for $PM_{2.5}$ element concentrations in $\mu g/g$ at site CUG for all 4 fractions obtained by sequential extraction.

	N	avg	stdev	lower QT	upper QT	median	10th PCT	90th PCT
Fraction 1								
mass	32	68	24	54	80	63	43	93
Mg	32	2960	1050	2120	3530	2910	1740	4070
Al	32	562	340	383	610	460	293	1260
K	32	26700	12700	18200	35200	24400	13200	45100
Ca	32	18500	10300	11500	20500 16300	9250	30500	
Ti	32	17	15	6.8	24	12	4.9	30
V	32	28	22	16	30	22	13	49
Cr	32	33	13	26	42	32	16	50
Mn	32	563	148	458	647	578	401	736
Fe	32	1000	595	675	1260	909	409	1500
Co	32	6.7	4.4	3.5	8.2	5.6	2.8	12
Ni	32	31	21	16	41	27	9.3	55
Cu	32	285	122	192	350	251	157	489
Zn	32	5660	1910	4800	6530	5910	3410	7180
As	32	204	84	159	232	202	117	278
Rb	32	117	50	79	154	110	56	186
Sr	32	99	92	49	112	80	47	147
Cd	32	47	19	38	58	45	22	66
Pb	32	717	656	264	938	428	191	2030
Fraction 2								

mass	32	68	24	54	80	63	43	93
Mg	32	2120	1350	1230	2630	1770	864	3870
Al	32	2040	970	1320	2620	1690	1230	3300
K	32	15400	10300	9370	16800	14000	7550	27900
Ca	31	13400	9130	6100	16700	11600	5000	25700
Ti	32	32	21	18	37	29	11	52
V	32	19	15	8.9	25	15	5.8	33
Cr	30	29	14	17	44	24	15	50
Mn	32	318	112	234	381	286	208	436
Fe	32	2330	1120	1850	2510	2150	1250	2780
Co	32	3.0	2.2	1.4	4.8	2.3	0.8	5.6
Ni	22	743	712	205	1180	407	122	1980
Cu	32	406	409	193	461	358	115	613
Zn	32	6570	5010	1430	10900	7260	882	12200
As	32	109	51	76	130	96	50	189
Rb	32	63	40	40	73	54	30	93
Sr	32	67	50	35	97	51	20	108
Cd	31	19	12	9.3	27	16	7.6	35
Pb	32	1030	619	428	1490	1020	381	1840
Fraction 3								
mass	32	68	24	54	80	63	43	93
Mg	32	2570	1980	1230	3260	1840	710	5780
Al	32	1650	1570	728	2160	1000	582	3390
K	31	10000	6930	3660	15200	8550	3000	19500
Ca	26	26900	58600	5830	14300	11400	2260	23900
Ti	32	268	165	162	300	228	124	471
V	32	16	9.8	8.0	22	14	6.1	32
Cr	23	50	32	24	72	51	15	102
Mn	32	345	121	255	421	324	215	479
Fe	31	2230	1660	579	3120	2620	222	4610
Co	32	5.1	2.9	2.9	6.5	4.7	2.4	8.7
Ni	27	288	314	87	434	154	7.6	722
Cu	32	190	117	104	246	167	79	326
Zn	32	6330	5290	746	10600	6090	243	12300
As	28	128	151	37	184	74	10	356
Rb	32	44	19	30	58	39	25	68
Sr	32	187	192	55	259	120	21	462
Cd	28	97	104	26	151	65	3.3	204
Pb	32	848	763	280	1434	428	223	2000
Fraction 4								
mass	32	68	24	54	80	63	43	93
Mg	32	1200	862	587	1620	1040	191	2230
Al	32	2380	1990	1170	2730	1720	823	4770
K	0							
Ca	25	15800	9700	9200	20600	16500	3260	31200
Ti	32	338	195	186	473	341	66	544
V	8	44	21	27	60	48	9.6	74
Cr	0							
Mn	11	93	102	27	115	61	3.9	172
Fe	32	7220	3900	4690	8680	5950	3290	12600
Co	14	3.9	4.1	1.8	5.6	2.1	1.2	7.0
Ni	5	446	856	48	114	61	29	1980
Cu	32	683	495	211	992	693	113	1185
Zn	23	5860	3160	4370	7390	5960	897	9940
As	22	33	27	9.9	43	28	6.4	77
Rb	0							
Sr	32	217	168	95	285	178	40	431
Cd	32	19	15	8.7	25	14	7.0	38
Pb	32	506	437	253	578	409	180	922
Sum 1–4								
mass	32	68	24	54	80	63	43	93
Mg	32	8840	4480	6000	10500	7560	5030	15200
Al	32	6630	3990	3830	8030	5370	3570	10900
K	32	61500	63000	35700	60500	47300	29500	111000
Ca	32	65600	55300	33300	76800	49200	23600	100000
Ti	32	656	283	454	822	581	362	1071
V	32	74	40	45	94	62	39	110
Cr	32	96	47	64	133	78	44	163
Mn	32	1260	362	1000	1460	1240	895	1590
Fe	32	12700	4610	9420	14100	11800	8480	21500
Co	32	17	8.5	9.9	23	14	8.0	27
Ni	32	854	1100	62	1370	379	36	2830
Cu	32	1560	795	861	1984	1458	573	2983

Zn	32	22800	14000	8730	32400	21900	6640	38700
As	32	448	207	291	626	404	215	701
Rb	32	224	90	172	275	210	132	324
Sr	32	570	330	358	685	493	245	1010
Cd	32	169	107	111	185	138	78	279
Pb	32	3110	1150	2530	3700	2960	1750	4490

Appendix C

Tables corresponding to chapter 6

Table C.1: Descriptive statistics (valid number, average, standard deviation, lower and upper quartiles, median, 10th and 90th percentile) for TSP mass and element concentrations at site CRAES from October 2007 to July 2008 (period C1, weekly sampling). Mass concentrations in $\mu g/m^3$, all element concentrations in ng/m^3.

	N	avg	stdev	lower QT	upper QT	median	10th PCT	90th PCT
mass	40	281	122	198	361	254	125	435
Li	42	12	5.8	7.0	16	11	5.2	19
Na	42	2870	1710	1290	3990	2530	1030	4500
Mg	42	5100	2550	2780	6640	5380	2180	8300
Al	42	12100	6830	6650	14800	11300	5210	18900
P	42	315	163	187	422	285	154	529
S	26	13700	8760	7970	15200	10400	6370	26400
K	42	5300	2490	3410	6540	5220	2550	8620
Ca	42	19500	9200	11700	25500	19500	8820	30300
Sc	42	2.3	1.3	1.3	2.9	2.1	1.0	3.9
Ti	42	765	451	426	1030	663	287	1320
V	42	21	11	13	26	20	9.9	34
Cr	42	27	12	17	35	25	15	41
Mn	42	304	151	176	385	278	135	506
Fe	42	9300	4860	5330	12200	8910	4290	15500
Co	42	5.0	2.7	2.6	6.6	4.6	2.1	7.7
Ni	42	17	9.2	8.6	23	15	7.7	28
Cu	42	104	59	58	135	80	50	166
Zn	42	908	591	524	1120	763	268	1700
Ga	42	12	8.5	5.3	16	9.4	4.0	24
As	42	32	22	15	44	27	12	54
Rb	42	30	18	18	37	27	11	49
Sr	42	101	48	56	147	99	42	160
Y	42	4.2	2.2	2.3	5.3	4.1	1.8	6.7
Zr	42	28	16	17	38	24	14	46
Nb	42	2.4	1.2	1.4	3.3	2.3	1.1	3.6
Mo	35	2.6	2.7	0.6	5.0	1.4	0.4	6.2
Ag	42	1.3	0.8	0.8	1.7	1.0	0.6	2.0
Cd	42	5.4	4.3	2.1	6.8	4.4	1.7	12
Sn	42	19	14	8.8	24	16	5.3	30
Sb	42	30	19	17	39	25	12	49
Cs	42	3.1	1.9	2.1	3.9	2.9	1.1	4.8
Ba	42	216	114	120	293	199	99	376
Pb	42	289	193	143	380	237	109	526

Table C.2: Descriptive statistics (valid number, average, standard deviation, lower and upper quartiles, median, 10th and 90th percentile) for TSP mass and element concentrations at site CRAES from July 2008 to September 2008 (period C2, 24-hourly sampling). Mass concentrations in μg/m^3, all element concentrations in ng/m^3.

	N	avg	stdev	lower QT	upper QT	median	10th PCT	90th PCT
mass	66	104	51	62	131	96	45	177
Li	64	2.2	1.6	0.8	3.4	2.1	0.2	4.3
Na	30	1040	761	229	1660	989	122	2140
Mg	64	794	525	322	1210	704	166	1460
Al	64	2110	1310	881	3090	1990	518	4090
S	66	9390	7590	3570	12600	7380	1750	22300
K	66	1310	696	743	1807	1240	402	2200
Ca	65	3390	1930	1770	4750	3230	738	6230
Sc	66	0.4	0.2	0.2	0.6	0.4	0.1	0.7
Ti	66	166	79	99	233	155	73	260
V	66	5.1	2.9	2.6	6.9	4.3	1.9	8.5
Cr	65	13	12	6	14	9.8	3	20
Mn	66	74	35	43	99	71	28	119
Fe	66	2190	1070	1240	3090	2000	893	3660
Co	66	1.3	0.7	0.8	1.6	1.2	0.5	2.2
Ni	56	4.1	3.1	1.7	5.6	3.7	1.0	7.7
Cu	66	34	23	18	45	29	11	58
Zn	65	258	181	110	389	216	44	525
Ga	66	2.5	1.3	1.3	3.4	2.4	1.0	4.5
As	66	11	7.1	4.7	16	10	3.0	21
Rb	66	7.0	3.8	3.7	10	6.6	2.3	12
Sr	66	24	14	12	32	20	8.5	40
Y	26	0.6	0.6	0.2	0.8	0.4	0.1	1.2
Zr	4	3.2	2.9	1.3	5.1	2.2	1.1	7.4
Nb	59	0.3	0.2	0.1	0.5	0.3	0.0	0.6
Ag	45	0.2	0.2	0.1	0.3	0.2	0.1	0.4
Cd	66	1.9	1.3	0.7	2.8	1.6	0.4	3.8
Sn	66	4.9	3.1	2.4	6.6	4.7	1.1	8.6
Sb	66	9.4	5.5	4.6	13	8.9	2.7	17
Cs	65	0.8	0.5	0.4	1.2	0.8	0.2	1.3
Ba	66	93	83	42	103	74	28	167
Pb	66	88	49	44	127	79	26	156

Table C.3: Descriptive statistics (valid number, average, standard deviation, lower and upper quartiles, median, 10th and 90th percentile) for TSP mass and element concentrations at site CRAES from October 2008 to February 2009 (period C3, weekly sampling). Mass concentrations in μg/m^3, all element concentrations in ng/m^3.

	N	avg	stdev	lower QT	upper QT	median	10th PCT	90th PCT
mass	21	238	74	187	277	235	145	290
Li	21	8.8	2.8	7.2	10	9.1	5.9	11
Na	21	2310	1080	1740	2800	2340	1100	3020
Mg	21	3290	1250	2320	3900	3300	1840	4600
Al	21	8480	3100	5990	10200	8500	4810	11200
P	21	247	85	181	294	255	145	337
S	21	7140	3470	4320	8380	6840	3500	12000
K	21	4220	2250	2960	4260	3880	2260	5910
Ca	21	12600	5590	8350	15100	12000	6930	19300
Sc	21	1.7	0.7	1.3	2.0	1.7	1.0	2.3
Ti	21	534	213	417	643	472	324	764
V	21	15	5.7	12	19	16	9	20
Cr	21	24	10	16	29	22	15	37
Mn	21	191	62	159	235	167	136	255
Fe	21	6660	2380	5160	8250	6360	3850	9100
Co	21	4.2	1.6	3.2	4.8	3.8	2.7	5.9
Ni	20	11	3.8	8.7	15	11	6.8	16
Cu	21	81	44	54	95	75	37	134
Zn	21	514	238	341	617	476	272	884
Ga	21	11	5.0	7.9	14	12	4.5	17
As	21	25	14	17	28	22	13	39
Rb	21	20	6.9	16	25	20	11	28
Sr	21	92	46	58	103	85	46	152
Y	21	3.1	1.2	2.3	3.6	3.0	1.9	3.9
Zr	18	18	5	16	21	19	12	25
Nb	18	2.0	0.8	1.3	2.3	1.9	1.2	2.8
Mo	9	1.1	0.5	0.6	1.4	1.1	0.3	1.8

Ag	18	0.9	0.2	0.8	1.1	0.9	0.6	1.2
Cd	16	4.1	1.7	2.7	5.3	4.1	2.0	6.9
Sn	21	14	6.9	8.5	19	12	6.9	23
Sb	21	22	9.4	14	28	22	11	32
Cs	21	2.1	0.7	1.7	2.5	2.2	1.2	3.0
Ba	21	219	215	115	215	159	87	294
Pb	21	194	64	147	239	196	111	274

Table C.4: Descriptive statistics (valid number, average, standard deviation, lower and upper quartiles, median, 10th and 90th percentile) for TSP mass and element concentrations at site CRAES for the whole sampling period C (Oct-07 – Feb-09, weekly and daily sampling). Mass concentrations in $\mu g/m^3$, all element concentrations in ng/m^3.

	N	avg	stdev	lower QT	upper QT	median	10th PCT	90th PCT
mass	128	182	117	92	249	157	52	360
Li	127	6.5	5.8	2.1	10	4.5	0.6	15
Na	93	2150	1550	1110	2810	1960	404	4240
Mg	127	2630	2510	693	3860	1720	279	6460
Al	127	6450	6190	1990	9680	4540	816	14400
P	64	296	147	184	370	268	145	520
S	113	9950	7590	4230	12600	8380	2460	21300
K	129	3080	2540	1220	4190	2150	592	6535
Ca	128	10200	9350	3210	15000	6290	1440	25400
Sc	129	1.2	1.2	0.4	1.8	0.7	0.2	2.8
Ti	129	421	387	154	559	276	90	959
V	129	12	10	4.2	17	8.7	2.3	26
Cr	128	19	13	10	25	16	5.1	37
Mn	129	168	139	69	230	130	40	381
Fe	129	5230	4420	1970	6910	3600	1130	11800
Co	129	3.0	2.5	1.2	4.1	2.2	0.7	6.6
Ni	118	9.9	8.4	3.7	14	7.7	1.2	22
Cu	129	65	52	29	83	50	15	135
Zn	128	513	472	178	648	392	91	1100
Ga	129	6.9	7.0	2.2	10	4.1	1.2	17
As	129	20	17	10	26	15	4.0	43
Rb	129	17	15	6.4	23	12	3.1	36
Sr	129	60	51	19	92	40	11	147
Y	89	2.9	2.2	0.9	4.2	2.6	0.3	5.5
Zr	64	24	15	16	31	21	10	42
Nb	119	1.3	1.3	0.3	2.1	0.7	0.1	3.3
Mo	44	2.3	2.5	0.6	3.7	1.3	0.4	5.9
Ag	105	0.8	0.7	0.2	1.0	0.6	0.1	1.7
Cd	124	3.4	3.1	1.2	4.4	2.5	0.6	6.8
Sn	129	11	11	4.3	15	7.2	1.9	24
Sb	129	18	15	8.7	23	14	3.9	39
Cs	128	1.8	1.6	0.8	2.5	1.3	0.3	3.6
Ba	129	153	137	65	201	114	35	314
Pb	129	171	149	75	216	134	41	344

Table C.5: Descriptive statistics (valid number, average, standard deviation, lower and upper quartiles, median, 10th and 90th percentile) for TSP mass and element concentrations at site CRAES from October 2007 to July 2008 (period C1, weekly sampling). Mass concentrations in $\mu g/m^3$, all element concentrations in $\mu g/g$.

	N	avg	stdev	lower QT	upper QT	median	10th PCT	90th PCT
mass	41	279	122	195	360	254	138	434
Li	41	43	9.4	37	47	41	34	52
Na	41	10100	3130	7450	12100	9720	6560	13500
Mg	41	18300	4430	15400	20800	17300	13100	25000
Al	41	43000	12500	34900	49800	41500	28100	55500
P	41	1140	280	924	1339	1138	851	1480
S	26	51800	24900	30000	69500	43800	23700	88000
K	41	19700	6560	16100	21400	17916	14400	24600
Ca	41	70900	16900	60600	84500	68800	48600	93900
Sc	41	8.2	2.6	6.4	9.9	7.7	5.4	11
Ti	41	2700	833	2100	3210	2520	1800	3904
V	41	77	19	66	91	75	54	98
Cr	41	101	24	87	114	100	76	136
Mn	41	1130	345	886	1350	1020	777	1490
Fe	41	33400	8310	29000	38500	32600	24800	42900
Co	41	18	5.3	15	22	17	12	25

Ni	41	62	19	52	73	57	38	91
Cu	41	389	151	291	419	346	263	652
Zn	41	3310	1390	2280	3990	3310	1650	4920
Ga	41	43	22	26	62	37	22	71
As	41	116	49	79	139	110	64	167
Rb	41	108	33	89	120	95	83	146
Sr	41	378	139	284	446	362	254	498
Y	41	15	4.1	12	17	15	11	22
Zr	41	103	30	87	115	99	72	132
Nb	41	8.9	2.7	7.0	10	8.6	6.2	12
Mo	34	10	8.7	2.1	18	6.8	1.5	22
Ag	41	4.9	1.9	3.5	6.1	4.6	2.7	7.6
Cd	41	19	11	12	23	16	9.6	34
Sn	41	65	31	44	79	59	37	95
Sb	41	109	46	76	135	102	61	163
Cs	41	11	3.7	9.2	13	10	7.7	15
Ba	41	871	741	564	985	752	465	1120
Pb	41	1060	425	742	1170	1080	574	1630

Table C.6: Descriptive statistics (valid number, average, standard deviation, lower and upper quartiles, median, 10th and 90th percentile) for TSP mass and element concentrations at site CRAES from July 2008 to September 2008 (period C2, 24-hourly sampling). Mass concentrations in μg/m^3, all element concentrations in μg/g.

	N	avg	stdev	lower QT	upper QT	median	10th PCT	90th PCT
mass	66	104	51	62	131	96	45	177
Li	64	19	8.9	13	25	18	5.7	29
Na	30	10400	7500	2710	16500	9210	1210	20600
Mg	64	7470	4000	3850	10040	7290	2670	13300
Al	64	19100	7450	13300	24600	18600	9840	27400
S	66	82000	38100	45500	114000	86300	31400	130000
K	66	12600	3760	9940	14400	12400	8520	17700
Ca	65	32300	13200	23000	41300	32600	15800	45800
Sc	66	3.9	1.1	3.1	4.7	3.9	2.6	5.2
Ti	66	1640	362	1400	1840	1630	1160	2130
V	66	48	13	39	54	46	35	69
Cr	65	126	111	65	144	91	50	238
Mn	66	729	188	607	839	683	513	1000
Fe	66	21600	5860	17400	25400	21000	13700	29400
Co	66	14	6.9	8.7	16	12	7.2	21
Ni	56	35	24	22	42	28	12	72
Cu	66	322	126	231	402	296	202	457
Zn	65	2390	1620	1430	2910	2140	863	3820
Ga	66	24	4.6	21	26	23	19	29
As	66	109	66	68	131	97	48	189
Rb	66	68	23	51	74	64	44	103
Sr	66	226	77	176	257	221	153	312
Y	25	4.3	3.8	1.9	4.7	3.1	0.8	10
Zr	4	48	50	10	85	34	8.5	115
Nb	59	2.7	1.3	2.0	3.6	2.8	0.7	4.3
Mo	0							
Ag	46	2.3	2.8	0.9	2.6	1.8	0.5	3.7
Cd	66	18	9.2	12	20	15	8.1	30
Sn	66	45	20	33	51	43	29	70
Sb	66	91	43	65	107	83	46	147
Cs	65	7.5	4.3	5.4	9.1	6.5	3.1	14
Ba	66	989	1140	505	931	673	385	2130
Pb	66	839	337	625	973	768	458	1450

Table C.7: Descriptive statistics (valid number, average, standard deviation, lower and upper quartiles, median, 10th and 90th percentile) for TSP mass and element concentrations at site CRAES from October 2008 to February 2009 (period C3, weekly sampling). Mass concentrations in μg/m^3, all element concentrations in μg/g.

	N	avg	stdev	lower QT	upper QT	median	10th PCT	90th PCT
mass	21	238	74	187	277	235	145	290
Li	21	37	6.6	34	41	37	30	45
Na	21	9430	2670	7990	10500	9060	7550	12000
Mg	21	13600	2080	12000	15000	13500	11500	15800
Al	21	35100	4980	33200	38700	35400	29600	39800
P	21	1030	161	962	1080	1030	881	1250

S	21	29500	10500	24300	35300	27600	18600	39700
K	21	17500	7100	14400	18000	15800	13800	18600
Ca	21	52500	12900	42100	62100	56300	36500	66400
Sc	21	7.0	1.2	6.3	7.8	7.1	5.9	8.4
Ti	21	2210	351	2050	2360	2240	1740	2640
V	21	64	9.8	63	69	64	56	71
Cr	21	100	19	86	109	91	83	132
Mn	21	810	153	743	877	798	612	1030
Fe	21	27700	3920	26100	30400	27300	24000	32200
Co	21	17	3.1	17	20	19	13	20
Ni	20	47	6.1	43	50	47	37	55
Cu	21	335	135	259	350	295	224	463
Zn	21	2090	597	1780	2260	1990	1450	2730
Ga	21	46	16	29	55	51	25	63
As	21	101	37	76	111	93	70	133
Rb	21	84	13	78	88	85	75	96
Sr	21	377	134	299	392	363	291	413
Y	21	13	2.7	11	14	13	10	15
Zr	18	84	14	73	92	84	69	101
Nb	18	8.7	1.7	7.8	10	8.9	6.8	10
Mo	9	4.9	2.3	3.3	5.8	4.1	2.1	9.1
Ag	18	3.9	0.9	3.3	4.3	3.9	2.9	5.2
Cd	16	17	6.2	13	18	15	11	26
Sn	21	58	18	44	68	55	39	84
Sb	21	90	24	77	106	87	56	111
Cs	21	9.0	2.0	8.0	10	8.6	7.7	11
Ba	21	876	752	580	741	681	555	1100
Pb	21	817	176	714	918	836	591	1020

Table C.8: Descriptive statistics (valid number, average, standard deviation, lower and upper quartiles, median, 10th and 90th percentile) for TSP mass and element concentrations at site CRAES for the whole sampling period C (Oct-07 – Feb-09, weekly and daily sampling). Mass concentrations in $\mu g/m^3$, all element concentrations in $\mu g/g$.

	N	avg	stdev	lower QT	upper QT	median	10th PCT	90th PCT
mass	128	182	117	92	249	157	52	360
Li	126	30	14	18	40	31	11	47
Na	92	10000	4890	7210	12300	9640	3970	16500
Mg	126	12000	6250	7290	15800	12000	3460	19500
Al	126	29500	14200	18400	38000	27800	11700	47900
P	63	1160	494	948	1240	1060	851	1430
S	113	65300	38100	31400	93500	59000	24700	121000
K	128	15700	6280	12100	18000	14600	9290	22000
Ca	127	48100	22500	32000	63500	44600	20200	80500
Sc	128	5.8	2.7	3.8	7.4	5.2	3.0	8.9
Ti	128	2070	731	1600	2370	1880	1270	3120
V	128	60	20	44	72	57	37	88
Cr	127	114	82	78	124	94	60	174
Mn	128	869	302	650	993	808	568	1320
Fe	128	26300	8360	20100	30500	26100	15800	36900
Co	128	16	6.3	11	20	15	8.1	23
Ni	117	47	24	28	57	46	18	82
Cu	128	346	138	255	413	309	213	539
Zn	127	2640	1500	1710	3330	2310	1140	4310
Ga	128	33	17	22	39	26	20	63
As	128	110	57	74	132	98	55	185
Rb	128	83	31	63	96	81	48	122
Sr	128	300	133	216	366	268	167	454
Y	87	11	5.9	6.0	15	12	2.3	18
Zr	63	94	31	78	111	90	69	125
Nb	118	5.7	3.6	2.8	8.7	4.7	1.5	10
Mo	43	9.0	8.0	2.3	16	5.8	1.5	21
Ag	105	3.6	2.5	1.9	4.6	3.3	0.9	6.4
Cd	123	18	9.5	12	21	16	8.8	30
Sn	128	54	25	38	67	48	29	84
Sb	128	97	42	70	115	87	57	147
Cs	127	9.0	4.1	6.4	11	8.5	3.9	15
Ba	128	933	964	555	915	685	422	1620
Pb	128	906	362	684	1090	840	532	1450

Table C.9: Descriptive statistics (valid number, average, standard deviation, lower and upper quartiles, median, 10th and 90th percentile) for day-time $PM_{2.5}$ mass and element concentrations at site CRAES from October 2007 to July 2008 (period C1, weekly sampling). Mass concentrations in $\mu g/m^3$, all element concentrations in ng/m^3.

DAY	N	avg	stdev	lower QT	upper QT	median	10th PCT	90th PCT
mass	38	97.4	58.6	50.1	141.4	89.6	32.1	158.5
Li	39	3.28	2.16	1.86	4.04	2.88	1.04	6.81
Na	39	919	686	420	1270	664	252	2200
Mg	39	991	905	346	1170	672	225	2260
Al	39	2670	2940	874	3060	1560	701	8470
P	39	94.1	57.4	58.8	120	77.1	34.3	167
S	–							
K	39	2440	1250	1400	3430	2330	870	3980
Ca	39	3840	3370	1650	4370	2670	1190	9340
Sc	39	0.48	0.51	0.17	0.58	0.31	0.10	1.00
Ti	39	153	149	69.3	167	92.9	48.1	340
V	39	5.17	4.22	2.30	6.11	3.82	1.76	9.92
Cr	38	9.73	7.88	4.94	11.4	7.81	2.59	16.8
Mn	39	119	63.1	80.1	144	113	41.9	199
Fe	39	2190	1850	1130	2640	1550	635	4140
Co	39	1.55	2.23	0.70	1.63	1.10	0.31	2.90
Ni	37	6.97	9.19	2.83	7.09	4.58	1.48	9.68
Cu	39	50.5	29.5	23.3	73.5	43.4	16.2	88.9
Zn	39	519	260	316	760	510	141	888
Ga	39	5.64	3.74	2.65	7.80	5.10	1.56	12.2
As	39	19.1	12.4	8.14	28.5	17.4	5.37	34.9
Rb	39	12.8	6.36	7.21	17.2	12.4	4.92	22.8
Sr	39	21.9	19.7	9.57	23.6	14.7	5.80	45.1
Y	38	0.77	0.89	0.19	0.99	0.43	0.07	1.79
Zr	39	16.7	9.10	10.5	20.5	15.3	6.49	30.1
Nb	39	0.68	0.54	0.31	0.90	0.58	0.17	1.52
Mo	18	6.68	6.81	1.99	8.62	3.88	1.77	17.9
Ag	39	0.80	0.35	0.50	1.04	0.76	0.33	1.37
Cd	39	3.21	2.03	1.44	4.68	2.77	1.00	6.46
Sn	39	11.7	6.84	5.55	17.6	10.4	4.03	22.3
Sb	39	16.4	10.5	7.14	26.6	11.9	4.98	33.7
Cs	39	1.65	0.73	0.93	2.25	1.57	0.72	2.52
Ba	33	41.2	35.0	16.3	60.9	31.9	6.13	93.4
Pb	39	182	95.8	106	270	163	59.7	334

Table C.10: Descriptive statistics (valid number, average, standard deviation, lower and upper quartiles, median, 10th and 90th percentile) for day-time $PM_{2.5}$ mass and element concentrations at site CRAES from July 2008 to September 2008 (period C2, 12-hourly sampling). Mass concentrations in $\mu g/m^3$, all element concentrations in ng/m^3.

DAY	N	avg	stdev	lower QT	upper QT	median	10th PCT	90th PCT
mass	56	58.4	37.1	36.8	77.3	52.5	19.6	88.8
Li	61	1.89	1.47	0.85	2.40	1.51	0.48	4.21
Na	52	5620	5540	2310	7760	4040	1380	10500
Mg	46	1090	1050	345	1400	867	162	2320
Al	63	1600	1150	638	2260	1500	323	3353
P	47	2700	2190	1020	3510	2400	308	6100
S	67	6560	6430	2080	8520	4420	1080	17200
K	49	986	831	446	1180	806	219	1990
Ca	34	3390	2610	1760	4160	2840	584	6620
Sc	59	0.06	0.04	0.02	0.08	0.05	0.01	0.13
Ti	55	16.2	15.1	6.33	22.3	10.0	4.60	39.1
V	67	1.31	1.13	0.52	1.86	0.85	0.30	2.95
Cr	64	11.7	12.9	4.15	13.9	7.55	1.78	29.2
Mn	67	24.7	14.1	15.0	32.5	20.6	9.70	43.6
Fe	67	375	204	225	505	336	159	635
Co	67	0.44	0.34	0.21	0.52	0.33	0.14	0.95
Ni	48	7.63	8.78	1.98	8.87	5.22	0.89	18.8
Cu	67	27.2	30.7	9.98	33.6	17.8	6.55	54.4
Zn	51	310	381	83.8	373	160	19.3	780
Ga	67	1.48	0.84	0.85	1.88	1.37	0.60	2.72
As	67	7.03	5.07	3.15	10.8	5.63	2.15	14.3
Rb	67	3.90	2.60	1.91	5.22	3.01	1.55	7.63
Sr	67	12.7	7.20	6.71	16.3	12.3	4.89	22.2
Y	30	0.72	1.09	0.18	0.86	0.34	0.05	1.86
Zr	39	32.6	22.3	13.0	49.4	30.3	5.85	58.3

Nb	15	0.69	0.36	0.38	1.04	0.59	0.26	1.15
Mo	57	16.4	16.7	6.37	21.0	11.9	2.21	36.2
Ag	56	0.77	1.04	0.29	0.96	0.59	0.14	1.29
Cd	67	1.15	0.91	0.42	1.65	0.97	0.18	2.44
Sn	64	2.91	2.15	1.30	3.85	2.52	0.53	5.40
Sb	5	2.66	3.23	0.09	3.31	1.90	0.08	7.91
Cs	67	0.90	0.46	0.53	1.26	0.77	0.39	1.50
Ba	56	267	386	50.9	342	111	22.6	785
Pb	61	48.9	37.8	16.6	70.2	41.4	11.1	99.4

Table C.11: Descriptive statistics (valid number, average, standard deviation, lower and upper quartiles, median, 10th and 90th percentile) for day-time $PM_{2.5}$ mass and element concentrations at site CRAES from October 2008 to February 2009 (period C3, weekly sampling). Mass concentrations in $\mu g/m^3$, all element concentrations in ng/m^3.

DAY	N	avg	stdev	lower QT	upper QT	median	10th PCT	90th PCT
mass	16	126	39.2	95.1	138	118	88.3	201
Li	16	4.38	1.34	3.56	4.98	4.41	2.53	6.85
Na	16	1130	479	759	1280	1010	723	1930
Mg	16	1350	631	951	1580	1210	803	2340
Al	16	3540	1670	2400	4345	3190	1850	6440
P	16	125	45.6	88.2	148	117	76.9	208
S	16	5150	2220	2890	6350	5180	2520	7310
K	16	2810	1180	2190	3350	2350	1690	4120
Ca	16	5280	2600	3830	6250	4720	2500	8610
Sc	16	0.71	0.34	0.53	0.82	0.64	0.35	1.38
Ti	16	222	107	161	251	194	123	438
V	16	6.82	2.92	4.58	7.65	6.26	3.89	13.2
Cr	16	11.7	4.39	9.05	15.1	11.2	6.75	18.3
Mn	16	111	35.1	94.1	116	101	81.1	184
Fe	16	2940	1470	2120	3140	2540	1770	5510
Co	16	1.93	0.72	1.45	2.05	1.76	1.33	2.90
Ni	16	6.20	2.38	4.58	8.23	6.15	2.56	9.68
Cu	16	44.9	24.3	29.5	44.9	40.3	26.7	80.6
Zn	16	394	167	263	537	356	192	634
Ga	16	8.39	2.77	6.64	9.50	8.50	5.17	11.6
As	16	17.6	7.76	12.0	21.2	17.0	9.61	29.9
Rb	16	11.9	4.01	8.77	14.6	10.6	7.76	18.1
Sr	16	40.2	18.2	25.9	51.7	36.8	18.8	70.8
Y	16	1.31	0.58	0.95	1.72	1.24	0.58	2.24
Zr	16	21.6	7.00	17.9	25.0	21.7	14.8	26.8
Nb	16	0.86	0.37	0.61	0.98	0.78	0.46	1.38
Mo	8	6.98	4.20	3.19	10.8	7.60	1.03	11.6
Ag	16	0.73	0.25	0.54	0.91	0.66	0.46	1.04
Cd	12	2.84	0.99	2.00	3.45	3.06	1.55	3.99
Sn	16	11.3	4.97	8.20	13.8	10.4	4.57	19.0
Sb	16	12.8	6.59	6.88	16.9	12.0	5.51	22.3
Cs	16	1.52	0.51	1.04	1.91	1.53	0.94	2.19
Ba	15	95.0	90.9	52.3	94.0	65.2	45.1	222
Pb	16	160	43.8	127	188	172	97.6	211

Table C.12: Descriptive statistics (valid number, average, standard deviation, lower and upper quartiles, median, 10th and 90th percentile) for night-time $PM_{2.5}$ mass and element concentrations at site CRAES from October 2007 to July 2008 (period C1, weekly sampling). Mass concentrations in $\mu g/m^3$, all element concentrations in ng/m^3.

NIGHT	N	avg	stdev	lower QT	upper QT	median	10th PCT	90th PCT
mass	36	103	49.3	63.1	138	89.6	46.8	176
Li	36	3.34	2.35	1.64	4.29	2.72	1.40	5.55
Na	36	845	526	490	1140	698	290	1570
Mg	36	864	894	355	1200	452	270	1990
Al	36	2360	3290	759	3370	913	528	5820
P	36	97.4	59.6	57.8	134	75.2	36.0	162
S	–							
K	36	2820	1820	1460	3680	2480	1190	4510
Ca	35	3030	2790	1120	4270	1650	964	7460
Sc	36	0.44	0.58	0.16	0.56	0.21	0.11	1.07
Ti	36	138	172	53.6	179	74.3	43.2	335
V	36	5.02	4.92	2.20	6.76	3.20	1.74	10.9
Cr	35	7.58	5.12	3.55	10.5	6.43	3.06	15.6
Mn	36	101	49.8	71.8	128	87.2	42.0	156

Fe	36	1940	1980	865	2400	1110	688	4270
Co	36	1.21	0.81	0.70	1.50	1.01	0.53	2.17
Ni	34	6.07	6.54	2.17	6.95	3.83	1.73	11.7
Cu	36	77.2	38.6	54.4	95.3	68.2	32.6	129
Zn	36	507	296	236	691	467	190	819
Ga	36	6.74	4.26	3.41	9.06	6.09	1.97	13.7
As	36	20.8	11.6	10.3	28.9	18.4	8.37	39.7
Rb	36	12.7	6.39	6.89	16.4	12.7	5.46	22.4
Sr	36	25.1	30.0	9.42	25.6	13.4	6.23	66.1
Y	31	0.91	0.92	0.33	1.17	0.57	0.18	1.93
Zr	34	17.5	8.31	12.6	22.0	15.7	9.44	27.5
Nb	36	0.62	0.44	0.31	0.76	0.53	0.15	1.18
Mo	22	7.82	5.80	4.35	10.4	6.49	1.84	13.5
Ag	36	0.81	0.35	0.55	1.03	0.72	0.41	1.38
Cd	36	4.05	2.87	1.69	5.16	3.53	1.36	8.08
Sn	36	14.6	7.15	8.42	18.3	13.8	6.65	27.2
Sb	36	31.7	14.2	22.2	40.7	29.6	12.6	48.9
Cs	36	1.63	0.75	0.94	2.14	1.65	0.77	2.64
Ba	33	76.3	144	23.2	60.3	39.8	10.8	121
Pb	36	213	110	122	274	189	84.2	386

Table C.13: Descriptive statistics (valid number, average, standard deviation, lower and upper quartiles, median, 10th and 90th percentile) for night-time $PM_{2.5}$ mass and element concentrations at site CRAES from July 2008 to September 2008 (period C2, 12-hourly sampling). Mass concentrations in $\mu g/m^3$, all element concentrations in ng/m^3.

NIGHT	N	avg	stdev	lower QT	upper QT	median	10th PCT	90th PCT
mass	47	53.7	38.9	27.2	67.3	43.8	16.6	124
Li	58	1.81	1.45	0.58	2.66	1.46	0.27	4.07
Na	47	8350	7600	3110	10800	6470	511	25200
Mg	46	1320	1230	420	1740	878	189	3800
Al	57	1900	1380	824	2470	1540	323	3990
P	51	3280	2300	1160	4590	2732	798	6430
S	65	6270	6110	2080	8710	3600	1210	16500
K	52	947	713	385	1350	836	197	1880
Ca	39	4170	3450	1360	8030	2580	492	9520
Sc	54	0.06	0.05	0.03	0.08	0.04	0.01	0.11
Ti	53	16.6	14.1	6.36	22.5	12.8	4.60	38.4
V	63	1.48	1.65	0.48	1.85	0.89	0.28	3.24
Cr	61	15.0	14.6	4.33	21.4	12.4	2.16	28.3
Mn	65	24.1	13.3	14.1	31.6	20.9	8.37	43.5
Fe	64	375	210	200	546	331	146	658
Co	65	0.67	0.75	0.27	0.80	0.43	0.14	1.35
Ni	49	9.15	9.58	2.37	11.9	5.62	0.87	29.4
Cu	65	44.1	48.3	15.0	44.1	29.2	8.67	103
Zn	54	365	502	67.7	436	171	13.8	915
Ga	65	1.62	0.85	1.03	1.91	1.49	0.76	2.75
As	65	7.10	4.73	3.75	9.49	6.05	2.26	14.9
Rb	65	3.73	2.48	1.91	5.40	2.93	1.44	7.22
Sr	65	14.6	12.1	6.29	18.1	11.7	4.62	25.5
Y	25	0.80	0.74	0.27	1.33	0.59	0.07	1.45
Zr	31	31.1	22.7	12.4	45.0	28.0	4.26	64.5
Nb	16	0.49	0.33	0.26	0.59	0.45	0.06	0.99
Mo	53	22.5	22.2	7.00	31.2	12.7	3.86	61.1
Ag	58	0.58	0.43	0.29	0.84	0.47	0.15	1.11
Cd	65	1.19	0.99	0.53	1.59	0.91	0.36	2.34
Sn	65	2.98	2.21	1.34	3.84	2.54	0.87	5.79
Sb	6	2.69	3.22	0.09	4.07	1.67	0.08	8.55
Cs	65	0.82	0.46	0.49	1.11	0.64	0.32	1.51
Ba	52	348	589	41.2	373	119	24.2	959
Pb	61	46.4	35.1	16.9	62.7	41.2	13.2	92.6

Table C.14: Descriptive statistics (valid number, average, standard deviation, lower and upper quartiles, median, 10th and 90th percentile) for night-time $PM_{2.5}$ mass and element concentrations at site CRAES from October 2008 to February 2009 (period C3, weekly sampling). Mass concentrations in $\mu g/m^3$, all element concentrations in ng/m^3.

NIGHT	N	avg	stdev	lower QT	upper QT	median	10th PCT	90th PCT
mass	21	102	46.9	71.6	119	96.2	46.0	149
Li	21	2.95	1.45	2.24	3.68	2.56	1.69	4.09
Na	21	835	413	607	993	738	456	1290

Mg	21	710	623	273	907	462	230	1400
Al	21	1790	1630	685	2240	1390	522	3800
P	21	87.9	42.7	60.9	112	79.3	51.5	133
S	21	5170	2400	2980	7330	4460	2760	8660
K	21	3080	3000	1870	2530	2000	1290	4820
Ca	21	2090	1640	1000	2650	1700	821	3330
Sc	21	0.34	0.32	0.13	0.38	0.24	0.10	0.68
Ti	21	111	97.3	46.4	116	76.5	43.5	206
V	21	3.89	2.83	1.91	4.53	2.94	1.73	7.58
Cr	21	14.6	8.66	7.78	20.0	11.3	6.88	28.6
Mn	21	74.5	27.2	62.0	85.7	72.7	44.8	92.7
Fe	21	1500	1110	903	1680	1130	710	2590
Co	21	1.24	0.68	0.91	1.65	1.07	0.39	1.88
Ni	20	4.34	1.67	3.16	5.76	3.89	2.52	6.61
Cu	21	60.2	44.7	31.7	65.0	48.2	25.6	146
Zn	21	350	140	245	411	334	201	549
Ga	21	7.70	3.66	4.22	10.3	7.83	3.27	12.3
As	21	17.9	7.42	12.0	21.1	16.7	10.7	25.6
Rb	21	8.87	3.66	5.59	10.5	8.79	5.07	11.1
Sr	21	35.7	51.9	9.03	30.6	12.3	7.52	80.5
Y	20	0.64	0.66	0.27	0.75	0.42	0.12	1.36
Zr	21	20.1	9.70	14.8	21.3	20.0	11.7	26.4
Nb	21	0.44	0.30	0.22	0.59	0.39	0.16	0.76
Mo	9	14.7	29.4	2.78	7.64	3.71	1.38	92.3
Ag	21	0.83	0.39	0.62	0.93	0.75	0.54	1.06
Cd	16	3.26	1.30	2.14	4.42	2.96	1.71	4.67
Sn	21	13.5	12.3	8.15	15.1	10.4	5.91	16.8
Sb	21	15.6	6.00	10.7	17.8	15.3	8.10	21.8
Cs	21	1.14	0.42	0.79	1.39	1.23	0.64	1.50
Ba	19	131	267	18.6	83.1	33.1	10.5	695
Pb	21	178	70.7	130	214	173	95.8	265

Table C.15: Descriptive statistics (valid number, average, standard deviation, lower and upper quartiles, median, 10th and 90th percentile) for day-time $PM_{2.5}$ mass and element concentrations at site CRAES from October 2007 to July 2008 (period C1, weekly sampling). Mass concentrations in $\mu g/m^3$, all element concentrations in $\mu g/g$.

DAY	N	avg	stdev	lower QT	upper QT	median	10th PCT	90th PCT
mass	39	96.4	58.2	50.1	141	86.4	32.1	158
Li	38	35.4	9.45	28.2	40.1	35.7	24.0	48.9
Na	38	9930	4560	7660	10000	8940	6380	13600
Mg	38	10200	4960	5630	14000	9820	4050	17900
Al	38	27000	17300	17900	34600	22700	7650	46500
P	38	1040	397	785	1120	952	660	1667
S	–							
K	38	29100	17000	23600	29900	26800	19400	35100
Ca	38	41800	24400	25700	49900	38500	16200	72200
Sc	38	4.73	2.53	2.62	6.22	4.26	1.54	8.71
Ti	38	1590	760	1020	1960	1425	650	2600
V	38	55.7	24.7	37.7	62.6	55.0	28.8	90.6
Cr	37	103	62.8	71.8	106	95.3	55.3	148
Mn	38	1450	873	957	1430	1150	872	2690
Fe	38	23600	12700	15400	29100	20400	12300	39100
Co	38	16.9	21.2	9.18	16.5	12.2	6.21	26.5
Ni	36	71.8	80.4	32.2	67.6	47.4	25.0	132
Cu	38	596	437	378	641	503	288	931
Zn	38	6480	4990	4170	7410	5710	3090	9070
Ga	38	64.2	32.4	34.9	87.3	55.2	32.2	117
As	38	225	165	151	248	202	94.2	314
Rb	38	157	113	114	158	139	99.6	197
Sr	38	240	136	151	283	223	106	453
Y	37	7.79	6.09	2.05	12.8	6.71	1.33	16.3
Zr	38	216	151	134	286	182	69.2	347
Nb	38	7.31	3.89	4.65	9.08	6.79	2.42	13.2
Mo	17	110	150	28.0	129	47.2	7.94	400
Ag	38	10.4	7.40	6.62	11.8	9.26	4.34	14.9
Cd	38	38.6	30.5	24.8	44.9	33.3	15.6	55.4
Sn	38	159	239	90.0	150	110	76.8	196
Sb	38	184	93.0	120	218	168	90.6	320
Cs	38	21.0	15.9	15.2	22.3	19.1	11.0	25.8
Ba	33	526	597	216	630	387	79.8	1030
Pb	38	2200	1290	1480	2600	2040	1200	3170

Table C.16: Descriptive statistics (valid number, average, standard deviation, lower and upper quartiles, median, 10th and 90th percentile) for day-time $PM_{2.5}$ mass and element concentrations at site CRAES from July 2008 to September 2008 (period C2, 12-hourly sampling). Mass concentrations in $\mu g/m^3$, all element concentrations in $\mu g/g$.

DAY	N	avg	stdev	lower QT	upper QT	median	10th PCT	90th PCT
mass	56	58.4	37.1	36.8	77.3	52.5	19.6	88.8
Li	52	45.1	62.6	14.4	57.8	24.3	8.53	86.8
Na	44	109000	98900	40500	139000	84400	21100	219000
Mg	40	17800	15295	7240	20600	13900	3040	42000
Al	53	28400	18828	17000	36900	23400	5430	62900
P	39	39500	26332	18000	58800	37400	7560	73900
S	56	110000	53500	68900	150000	118000	33500	181000
K	42	16500	12400	8790	19600	14300	5230	31500
Ca	28	52900	45500	35200	60100	48300	8470	84600
Sc	51	1.47	1.79	0.45	1.63	1.06	0.33	3.00
Ti	50	268	179	122	331	250	73.3	527
V	56	25.1	16.8	14.7	29.9	21.3	9.16	44.2
Cr	56	222	239	84.4	264	148	47.0	430
Mn	56	505	365	281	583	401	223	792
Fe	56	7730	5560	4280	8730	6240	3370	14000
Co	56	8.48	7.10	4.20	9.34	5.51	2.80	21.0
Ni	40	114	108	32.7	144	69.6	18.2	287
Cu	56	459	436	224	509	347	163	756
Zn	45	4630	7070	1195	4400	2820	492	9730
Ga	56	28.5	13.7	19.8	32.9	24.0	15.9	46.9
As	56	140	94.0	79.9	162	121	54.8	209
Rb	56	78.7	58.6	46.5	91.1	63.5	33.4	141
Sr	56	251	193	127	307	198	86.7	507
Y	26	11.2	11.1	4.46	11.4	8.94	1.58	21.0
Zr	33	909	1310	191	1250	497	73.1	1680
Nb	14	31.1	52.5	5.23	36.4	7.80	3.99	85.3
Mo	47	280	234	114	378	202	62.9	660
Ag	49	16.8	23.8	5.56	16.0	10.5	1.29	39.4
Cd	56	25.3	28.5	11.8	28.9	16.9	6.58	40.4
Sn	54	62.1	57.3	36.3	74.4	46.2	13.8	99.0
Sb	5	14.8	14.1	2.04	23.6	12.7	1.33	34.1
Cs	56	18.9	15.2	11.2	21.7	15.7	6.77	28.7
Ba	48	4720	8040	811	5270	1700	291	10200
Pb	54	866	579	535	1130	743	215	1610

Table C.17: Descriptive statistics (valid number, average, standard deviation, lower and upper quartiles, median, 10th and 90th percentile) for day-time $PM_{2.5}$ mass and element concentrations at site CRAES from October 2008 to February 2009 (period C3, weekly sampling). Mass concentrations in $\mu g/m^3$, all element concentrations in $\mu g/g$.

DAY	N	avg	stdev	lower QT	upper QT	median	10th PCT	90th PCT
mass	16	126	39.2	95.1	138	118	88.3	201
Li	16	35.2	6.19	30.8	38.3	34.1	27.9	43.4
Na	16	8860	1490	7720	10300	8530	7430	11100
Mg	16	10500	2620	8480	12200	10500	7850	13900
Al	16	27400	6880	23100	32500	26400	16800	37600
P	16	987	134	843	1090	995	824	1150
S	16	41400	14500	27200	52700	40900	24700	61300
K	16	22600	7420	18300	26700	20200	16400	29300
Ca	16	41300	14200	32200	51100	41400	25500	58900
Sc	16	5.51	1.35	4.81	6.46	5.59	3.22	7.47
Ti	16	1720	453	1460	2020	1680	1120	2370
V	16	53.4	10.1	46.1	60.4	52.1	40.9	67.0
Cr	16	93.8	33.7	78.6	99.2	87.1	70.8	116
Mn	16	892	171	810	973	862	682	1100
Fe	16	22700	5380	18900	25700	22400	16300	30100
Co	16	15.3	2.26	14.3	16.0	15.1	12.3	19.2
Ni	16	50.4	18.6	39.9	56.3	47.9	27.7	80.0
Cu	16	353	125	279	414	329	221	587
Zn	16	3160	1060	2240	3900	3020	2050	4770
Ga	16	68.0	18.1	56.5	76.4	63.6	50.1	86.7
As	16	143	61.7	105	164	126	82.5	191
Rb	16	94.7	17.5	79.8	109	93.3	73.7	115
Sr	16	312	84.7	249	360	313	204	439
Y	16	10.4	3.33	7.97	13.2	11.0	5.44	14.6
Zr	16	179	50.7	135	220	193	88.5	229

Nb	16	6.70	1.38	5.86	7.26	6.90	4.60	9.01
Mo	8	68.1	50.3	17.4	109	76.4	7.50	132
Ag	16	6.05	2.12	4.79	7.53	5.25	3.78	9.22
Cd	12	24.4	9.84	16.1	34.4	22.4	14.3	37.9
Sn	16	89.4	28.6	69.4	113	90.2	39.2	119
Sb	16	103	48.9	66.5	121	98.1	52.2	177
Cs	16	12.2	3.08	9.84	14.1	12.7	8.07	15.5
Ba	15	739	648	411	655	574	334	1720
Pb	16	1300	310	1070	1530	1250	951	1800

Table C.18: Descriptive statistics (valid number, average, standard deviation, lower and upper quartiles, median, 10th and 90th percentile) for night-time $PM_{2.5}$ mass and element concentrations at site CRAES from October 2007 to July 2008 (period C1, weekly sampling). Mass concentrations in $\mu g/m^3$, all element concentrations in $\mu g/g$.

NIGHT	N	avg	stdev	lower QT	upper QT	median	10th PCT	90th PCT
mass	36	103	49.3	63.1	138	89.6	46.8	176
Li	35	32.3	8.71	25.1	38.7	30.7	23.5	42.6
Na	35	8270	2170	6590	9670	8130	5860	11000
Mg	35	8160	4770	4060	10700	7350	2920	14900
Al	35	20700	15900	8280	28600	15800	5620	36700
P	35	965	254	774	1090	960	685	1240
S	–							
K	35	29000	14500	23200	28800	25200	19900	33100
Ca	35	28700	17000	13600	45300	24400	10100	51700
Sc	35	3.86	2.72	1.84	5.13	2.99	1.40	6.64
Ti	35	1240	803	653	1680	988	495	2080
V	35	47.3	23.1	27.2	61.7	46.8	21.8	78.7
Cr	35	74.4	38.3	49.2	86.8	68.6	38.4	103
Mn	35	1060	335	878	1080	976	776	1570
Fe	35	17900	8780	10200	22600	15500	9260	26500
Co	35	12.1	4.51	9.18	13.7	11.6	6.53	18.5
Ni	33	69.5	103	28.2	63.3	40.4	14.6	148
Cu	35	825	335	573	994	797	429	1280
Zn	35	5320	2340	4060	6200	5050	3000	7700
Ga	35	71.2	39.3	41.1	100	53.7	34.5	131
As	35	224	108	160	259	212	99.8	357
Rb	35	129	39.0	103	145	127	97.9	163
Sr	35	260	314	103	297	190	74.2	361
Y	30	8.57	5.77	3.60	14.2	7.40	2.82	14.8
Zr	33	191	87.5	128	260	160	106	336
Nb	35	5.98	2.83	3.70	7.75	5.93	2.52	9.51
Mo	22	81.1	54.7	47.5	133	63.7	19.9	168
Ag	35	9.04	4.51	7.24	9.88	8.20	4.48	12.6
Cd	35	44.9	43.5	26.1	51.6	38.5	20.6	56.0
Sn	35	157	62.2	113	189	153	87.2	228
Sb	35	340	130	235	441	337	178	504
Cs	35	17.2	6.09	13.5	19.5	16.9	11.1	21.6
Ba	32	858	1495	187	752	401	97.7	1480
Pb	35	2280	837	1790	2700	2310	1260	3170

Table C.19: Descriptive statistics (valid number, average, standard deviation, lower and upper quartiles, median, 10th and 90th percentile) for night-time $PM_{2.5}$ mass and element concentrations at site CRAES from July 2008 to September 2008 (period C2, 12-hourly sampling). Mass concentrations in $\mu g/m^3$, all element concentrations in $\mu g/g$.

NIGHT	N	avg	stdev	lower QT	upper QT	median	10th PCT	90th PCT
mass	48	57.1	45.1	27.4	68.9	43.9	16.6	127
Li	42	36.0	32.8	14.1	50.6	26.0	5.56	86.8
Na	36	117000	104000	42700	149000	92900	12200	267200
Mg	33	24000	24600	8000	35500	12500	4000	59600
Al	42	36500	43100	11300	51500	25200	5640	67300
P	34	60000	53400	25000	73900	41500	15000	141000
S	48	121000	114000	53300	130000	96800	33500	225000
K	37	16600	11800	7550	25100	13300	3660	37300
Ca	27	57900	59500	17000	74100	41900	8760	129000
Sc	39	1.41	2.51	0.29	1.52	0.75	0.14	3.17
Ti	38	279	258	156	325	216	102	431
V	47	25.2	19.7	12.3	32.3	20.3	6.51	56.9
Cr	46	283	321	65.0	367	168	38.9	787
Mn	48	501	353	276	556	418	155	1050

Fe	47	7210	4520	4120	8490	6620	2220	13800
Co	48	12.2	13.3	4.82	14.6	7.12	2.06	31.7
Ni	35	119	110	27.8	179	103	13.1	251
Cu	48	817	765	309	1060	548	163	2020
Zn	41	5590	7920	1110	4980	2710	417	14700
Ga	48	33.8	23.9	20.9	37.0	25.2	13.2	66.4
As	48	148	123	82.3	162	122	48.3	267
Rb	48	78.1	63.9	43.4	95.5	65.1	26.3	129
Sr	48	265	237	133	300	203	76.3	474
Y	19	13.9	20.1	2.58	11.3	8.08	0.92	59.5
Zr	23	659	674	257	900	427	78.6	1680
Nb	13	10.2	8.94	2.17	14.6	8.14	0.93	20.0
Mo	38	279	268	105	355	206	38.4	642
Ag	43	13.2	13.2	3.96	16.4	8.96	1.27	32.5
Cd	48	23.7	20.9	12.9	26.3	17.9	8.64	46.4
Sn	48	59.8	49.6	28.5	75.7	48.4	14.6	113
Sb	6	21.3	25.6	2.04	30.3	13.5	1.33	67.3
Cs	48	17.2	12.0	8.45	21.2	14.4	5.37	30.6
Ba	36	5609	9479	772	5761	2030	299	17099
Pb	46	958	821	499	1158	757	294	1663

Table C.20: Descriptive statistics (valid number, average, standard deviation, lower and upper quartiles, median, 10th and 90th percentile) for night-time $PM_{2.5}$ mass and element concentrations at site CRAES from October 2008 to February 2009 (period C3, weekly sampling). Mass concentrations in $\mu g/m^3$, all element concentrations in $\mu g/g$.

NIGHT	N	avg	stdev	lower QT	upper QT	median	10th PCT	90th PCT
mass	21	102	46.9	71.6	119	96.2	46.0	149
Li	21	29.3	9.36	21.5	33.6	27.9	20.3	40.3
Na	21	8280	1810	6530	9350	8610	5800	11000
Mg	21	6340	3000	4370	8390	5430	3560	10800
Al	21	16300	9090	10500	20400	13700	6300	29900
P	21	887	316	655	975	851	568	1100
S	21	52400	17400	39000	67300	45000	36600	75200
K	21	29100	17500	19500	31100	21500	18700	37900
Ca	21	20200	8590	14900	26500	19400	8620	31400
Sc	21	3.13	1.66	1.98	3.95	2.89	1.30	5.07
Ti	21	1030	505	650	1350	940	463	1450
V	21	36.6	13.0	30.1	45.2	36.4	19.5	45.9
Cr	21	176	153	83.9	237	109	68.4	347
Mn	21	779	211	633	891	746	568	1030
Fe	21	14400	5950	9850	16500	13000	9480	22200
Co	21	11.9	3.37	9.00	14.0	12.7	7.51	15.9
Ni	20	43.8	13.4	36.6	51.3	43.3	29.0	61.2
Cu	21	668	647	372	707	444	286	1070
Zn	21	3690	1210	2800	4640	4010	2210	4840
Ga	21	75.9	24.1	56.9	96.2	73.0	44.8	100
As	21	193	77.7	144	249	172	105	301
Rb	21	90.3	20.0	74.9	100	84.5	70.6	113
Sr	21	292	319	151	309	174	81.5	417
Y	20	6.53	6.86	2.66	7.11	4.63	1.53	12.2
Zr	21	242	209	165	298	181	97.2	357
Nb	21	4.34	1.99	2.78	5.29	4.11	2.24	7.07
Mo	9	281	668	29.2	80.8	57.7	11.7	2060
Ag	21	9.11	4.59	6.39	10.9	6.98	5.03	13.6
Cd	16	36.3	13.4	27.9	41.9	33.5	17.0	61.1
Sn	21	181	294	78.7	155	109	68.2	220
Sb	21	175	80.9	128	217	178	64.8	261
Cs	21	11.7	3.01	9.71	12.7	11.1	8.96	16.5
Ba	19	1000	1730	240	553	427	158	5590
Pb	21	1840	519	1460	2070	1820	1180	2290

Table C.21: Descriptive statistics (valid number, average, standard deviation, lower and upper quartiles, median, 10th and 90th percentile) for PM_1 mass and element concentrations at site CRAES from July 2008 to September 2008 (period C2, 24-hourly sampling). Mass concentrations in $\mu g/m^3$, all element concentrations in ng/m^3.

	N	avg	stdev	lower QT	upper QT	median	10th PCT	90th PCT
mass	49	34.1	17.2	20.8	45.1	31.4	14.5	63.6
Li	52	1.18	0.81	0.69	1.38	1.04	0.43	2.01
Na	51	5740	2710	4010	8520	5560	1730	9190

Mg	41	531	384	249	732	409	113	1130
Al	53	1090	451	794	1240	1020	630	1580
P	47	1290	829	646	1650	1280	238	2180
S	54	4420	3600	1660	5734	3000	1140	9850
K	52	585	344	394	702	536	216	933
Ca	54	4030	977	3400	4760	3880	2920	5540
Sc	36	0.05	0.04	0.02	0.06	0.04	0.01	0.09
Ti	53	12.4	8.05	8.06	15.6	10.4	4.15	24.0
V	54	0.97	0.72	0.53	1.29	0.73	0.30	1.82
Cr	51	10.2	6.71	5.86	13.2	8.21	4.10	19.9
Mn	54	16.7	7.27	10.3	22.2	15.8	7.79	26.9
Fe	54	205	96.9	147	255	188	97.0	333
Co	54	0.79	2.79	0.21	0.51	0.29	0.19	0.85
Ni	54	9.33	11.1	6.46	9.78	7.68	4.85	11.5
Cu	54	20.2	13.9	9.28	25.8	17.2	7.00	32.7
Zn	49	273	214	166	281	197	121	503
Ga	54	1.51	0.67	0.98	1.94	1.44	0.67	2.69
As	54	6.09	2.95	3.51	8.13	5.52	2.93	9.94
Rb	54	2.36	1.60	1.23	3.14	2.13	0.93	3.94
Sr	54	10.8	5.48	7.77	11.6	9.12	6.88	15.1
Mo	54	28.3	7.07	23.5	33.4	29.3	18.7	36.5
Cd	54	0.84	0.49	0.44	1.31	0.69	0.28	1.48
Sn	54	2.67	1.75	1.42	3.54	2.19	0.90	4.73
Sb	45	5.87	3.46	3.17	8.42	5.81	1.33	10.4
Cs	54	0.66	0.31	0.44	0.79	0.64	0.30	1.10
Ba	49	136	235	29.5	86.6	55.4	19.8	433
Pb	53	33.5	23.9	13.8	52.4	30.3	6.87	61.1

Table C.22: Descriptive statistics (valid number, average, standard deviation, lower and upper quartiles, median, 10th and 90th percentile) for PM_1 mass and element concentrations at site CRAES from July 2008 to September 2008 (period C2, 24-hourly sampling). Mass concentrations in $\mu g/m^3$, all element concentrations in $\mu g/g$.

	N	avg	stdev	lower QT	upper QT	median	10th PCT	90th PCT
mass	49	34.1	17.2	20.8	45.1	31.4	14.5	63.6
Li	47	45.9	50.7	15.6	57.1	30.6	10.2	90.6
Na	46	249000	200000	96600	356000	194000	44200	561000
Mg	37	22400	20300	7200	34900	17100	4490	52500
Al	48	45700	44500	19800	55800	31800	12700	99200
P	42	56600	75800	20300	63800	35500	8490	120000
S	49	128000	77700	85600	170000	105792	55400	206000
K	47	21600	18600	10500	22900	17033	9410	42500
Ca	49	181000	202000	81000	200000	130000	60500	246000
Sc	36	2.27	4.01	0.38	2.61	0.88	0.17	3.56
Ti	48	588	912	185	618	300	83.1	1310
V	49	34.8	32.2	15.5	37.9	26.1	11.5	76.9
Cr	46	419	391	141	511	306	92.6	940
Mn	49	657	581	334	810	422	268	1160
Fe	49	8360	7960	3420	9880	5030	2410	20400
Co	49	25.6	64.7	5.86	21.0	10.9	3.44	52.1
Ni	49	453	892	140	404	281	102	551
Cu	49	744	692	434	903	550	201	1230
Zn	45	12100	15900	4300	10100	6700	3180	26900
Ga	49	54.8	38.7	32.5	60.0	41.3	27.7	92.2
As	49	218	145	127	227	164	106	468
Rb	49	82.5	65.3	41.5	103	59.5	29.8	158
Sr	49	477	515	214	479	345	130	1030
Mo	49	1240	1230	571	1440	917	409	2550
Cd	49	29.7	21.1	16.3	34.9	22.8	12.5	60.6
Sn	49	99.5	101	47.8	114	71.5	31.7	193
Sb	42	255	379	77.9	243	151	54.1	477
Cs	49	29.1	34.9	11.8	28.7	17.7	8.05	50.3
Ba	44	7450	15900	840	3830	1620	567	17400
Pb	48	1080	729	592	1420	888	292	1970

Table C.23: Descriptive statistics (valid number, average, standard deviation, lower and upper quartiles, median, 10th and 90th percentile) for TSP mass and watersoluble ion concentrations at site CRAES from July 2008 to September 2008 (period C2, 24-hourly sampling). Mass concentrations in μg/m^3, all ion concentrations in ng/^{3}.

	N	avg	stdev	lower QT	upper QT	median	10th PCT	90th PCT
mass	66	104	51	62	131	96	45	177
Li	65	1.7	0.6	1.2	2.1	1.6	1.1	2.5
Na	65	7560	555	7170	7930	7530	6830	8240
Mg	65	460	232	259	589	431	221	736
Al	65	162	96	76	225	142	58	278
P	65	5010	458	4740	5260	5010	4470	5720
S	65	10600	9470	3720	13800	8550	2040	26600
K	65	983	571	492	1320	889	300	1650
Ca	65	3490	1540	2260	4580	3360	1580	5830
Sc	65	0.04	0.02	0.03	0.05	0.04	0.02	0.06
Ti	65	5.2	2.8	2.9	6.9	4.7	2.3	9.7
V	65	2.1	1.6	0.9	2.8	1.6	0.7	4.1
Cr	65	6.0	3.5	4.6	6.5	5.3	3.7	8.0
Mn	65	41	19	24	55	37	16	69
Fe	65	209	108	111	290	191	86	356
Co	65	0.6	0.3	0.4	0.7	0.5	0.3	0.9
Ni	65	4.7	1.7	3.7	5.5	4.6	2.5	6.6
Cu	65	21	16	9.4	25	16	6.0	37
Zn	65	221	148	77	341	193	58	434
Ga	65	0.5	0.4	0.2	0.7	0.3	0.1	1.2
As	65	8.4	4.6	4.6	12	7.4	3.3	15
Rb	65	4.4	2.7	2.1	6.1	4.4	1.3	7.6
Sr	65	12	9.1	6.2	14	10	4.5	23
Y	65	0.2	0.1	0.1	0.2	0.2	0.1	0.3
Cd	65	1.7	1.2	0.6	2.5	1.7	0.4	3.1
Sn	65	2.0	1.4	0.9	2.7	1.7	0.5	4.0
Sb	65	3.4	4.9	1.5	3.9	2.4	0.7	5.8
Cs	65	0.6	0.4	0.2	0.8	0.6	0.1	1.0
Ba	65	16	17	7.9	22	12	4.8	26
Pb	65	44	25	22	62	39	15	78
Chloride	45	299	378	131	350	196	83	491
Nitrate	64	2880	2020	1250	3870	2070	997	6020
Phosphate	65	7530	1090	6890	8120	7640	5880	8640
Sulphate	65	28200	23000	10600	37400	22800	5650	70600

Table C.24: Descriptive statistics (valid number, average, standard deviation, lower and upper quartiles, median, 10th and 90th percentile) for TSP mass and watersoluble ion concentrations at site CRAES from July 2008 to September 2008 (period C2, 24-hourly sampling). Mass concentrations in μg/m^3, all ion concentrations in μg/g.

	N	avg	stdev	lower QT	upper QT	median	10th PCT	90th PCT
mass	66	104	51	62	131	96	45	177
Li	65	18	5	14	21	17	13	25
Na	65	91900	44600	57100	121000	77400	41700	169000
Mg	65	4530	883	3940	4900	4480	3590	5720
Al	65	1510	488	1214	1825	1494	891	2080
P	65	62100	32800	36400	85800	50700	26500	119000
S	65	85700	39400	49100	120000	85900	33700	135000
K	65	9520	3849	6946	11100	8600	5910	14300
Ca	65	35000	6920	30400	39300	35800	25200	43400
Sc	65	0.5	0.3	0.3	0.5	0.4	0.3	0.7
Ti	65	50	11	42	56	50	38	64
V	65	20	11	12	21	16	10	37
Cr	65	68	38	39	83	56	31	121
Mn	65	401	103	326	442	385	284	584
Fe	65	2020	541	1620	2370	1960	1370	2670
Co	65	6.0	3.2	3.9	7.1	5.2	3.3	10
Ni	65	55	29	32	73	45	25	100
Cu	65	189	90	122	225	170	106	303
Zn	65	2010	938	1330	2400	1810	1030	3390
Ga	65	4.3	2.1	2.6	5.4	3.8	2.2	7.4
As	65	81	26	64	93	77	53	119
Rb	65	43	24	29	49	36	22	75
Sr	65	115	45	96	121	106	86	140
Y	65	1.7	0.4	1.5	2.0	1.7	1.4	2.2
Cd	65	16	10	11	19	14	7	29

Sn	65	18	8	13	21	16	9	28
Sb	65	35	62	18	33	23	13	50
Cs	65	5.3	3.8	3.4	5.7	4.5	1.7	11
Ba	65	152	112	105	163	128	85	243
Pb	65	429	190	334	473	373	243	691
Chloride	45	3510	3160	1130	5420	2390	649	8170
Nitrate	64	33200	21600	11000	51800	30700	7480	60400
Phosphate	65	93500	50600	57800	121000	77800	36100	182000
Sulphate	65	246000	110000	140000	341000	243000	95500	397000

Table C.25: Descriptive statistics (valid number, average, standard deviation, lower and upper quartiles, median, 10th and 90th percentile) for BC concentrations in $\mu g/m^3$ from $PM_{2.5}$ and PM_1 samples at site CRAES during sampling period C (24th of July – 28th of September 2008). Sampling intervals of 12 hours for $PM_{2.5}$ and 24-hours for PM_1.

	N	avg	stdev	lower QT	upper QT	median	10th PCT	90th PCT
CRAES								
$PM_{2.5}$ (all)	132	2.26	1.15	1.39	2.97	2.10	0.91	3.74
$PM_{2.5}$ (day)	66	2.11	1.11	1.29	2.65	1.96	0.86	3.59
$PM_{2.5}$ (night)	66	2.41	1.18	1.41	3.16	2.31	0.97	3.86
PM_1	66	1.67	0.88	1.06	2.18	1.60	0.47	2.81
CUG								
$PM_{2.5}$	67	2.33	1.18	1.43	3.13	2.19	0.83	3.98

Table C.26: Descriptive statistics (valid number, average, standard deviation, lower and upper quartiles, median, 10th and 90th percentile) for APM mass concentrations at site CUG from the 21th of July to the 26th of September 2008 (24-hourly sampling). Mass concentrations in $\mu g/m^3$ for the different size classes for total, transparent, and opaque particles, respectively.

Size class	N	avg	stdev	lower QT	upper QT	median	10th PCT	90th PCT
Total								
2.5 – 5	67	23.5	12.7	14.8	28.0	21.7	11.8	35.8
5 – 10	67	26.3	13.8	16.5	32.4	23.4	11.8	45.1
10 – 20	67	20.9	9.66	14.5	24.6	18.8	10.1	31.1
20 – 40	67	7.84	3.89	4.83	11.0	7.37	3.28	12.7
40 – 80	67	1.45	1.93	0.00	1.74	1.07	0.00	4.99
2.5 – 80	67	79.9	36.9	53.8	96.0	74.3	40.6	124
2.5 – 10	67	49.8	26.1	31.5	59.2	45.8	23.6	76.8
Transparent								
2.5 – 5	67	20.5	11.1	12.8	24.6	19.4	9.72	30.2
5 – 10	67	20.8	11.1	12.7	26.2	17.9	9.66	35.0
10 – 20	67	16.9	7.98	11.2	20.7	15.7	9.01	26.7
20 – 40	67	6.48	3.37	3.83	8.46	6.11	2.20	11.5
40 – 80	67	1.25	1.65	0.00	1.61	1.04	0.00	3.05
2.5 – 80	67	65.9	30.6	42.8	81.6	62.6	31.1	97.6
2.5 – 10	67	41.3	21.8	26.5	51.0	36.8	18.4	62.4
Opaque								
2.5 – 5	67	2.95	2.04	1.76	3.56	2.64	1.13	5.04
5 – 10	67	5.51	3.93	3.06	7.15	4.35	1.96	11.8
10 – 20	67	4.00	2.77	2.09	4.93	3.09	1.41	7.54
20 – 40	67	1.35	1.25	0.00	2.32	1.22	0.00	3.31
40 – 80	67	0.20	0.70	0.00	0.00	0.00	0.00	0.00
2.5 – 80	67	14.0	8.84	7.98	18.4	11.1	6.70	25.0
2.5 – 10	67	8.46	5.77	4.74	10.6	6.65	3.37	15.9

Table C.27: Factor (Fa) scores of the seven extracted factors (Principal Component Analysis, substitution of missing data by average value, Varimax standardized rotation) for element concentrations of TSP samples (data from 24-hourly samples in $\mu g/g$) together with some meteorological parameters (visibility, dew point, temperature, wind speed, and predominant wind direction). Scores $> \pm 1$ are marked in bold.

	Fa1	Fa2	Fa3	Fa4	Fa5	Fa6	Fa7
24/07/2008	-0.98	-0.60	**1.68**	**1.15**	-0.04	-0.26	**1.03**
25/07/2008	-0.89	0.35	**1.92**	-0.32	0.22	-0.39	0.76
26/07/2008	-0.78	-0.55	**1.67**	0.37	-0.15	-0.24	0.92
27/07/2008	-0.41	0.66	**2.04**	0.08	-0.33	0.66	0.78
28/07/2008	-0.76	-0.49	**1.45**	-0.27	-0.22	-0.21	0.95
29/07/2008	-0.72	**-1.04**	0.16	**1.40**	0.31	**5.71**	-0.62
30/07/2008	**2.51**	**-1.37**	0.37	**-1.60**	**-1.07**	0.66	**1.65**
31/07/2008	**1.48**	**-1.60**	0.23	**-1.51**	**-1.32**	**2.19**	0.08
01/08/2008	**-1.18**	**-1.03**	-0.65	-0.59	**1.93**	0.93	0.47

02/08/2008	**-1.46**	0.04	-0.64	**-1.71**	0.23	0.18	**-1.03**
03/08/2008	-0.62	0.13	**1.07**	-0.04	0.51	0.83	**1.24**
04/08/2008	-0.31	0.13	**1.13**	-0.64	0.29	-0.39	**1.01**
05/08/2008	-0.50	0.12	**1.02**	-0.30	0.31	0.14	-0.26
06/08/2008	-0.28	**1.45**	**1.66**	-0.41	-0.14	0.78	0.26
07/08/2008	-0.84	-0.64	**1.77**	-0.43	-0.31	-0.66	**1.26**
08/08/2008	-0.52	0.56	**1.58**	-0.30	**2.08**	**-1.13**	**1.30**
09/08/2008	0.82	0.67	**1.36**	-0.98	-0.51	-0.32	0.22
10/08/2008	**3.18**	-0.69	-0.23	-0.29	0.08	**-1.45**	-0.41
11/08/2008	**2.11**	**-1.63**	-0.95	-0.61	**5.42**	-0.30	-0.87
12/08/2008	0.53	**-1.55**	0.44	-0.99	-0.85	0.18	-0.71
13/08/2008	0.20	-0.61	0.43	-0.60	-0.75	-0.74	**-1.49**
14/08/2008	-0.52	**-1.37**	-0.75	-0.77	-0.47	**-1.22**	-0.62
15/08/2008	**-1.07**	0.01	-0.79	-0.79	0.10	0.56	-0.77
16/08/2008	**-1.12**	-0.76	-0.72	0.60	-0.34	0.46	-0.90
17/08/2008	0.10	**-1.18**	-0.95	-0.92	-0.64	0.05	-0.54
18/08/2008	-0.16	-0.51	-0.55	**-1.16**	0.08	-0.59	**-1.10**
19/08/2008	**-1.02**	-0.86	-0.09	**-1.03**	0.09	0.30	**-1.14**
20/08/2008	**2.29**	**1.84**	0.61	0.85	**1.34**	0.89	0.08
21/08/2008	-0.16	**-1.37**	-0.05	**1.13**	-0.50	-0.41	**-0.11**
22/08/2008	-0.15	0.00	-0.86	**-1.15**	-0.27	0.60	-0.79
23/08/2008	-0.63	**1.16**	-0.65	**-1.09**	-0.01	0.99	-0.03
24/08/2008	-0.84	-0.78	0.23	-0.69	-0.26	-0.91	0.22
25/08/2008	-0.75	-0.57	0.18	0.67	**1.57**	-0.48	0.74
26/08/2008	-0.25	**-1.79**	0.29	0.42	-0.70	-0.12	0.00
27/08/2008	-0.42	-0.43	0.86	-0.66	-0.63	-0.54	**-1.09**
28/08/2008	-0.52	0.39	**1.33**	-0.23	**1.00**	**-1.19**	**-1.04**
29/08/2008	0.14	**-1.06**	0.67	**1.96**	-0.26	-0.81	-0.60
30/08/2008	0.14	-0.05	**-1.32**	-0.28	0.18	0.49	**2.50**
31/08/2008	-0.89	-0.21	**-1.23**	-0.17	**1.77**	-0.21	-0.66
01/09/2008	-0.41	0.39	-0.98	-0.83	-0.26	**1.06**	-0.12
02/09/2008	0.53	**1.33**	-0.10	-0.64	-0.36	0.43	0.44
03/09/2008	**1.23**	0.40	-0.37	0.21	**1.76**	0.63	**1.07**
04/09/2008	-0.53	0.36	-0.16	0.75	0.12	-0.22	-0.20
05/09/2008	-0.26	0.45	-0.13	-0.10	-0.39	-0.33	-0.75
06/09/2008	0.91	0.98	0.71	0.05	-0.72	-0.58	-0.51
07/09/2008	-0.58	-0.35	-0.07	**1.54**	-0.05	0.08	**-1.66**
08/09/2008	**2.27**	0.56	0.21	**1.39**	-0.39	-0.09	0.50
09/09/2008	**1.35**	**-1.96**	-0.16	**1.76**	-0.94	-0.88	0.89
10/09/2008	-0.20	0.99	-0.80	-0.11	-0.64	0.30	**1.22**
11/09/2008	-0.22	**1.92**	-0.49	-0.90	-0.44	0.82	-0.79
12/09/2008	**1.06**	**1.15**	-0.90	-0.62	-0.83	-0.36	-0.38
13/09/2008	0.63	**2.16**	-0.53	-0.25	-0.21	-0.25	**-1.02**
14/09/2008	0.22	0.75	0.11	0.54	0.40	-0.58	-0.68
15/09/2008	-0.25	0.39	-0.14	-0.10	-0.41	-0.75	**-1.77**
16/09/2008	**1.43**	0.94	-0.16	0.37	-0.21	0.47	-0.12
17/09/2008	-0.06	0.26	0.20	**1.19**	-0.47	0.78	-0.61
18/09/2008	0.16	0.89	-0.31	-0.14	-0.33	-0.18	-0.59
19/09/2008	-0.35	0.90	0.00	**1.70**	0.92	-0.29	-0.17
20/09/2008	0.13	**1.08**	0.39	0.07	-0.46	**-1.09**	**-2.02**
21/09/2008	0.93	-0.57	0.22	**2.04**	-0.98	-0.15	-0.92
22/09/2008	-0.56	-0.58	**-2.08**	**-1.02**	**-1.10**	**-1.37**	**1.67**
23/09/2008	**-1.51**	0.76	**-1.81**	**3.01**	0.10	-0.67	0.43
24/09/2008	-0.34	**-1.01**	**-1.96**	**1.95**	-0.70	-0.36	0.54
25/09/2008	-0.95	0.37	**-2.05**	-0.97	0.02	**-1.06**	**2.73**
26/09/2008	-0.41	**2.31**	**-1.16**	0.36	-0.27	-0.08	0.81
27/09/2008	**1.01**	0.36	**-1.23**	0.65	-0.91	0.66	**1.34**

Table C.28: Factor (Fa) scores of the eight extracted factors (Principal Component Analysis, case by case exclusion of missing data, Varimax standardized rotation) for watersoluble ion concentrations of TSP samples (data from 24-hourly samples in μg/g) together with some meteorological parameters (visibility, dew point, temperature, wind speed, and predominant wind direction). Scores $> \pm 1$ are marked in bold.

	Fa1	Fa2	Fa3	Fa4	Fa5	Fa6	Fa7	Fa8
	Fa1	Fa2	Fa3	Fa4	Fa5	Fa6	Fa7	Fa8
24/07/2008	-0.87	-0.79	**1.50**	-0.13	0.55	-0.09	**-1.37**	0.62
25/07/2008	-0.77	**-1.24**	0.60	-0.15	0.03	0.39	-0.83	**1.18**
26/07/2008	-0.58	**-1.23**	0.50	**-1.05**	0.87	-0.14	**-1.13**	**1.01**
27/07/2008	-0.45	**-1.13**	0.41	**-1.20**	**1.35**	-0.09	**-1.06**	**1.43**
28/07/2008	-0.47	-0.91	0.50	-0.36	0.75	0.12	**-1.00**	**1.23**
29/07/2008	0.07	**1.76**	0.33	**2.10**	-0.54	-0.39	-0.32	0.43
30/07/2008	**3.22**	-0.47	**-1.09**	0.90	0.27	0.35	-0.87	**1.27**
31/07/2008	**1.90**	**1.25**	**-1.32**	**-1.25**	0.75	0.35	0.24	**1.94**

01/08/2008	-0.25	0.95	-0.92	0.53	-0.48	0.74	-0.19	0.61
02/08/2008	**-1.19**	0.39	**-1.23**	0.34	0.20	0.57	0.49	0.50
03/08/2008	-0.11	-0.57	-0.46	-0.05	**-1.66**	0.15	**-1.02**	**1.06**
04/08/2008	0.00	-0.80	-0.04	-0.27	-0.98	0.13	-0.95	0.76
05/08/2008	-0.30	-0.67	0.00	0.40	0.35	0.41	-0.02	**1.00**
06/08/2008	-0.26	-0.77	0.95	**1.52**	0.98	0.44	-0.78	**1.24**
07/08/2008	-0.77	-0.94	0.82	**-1.49**	0.94	0.31	**-1.22**	**1.46**
08/08/2008	-0.38	-0.40	0.06	**-1.56**	**-6.85**	0.02	**-1.11**	0.76
09/08/2008	0.50	-0.53	**1.18**	-0.13	0.59	0.60	0.03	0.95
10/08/2008	**2.91**	**1.19**	**1.59**	-0.53	0.04	0.69	0.24	-0.66
11/08/2008	**2.68**	**2.15**	-0.61	-0.77	-0.53	0.58	**1.34**	0.77
12/08/2008	0.62	-0.11	-0.50	-0.66	0.24	0.10	**1.29**	**1.12**
13/08/2008	-0.10	-0.26	0.90	0.78	-0.32	0.95	**2.36**	0.41
14/08/2008	-0.19	**1.93**	0.46	-0.25	-0.24	0.47	0.71	0.10
15/08/2008	**-1.32**	**2.07**	-0.55	-0.41	0.51	0.19	-0.49	0.48
16/08/2008	**-1.02**	0.29	-0.75	0.28	0.37	0.10	0.59	-0.27
17/08/2008	-0.16	**1.54**	-0.20	0.35	0.27	0.02	0.14	0.12
18/08/2008	-0.23	**1.45**	-0.47	-0.37	0.42	-0.16	0.73	**1.03**
19/08/2008	-0.75	0.53	-0.57	0.23	0.02	0.37	**1.24**	0.92
20/08/2008	**1.49**	0.46	**1.81**	**2.94**	-0.17	-0.17	-0.47	0.03
21/08/2008	-0.21	**2.53**	**1.90**	-0.52	-0.14	0.09	-0.15	0.67
22/08/2008	0.05	0.56	-0.95	0.40	0.47	0.47	0.33	0.41
23/08/2008	-0.36	0.08	**-1.52**	0.34	0.11	0.59	-0.39	0.11
24/08/2008	-0.78	-0.58	-0.55	**-1.02**	0.40	-0.03	-0.23	0.51
25/08/2008	-0.47	-0.19	0.65	0.86	-0.36	-0.08	-0.47	0.39
26/08/2008	-0.51	0.51	0.13	0.65	-0.09	**-7.10**	0.64	0.82
27/08/2008	-0.78	-0.63	0.07	-0.89	0.22	0.19	**1.63**	0.93
28/08/2008	-0.65	**-1.12**	**1.11**	**2.18**	-0.39	0.74	**1.32**	0.82
29/08/2008	-0.50	0.02	**2.43**	-0.67	0.18	0.07	0.52	-0.69
30/08/2008	-0.18	**1.32**	-0.26	0.52	0.60	-0.27	**-2.47**	-0.53
31/08/2008	-0.95	**1.30**	-0.78	0.18	-0.01	-0.08	0.44	-0.50
01/09/2008	-0.50	0.53	**-1.13**	0.37	0.56	0.08	-0.18	-0.24
02/09/2008	0.64	-0.97	**-1.05**	**1.38**	0.55	-0.12	-0.76	-0.18
03/09/2008	**1.26**	-0.43	-0.55	**1.07**	0.05	-0.58	**-1.53**	-0.45
04/09/2008	-0.71	-0.14	-0.20	0.32	0.04	-0.02	0.06	-0.13
05/09/2008	-0.35	-0.48	-0.39	**1.01**	0.01	0.13	0.38	-0.11
06/09/2008	0.22	**-1.05**	**1.58**	**1.14**	-0.12	0.75	0.50	-0.80
07/09/2008	-0.90	0.36	0.15	**-1.02**	-0.12	-0.41	**1.97**	-0.22
08/09/2008	**1.61**	-0.48	0.64	0.90	-0.60	-0.70	-0.25	-0.43
09/09/2008	**1.15**	0.50	**1.72**	**-3.41**	0.68	-0.48	-0.57	-0.90
10/09/2008	-0.26	-0.32	**-1.11**	-0.73	-0.09	0.03	-0.95	-0.88
11/09/2008	-0.39	-0.79	**-1.50**	-0.61	0.09	0.31	0.48	-0.75
12/09/2008	**1.04**	-1.18	**-1.91**	**-1.46**	0.77	**-1.21**	0.13	-0.97
13/09/2008	0.54	-0.95	**-1.61**	0.31	0.48	-0.12	0.58	-0.63
14/09/2008	0.28	-0.56	0.09	-0.15	-0.15	-0.06	0.27	-0.74
15/09/2008	-0.47	-0.62	-0.23	-0.41	-0.40	0.61	**2.11**	-0.80
16/09/2008	**2.04**	-0.79	-0.41	-0.32	-0.45	-0.46	-0.30	**-1.41**
17/09/2008	-0.13	-0.49	0.15	0.43	-0.33	**-1.19**	**1.24**	-0.51
18/09/2008	0.12	-1.21	**-1.41**	-0.44	0.07	-0.21	0.49	-0.80
19/09/2008	-0.25	-0.86	-0.18	0.45	-0.53	-0.05	0.35	-0.66
20/09/2008	-0.17	-1.38	0.32	0.17	-0.18	0.72	**2.39**	-0.94
21/09/2008	0.60	-0.41	**2.43**	**-1.32**	0.85	-0.35	0.39	**-1.63**
22/09/2008	-0.72	1.18	-0.65	-0.90	0.35	0.62	**-1.22**	**-1.45**
23/09/2008	**-1.53**	0.67	0.33	0.10	0.07	-0.30	-0.64	**-2.34**
24/09/2008	-0.42	0.58	0.49	0.06	-0.24	0.02	-0.05	**-2.55**
25/09/2008	**-1.45**	1.36	-0.17	0.88	0.03	**1.06**	**-1.71**	**-1.67**
26/09/2008								
27/09/2008	0.87	-1.04	-0.54	0.41	-0.10	0.34	-0.94	**-2.25**

Appendix D

Tables corresponding to chapter 7

Table D.1: Descriptive statistics (valid number, average, standard deviation, lower and upper quartiles, median, 10th and 90th percentile) for APM mass concentrations at site CUG from the 25th of April 2005 to the 30th of August 2007 (sampling in three and four day-intervals). Mass concentrations in $\mu g/m^3$ for the different size classes for all, transparent, and opaque particles, respectively.

Particle size	N	avg	stdev	lower QT	upper QT	median	10th PCT	90th PCT
Total								
2.5 – 5	243	50.0	23.4	34.8	59.8	44.1	27.2	77.2
5 – 10	243	62.6	25.6	44.7	75.5	58.9	36.2	91.9
10 – 20	243	61.7	27.8	42.6	76.9	55.0	31.9	94.4
20 – 40	243	35.5	26.2	18.6	43.6	26.6	14.0	68.1
40 – 80	243	8.7	12.8	2.1	8.1	4.5	1.2	23.2
2.5 – 80	243	219	104	148	267	193	117	329
2.5 – 10	243	113	48.3	79.1	135	104	64.0	169
Transparent								
2.5 – 5	243	44.5	22.0	30.6	54.3	39.4	23.8	68.6
5 – 10	243	55.0	24.6	37.8	66.8	50.9	30.4	82.4
10 – 20	243	55.0	26.9	36.2	68.4	48.9	28.4	85.6
20 – 40	243	32.1	25.2	16.2	39.0	23.3	11.9	60.8
40 – 80	243	7.8	12.2	1.7	7.5	3.9	1.1	19.3
2.5 – 80	243	194	100	128	236	169	99.7	292
2.5 – 10	243	99.6	45.8	68.4	120	92.7	54.2	151
Opaque								
2.5 – 5	243	5.4	2.5	3.7	6.6	4.8	2.8	8.9
5 – 10	243	7.6	2.9	5.4	9.0	7.0	4.3	11.6
10 – 20	243	6.7	2.9	4.5	8.6	6.4	3.3	10.5
20 – 40	243	3.5	2.3	1.8	4.5	3.1	1.1	6.1
40 – 80	243	0.8	1.1	0.0	1.1	0.5	0.0	2.0
2.5 – 80	243	24.1	9.6	16.5	29.8	22.1	12.7	36.7
2.5 – 10	243	13.0	5.3	9.3	16.0	11.9	7.1	20.6

List of Figures

List of Tables

Bibliography

ADAMSON, I., PRIEDITIS, H., HEDGECOCK, C. & VINCENT, R. 2000 Zinc is the toxic factor in the lung response to an atmospheric particulate sample. *Toxicology and Applied Pharmacology* **166**, 111–119.

AITIO, A. 2008 Research needs for environmental health risk assessment. *Journal of Toxicology and Environmental Health, Part A* **71**, 1254–1258.

AL-MASRI, M., AL-KHARFAN, K. & AL-SHAMALI, K. 2006 Speciation of Pb, Cu and Zn determined by sequential extraction for identification of air pollution sources in Syria. *Atmospheric Environment* **40**, 753–761.

AMDUR, M., BAYLES, J., UGRO, V. & UNDERHILL, D. 1978 Comparative irritant potency of sulfate salts. *Environmental Research* **16**, 1–8.

ANDERSON, H. 2009 Air pollution and mortality: a history. *Atmospheric Environment* **43**, 142–152.

ANDRADE, F., ORSINI, C. & MAENHAUT, W. 1994 Relation between aerosol sources and meteorological parameters for inhalable atmospheric particles in Sao Paulo City, Brazil. *Atmospheric Environment* **28**, 2307–2315.

ARIMOTO, R., DUCE, R., SAVOIE, D., PROSPERO, J., TALBOT, R., CULLEN, J., TOMZA, U., LEWIS, N. & RAY, B. 1996 Relationship among aerosol constituents from Asia and the North Pacific during PEM-West A. *Journal of Geophysical Research* **101**, 2011–2023.

ATKINSON, R., BREMNER, S., STEPHEN, A., ANDERSON, H., STRACHAN, D., BLAND, J. & DE LEON, A. 1999 Short-term associations between emergency hospital admissions for respiratory and cardiovascular disease and outdoor air pollution in London. *Archives of Environmental Health* **54**, 398–411.

AYRAULT, S., SENHOU, A., MOSKURA, M. & GAUDRY, A. 2010 Atmospheric trace element concentrations in total suspended particles near Paris, France. *Atmospheric Environment* **44**, 3700–3707.

BALLACH, J., HINTZENBERGER, R., SCHULTZ, E. & JEASCHKE, W. 2001 Development of an improved optical transmission technique for black carbon (BC) analysis. *Atmospheric Environment* **35**, 2089–2100.

BARTRA, J., MULLOL, J., DEL CUVILLO, A., DÁVILA, I., FERRER, M., JÁUREGUI, I., MONTORO, J., SASTRE, J. & VALERO, A. 2007 Air pollution and allergens. *Journal of Investigational Allergology and Clinical Immunology* **17**, 3–8, suppl. 2.

BEIJING MUNICIPAL BUREAU OF STATISTICS 2009 Beijing Statistical Yearbook 2008. http://www.bjstats.gov.cn.

BILOS, C., COLOMBO, J., SKORUPKA, C. & RODRIGUEZ PRESA, M. 2001 Sources, distribution and variability of airborne trace metals in La Plata City area, Argentina. *Environmental Pollution* **111**, 149–158.

BLUM, J. 2006 Dust agglomeration. *Advances in Physics* **55**, 881–947.

BORM, P., SCHINS, R. & ALBRECHT, C. 2004 Inhaled particles and lung cancer, part b: paradigms and risk assessment. *International Journal of Cancer* **110**, 3–14.

BOWES, S. & SWIFT, D. 1989 Deposition of inhaled particles in the oral airway during oronasal breathing. *Aerosol Science and Technology* **11**, 157–167.

BRANIS, M. & VETVICKA, J. 2010 Pm10, ambient temperature and relative humidity during the xxix Summer Olympic Games in Beijing: Were the athletes at risk? *Aerosol and Air Quality Research* **10**, 102–110.

BUZORIUS, G., HÄMERI, K., PEKKANEN, J. & KULMALA, M. 1999 Spatial variation of aerosol number concentration in Helsinki city. *Atmospheric Environment* **33**, 553–565.

CANEPARI, S., PERRINO, C., OLIVIERI, F. & ASTOLFI, M. 2008 Characterisation of the traffic sources of PM through size-segregated sampling, sequential leaching and ICP analysis. *Atmospheric Environment* **42**, 8161–8175.

CHEN, B., HONG, C. & KAN, H. 2004 Exposures and health outcomes from outdoor air pollutants in China. *Toxicology* **198**, 291–300.

CHEN, J., TAN, M., LI, Y., ZHENG, J., ZHANG, Y., SHAN, Z., ZHANG, G. & LI, Y. 2008 Characteristics of trace elements and lead isotope ratios in PM2.5 from four sites in Shanghai. *Journal of Hazardous Materials* **156**, 36–43.

CHEN, L. & LIPPMANN, M. 2009 Effects of metals within ambient air particulate matter (PM) on human health. *Inhalation Toxicology* **21**, 1–31.

CHENG, Y. 2000 *Concise regional geology of China*. Geological Publishing House, Beijing, China, 430 p.

CHESTER, R., LIN, F. & MURPHY, K. 1989 A three stage sequential leaching scheme for the characterisation of the sources and environmental mobility of trace metals in the marine aerosol. *Environmental Technology Letters* **10**, 887–900.

CHESTER, R., NIMMO, M. & PRESTON, M. 1999 The trace metal chemistry of atmospheric dry deposition samples collected at Cap Ferrat: a coastal site in the Western Mediterranean. *Marine Chemistry* **68**, 15–30.

CHILLRUD, S., EPSTEIN, D., ROSS, J., SAX, S., PEDERSON, D., SPENGLER, J. & KINNEY, P. 2004 Elevated airborne exposures of teenagers to manganese, chromium, and iron from steel dust and New York City's subway system. *Environmental Science and Technology* **38**, 732–737.

DAVIDSON, A. 1994 The Los Angeles aerosol characterization and source apportionment study: a meteorological–air quality analysis. *Aerosol Science and Technology* **21**, 269–282.

DIABATÉ, S., GÜNTHER, R., VÖLKEL, K., THIELE, D. & WOTTRICH, R. 2004 Gesundheitseffekte durch inhalierbare Feinststäube aus technischen Verbrennungsanlagen: In vitro Untersuchungen zur Wirkung feiner und ultrafeiner Partikel auf kultivierte Lungenzellen. *Forschungsbericht FZKA-BWPLUS* pp. –.

DIETZE, V., FRICKER, M., GOLTZSCHE, M. & SCHULTZ, E. 2006 Luftqualitätsmessung in deutschen Kurorten. Teil 1: Methodik und Absicherung. *Gefahrstoffe – Reinhaltung der Luft* **66**, 45–53.

DOCKERY, D., CUNNINGHAM, J., DAMOKOSH, A., NEAS, L., SPENGLER, J., KOUTRAKIS, P., WARE, J., RAIZENNE, M. & SPEIZER, F. 1996 Health effects of acid aerosols on North American children: Respiratory symptoms. *Environmental Health Perspectives* **104**, 500–505.

DOCKERY, D., POPE, C., XU, X., SPENGLER, J., WARE, J., FAY, M., FERRIS, B. & SPEIZER, F. 1993 An association between air pollution and mortality in six U.S. cities. *The New England Journal of Medicine* **329**, 1753–1759.

DOMRÖS, M. & PENG, G. 1988 *The climate of China*. Springer, Berlin, Heidelberg, New York, 360 p.

DONALDSON, K., BROWN, G., BROWN, D., BOLTON, R. & DAVIS, J. 1989 Inflammation generating potential of long and short fibre amosite asbestos samples. *British Journal of Industrial Medicine* **46**, 271–276.

DREHER, K., JASKOT, R., LEHMANN, J., RICHARDS, J., MCGEE, J., GHIO, A. & COSTA, D. 1994 Soluble transition metals mediate residual oil flyash induced acute lung injury. *Journal of Toxicology and Environmental Health* **50**, 258–305.

DUAN, F., HE, K., MA, Y., YANG, F., YU, X., CADLE, S., CHAN, T. & MULAWA, P. 2006 Concentration and chemical characteristics of PM2.5 in Beijing, China: 2001-2002. *Science of the Total Environment* **355**, 264–275.

FAN, S., TIAN, G., LI, G., HUANG, Y., QIN, J. & CHENG, S. 2009 Road fugitive dust emission characteristics in Beijing during Olympics Game 2008 in Beijing, China. *Atmospheric Environment* **43**, 6003–6010.

FANG, G.-C., CHANG, C.-N., CHU, C.-C., WU, Y.-S., FU, P.-C., YANG, I.-L. & CHEN, M.-H. 2003 Characterization of particulate, metallic elements of TSP, PM2.5 and PM2.5-10 aerosols at a farm sampling site in Taiwan, Taichung. *The Science of the Total Environment* **308**, 157–166.

FANG, M., CHAN, C. K. & YAO, X. 2009 Managing air quality in a rapidly developing nation: China. *Atmospheric Environment* **43**, 79–86.

FENGER, J. 1999 Urban air quality. *Atmospheric Environment* **33**, 4877–4900.

FENGER, J. 2009 Air pollution in the last 50 years – from local to global. *Atmospheric Environment* **43**, 13–22.

FERNÁNDEZ, A., TERNERO, M., BARRAGÁN, F. & JIMÉNEZ, J. 2000 An approach to characterization of sources of urban airborne particles through heavy metal speciation. *Chemosphere – Global Change Science* **2**, 123–136.

FERNÁNDEZ ESPINOSA, A. & TERNERO RODRÍGUEZ, M. 2004 Study of traffic pollution by metals in Seville (Spain) by physical and chemical speciation methods. *Analytical and Bioanalytical Chemistry* **379**, 684–699.

FERNÁNDEZ ESPINOSA, A., TERNERO RODRÍGUEZ, M., BARRAGÁN DE LA ROSA, F. & JIMÉNEZ SÁNCHEZ, J. 2002 A chemical speciation of trace metals for fine urban particles. *Atmospheric Environment* **36**, 773–780.

FRICKER, M. & SCHULTZ, E. 2002 Rubestimmung nach der Öl-Immersionsmethode unter niedrig belasteten, ländlichen Bedingungen. *Gefahrstoffe Reinhaltung der Luft* **62**, 385–390.

FUJIWARA, F., DOS SANTOS, M., MARRERO, J., POLLA, G., GÓMEZ, D., DAWIDOWSKIA, L. & SMICHOWSKI, P. 2006 Fractionation of eleven elements by chemical bonding from airborne particulate matter collected in an industrial city in Argentina. *Journal of Environmental Monitoring* **8**, 913–922.

GARCÍA, R., TORRES, M. & BÁEZ, A. 2008 Determination of trace elements in total suspended particles at the Southwest of Mexico City from 2003 to 2004. *Chemistry and Ecology* **24**, 157–167.

GIERÉ, R. & QUEROL, X. 2010 Solid particulate matter in the atmosphere. *Elements* **6**, 215–222.

GIETL, J. & KLEMM, O. 2009 Analysis of traffic and meteorology on airborne particulate matter in Münster, Northwest Germany. *Journal of the Air and Waste Management Association* **59**, 809–818.

GIETL, J., LAWRENCE, R., THORPE, A. & HARRISON, R. 2010 Identification of brake wear particles and derivation of a quantitative tracer for brake dust at a major road. *Atmospheric Environment* **44**, 141–146.

GILMOUR, P., ZIESENIS, A., MORRISON, E., VICKERS, M., DROST, E., FORD, I., KARG, E., MOSSA, C., SCHROEPPEL, A., FERRON, G., HEYDER, J., GREAVES, M., MACNEE, W. & DONALDSON, K. 2004 Pulmonary and systemic effects of short-term inhalation exposure to ultrafine carbon black particles. *Toxicology and Applied Pharmacology* **195**, 35–44.

GÓMEZ, D., GINÉ, M., SÁNCHEZ BELLATO, A. & SMICHOWSKI, P. 2005 Antimony: a traffic-related element in the atmosphere of Buenos Aires, Argentina. *Journal of Environmental Monitoring* **7**, 1162–1168.

GONG, X., WU, T., QIAO, Y. & XU, M. 2010 In situ leaching of trace elements in a coal ash dump and time dependence laboratory evaluation. *Energy Fuels* **24**, 84–90.

GROBÉTY, B., GIERÉ, R., DIETZE, V. & STILLE, P. 2010 Airborne particles in the urban environment. *Elements* **6**, 229–234.

GUINOT, B., ROGER, J.-C., CACHIER, H., PUCAI, W., JIANHUI, B. & TONG, Y. 2006 Impact of vertical atmospheric structure on Beijing aerosol distribution. *Atmospheric Environment* **40**, 5167–5180.

GUO, Y., JIA, Y., PAN, X., LIU, L. & WICHMANN, H.-E. 2009 The association between fine particulate air pollution and hospital emergency room visits for cardiovascular diseases in Beijing, China. *Science of the Total Environment* **407**, 4826–4830.

GWIAZDA, R., LEE, D., SHERIDAN, J. & SMITH, D. 2002 Low cumulative manganese exposure affects striatal GABA but not Dopamine. *Neurotoxicology* **23**, 69–76.

HAO, J., HE, D., WU, Y., FU, L. & HE, K. 2000 A study of the emission and concentration distribution of vehicular pollutants in the urban area of Beijing. *Atmospheric Environment* **34**, 453–465.

HAO, Y., GUO, Z., YANG, Z., FANG, M. & FENG, J. 2007 Seasonal variations and sources of various elements in the atmospheric aerosols in Qingdao, China. *Atmospheric Research* **85**, 27–37.

HARRISON, Y. & YIN, J. 2000 Particulate matter in the atmosphere: which particle properties are important for its effects on health? *The Science of the Total Environment* **249**, 85–101.

HASHIMOTO, Y., SEKINE, Y. & OTOSHI, T. 1992 Atmospheric aluminium from human activities. *Atmospheric Environment* **26**, 295–300.

HE, K., YANG, F., MA, Y., ZHANG, Q., YAO, X., CHAN, C., CADLE, S., CHAN, T. & MULAWA, P. 2001 The characteristics of PM2.5 in Beijing, China. *Atmospheric Environment* **35**, 4959–4970.

HEISER, U., NORRA, S., STÜBEN, D. & WAGNER, M. 1999 Sequentielle Schwermetallextraktion aus Staubniederschlägen und Straßensedimenten – Teil II: Sequentielle Schwermetallextraktion von städtischen Stäuben. *Umweltwissenschaften und Schadstoffforschung* **11**, 73–78, in German with English abstract.

HIEN, P., BINH, N., TRUONG, Y., NGO, N. & SIEU, L. 2001 Comparative receptor modelling study of TSP, PM2 and PM2-10 in Ho Chi Minh City. *Atmospheric Environment* **35**, 2669–2678.

HIGHWOOD, E. & KINNERSLEY, R. 2006 When smoke gets in our eyes: the multiple impacts of atmospheric black carbon on climate, air quality and health. *Environment International* **32**, 560–566.

HILL, S. 1997 Speciation of trace metals in the environment. *Chemical Society Reviews* **26**, 291–298.

HIRNER, A. 1992 Trace element speciation in soils and sediments using sequential extraction methods. *International Journal of Environmental Analytical Chemistry* **46**, 77–85.

HITZENBERGER, R., PETZOLD, A., BAUER, H., CTYROKY, P., POURESMAEIL, P., LASKUS, L. & PUXBAUM, H. 2006 Intercomparison of thermal and optical measurement methods for elemental carbon and black carbon at an urban location. *Environmental Science and Technology* **40**, 6377–6383.

HLAVAY, J., POLYÁAK, K., BÓDOG, I., MOLNÁR, A. & MÉSZÁROS, E. 1996 Distribution of trace elements in filter-collected aerosol samples. *Fresenius Journal of Analytical Chemistry* **354**, 227–232.

HU, M., HE, L., ZHANG, Y., WANG, M., KIM, Y. & MOON, K. 2002 Seasonal variation of ionic species in fine particles at Qingdao, China. *Atmospheric Environment* **36**, 5853–5859.

HU, X., LIU, S., WANG, Y. & LI, J. 2004 Numerical simulation of wind and temperature fields over Beijing area in summer. *Acta Meteorologica Sinica* **19**, 120–127.

IARC 2008 *Agents reviewed by the IARC Monographs, Volumes 1-99*.

JACOBSON, M. 2001 Strong radiative heating due to the mixing state of black carbon in atmospheric aerosols. *Nature* **409**, 695–697.

JANSEN, K., LARSON, T., KOENIG, J., MAR, T., FIELDS, C., STEWART, J. & LIPPMANN, M. 2005 Associations between health effects and particulate matter and black carbon in subjects with respiratory disease. *Environmental Health Perspectives* **113**, 1741–1746.

KASAMATSU, J., SHIMA, M., YAMAZAKI, S., TAMURA, K. & SUN, G. 2006 Effects of winter air pollution on pulmonary function of school children in Shenyang, China. *International Journal of Hygiene and Environmental Health* **209**, 435–444.

KENNEDY, N. & HINDS, W. 2002 Inhalability of large solid particles. *Journal of Aerosol Science* **33**, 237–255.

KIM, K.-H., MA, C.-J. & OKUDA, T. 2009 An analysis of long-term changes in airborne toxic metals in South Korea's two largest cities from 1991 to 2004. *Environmental Science and Pollution Research* **16**, 565–572.

KODAVANTI, U., JASKOT, R., COSTA, D. & DREHER, K. 1997 Pulmonary proinflammatory gene induction following acute exposure to residual oil fly ash: roles of particle-associated metals. *Inhalation Toxicology* **9**, 679–701.

KUANG, C., NEUMANN, T., NORRA, S. & STÜBEN, D. 2004 Land use-related chemical composition of street sediments in Beijing. *Environmental Science and Pollution Research* **11**, 73–83.

LEE, C., LI, X.-D., ZHANG, G., LI, J., DING, A.-J. & WANG, T. 2007 Heavy metals and Pb isotopic composition of aerosols in urban and suburban areas of Hong Kong and Guangzhou, South China – Evidence of the long-range transport of air contaminants. *Atmospheric Environment* **41**, 432–447.

LI, X., BROWN, D., SMITH, S., MACNEE, W. & DONALDSON, K. 1999 Short-term inflammatory responses following intratracheal instillation of fine and ultrafine carbon black in rats. *Inhalation Toxicology* **11**, 709–731.

LI, X., YUE, W., IIDA, A., LI, Y. & ZHANG, G. 2007 A study of the origin of individual PM2.5 particles in Shanghai air with synchrotron X-ray fluorescence microprobe. *Nuclear Instruments and Methods in Physics Research B* **260**, 336–342.

LI, Y., WANG, W., KAN, H., XU, X. & CHEN, B. 2010 Air quality and outpatient visits for asthma in adults during the 2008 Summer Olympic Games in Beijing. *Science of the Total Environment* **408**, 1226–1227.

LIM, J.-H., SABIN, L., SCHIFF, K. & STOLZENBACH, K. 2006 Concentration, size distribution, and dry deposition rate of particle-associated metals in the Los Angeles region. *Atmospheric Environment* **40**, 7810–7823.

LIN, J.-M., FANG, G.-C., HOLSEN, T. & NOLL, K. 1993 A comparison of dry deposition modeled from size distribution data and measured with a smooth surface for total particle mass, lead and calcium in Chicago. *Atmospheric Environment* **27**, 1131–1138.

LINGE, K. 2008 Methods for investigating trace element binding in sediments. *Critical Reviews in Environmental Science and Technology* **38**, 165–196.

LIPPMANN, M., ITO, K., HWANG, J., MACIEJCZYK, P. & CHEN, L. 2006 Cardiovascular effects of nickel in ambient air. *Environmental Health Perspectives* **114**, 1662–1669.

LIU, H., YUJUN, J., LIANG, B., ZHU, F., ZHANG, B. & SANG, J. 2009 Studies on wind environment around high buildings in urban areas. *Science in China Series D – Earth Science* **48**, 102–115, suppl. II.

LIU, L. & ZHANG, J. 2009 Ambient air pollution and children's lung function in China. *Environment International* **35**, 178–186.

LIU, X., GUO, J. & SUN, Z. 2008 Traffic operation with comments during Beijing Olympic Games. *Journal of Transportation Systems Engineering and Information Technology* **8**, 16–24.

LU, S., SHAO, L., WU, M., JONES, T., MEROLLA, L. & RICHARD, R. 2006 Correlation between plasmid DNA damage induced by PM_{10} and trace metals in inhalable particulate matters in Beijing air. *Science in China Series D: Earth Sciences* **49**, 1323–1331.

LU, S., YAO, Z., CHEN, X., WU, M., SHENG, G., FU, J. & DALY, P. 2008 The relationship between physicochemical characterization and the potential toxicity of fine particulates ($PM_{2.5}$) in Shanghai atmosphere. *Atmospheric Environment* **42**, 7205–7214.

MARRERO, J., JIMÉNEZ REBAGLIATI, R., GÓMEZ, D. & SMICHOWSKI, P. 2005 A study of uniformity of elements deposition on glass fiber filters after collection of airborne particulate matter (pm-10), using a high-volume sampler. *Talanta* **68**, 442–447.

MAYNARD, D. 2000 Overview of methods for analysing single ultrafine particles. *Philosophical Transactions of the Royal Society A* **358**, 2593–2610.

MCALISTER, J., SMITH, B. & TÖRÖK, A. 2008 Transition metals and water-soluble ions in deposits on a building and their potential catalysis of stone decay. *Atmospheric Environment* **42**, 7657–7668.

MCCLELLAN, R. 2002 Setting ambient air quality standards for particulate matter. *Toxicology* **181–182**, 329–347.

MENON, S., HANSEN, J., NAZARENKO, L. & LUO, Y. 2002 Climate effects of black carbon aerosols in China and India. *Science* **297**, 2250–2253.

MONACI, F., MONI, F., LANCIOTTI, E., GRECHI, D. & BARGAGLI, R. 2000 Biomonitoring of airborne metals in urban environments: new tracers of vehicle emission, in place of lead. *Environmental Pollution* **107**, 321–327.

MORENO, T., QUEROL, X., ALASTUEY, A., AMATO, F., PEY, J., PANDOLFI, M., KUENZLI, N., BOUSO, L., RIVERA, M. & GIBBONS, W. 2010 Effect of fireworks events on urban background trace metal aerosol concentrations: Is the cocktail worth the show? *Journal of Hazardous Materials* **183**, 945–949.

MORENO, T., QUEROL, X., ALASTUEY, A., MINGUILLÓN, M., PEY, J., RODRIGUEZ, S., MIRÓ, J., FELIS, C. & GIBBONS, W. 2007 Recreational atmospheric pollution episodes: Inhalable metalliferous particles from firework displays. *Atmospheric Environment* **41**, 913–922.

MORENO, T., QUEROL, X., ALASTUEY, A., VIANA, M., SALVADOR, P., SÁNCHEZ DE LA CAMPA, A., ARTINANO, B., DE LA ROSA, J. & GIBBONS, W. 2006 Variations in atmospheric PM trace metal content in Spanish towns: Illustrating the chemical complexity of the inorganic urban aerosol cocktail. *Atmospheric Environment* **40**, 6791–6803.

MUSCHAK, W. 1990 Biomonitoring of airborne metals in urban environments: new tracers of vehicle emission, in place of lead. *Science of the Total Environment* **93**, 419–431.

NATH, B., NORRA, S., CHATTERJEE, D. & STÜBEN, D. 2007 Fingerprinting of land use-related chemical patterns in street sediments from Kolkata, India. *Environmental Forensics* **8**, 313–328.

NATIONAL BUREAU OF STATISTICS OF CHINA 2009 Statistical Yearbook 2008. http://www.stats.gov.cn/english/statisticaldata/yearlydata/.

NAWROT, T., PLUSQUIN, M., HOGERVORST, J., ROELS, H., CELIS, H., THIJS, L., VANGRONSVELD, J., VAN HECKE, E. & STAESSEN, J. 2006 Environmental exposure to cadmium and risk of cancer: a prospective population-based study. *Lancet* **7**, 119–126.

NEMMAR, A., HOET, P., VANQUICKENBORNE, B., DINSDALE, D., THOMEER, M., HOYLAERTS, M., VANBILLOEN, H., MORTELMANS, L. & NEMERY, B. 2002 Passage of inhaled particles into the blood circulation in humans. *Circulation* **105**, 411–414.

NICOLÁS, J., YUBERO, E., PASTOR, C., CRESPO, J. & CARRATALÁ, A. 2009 Influence of meteorological variability upon aerosol mass size distribution. *Atmospheric Research* **94**, 330–337.

NING, D., ZHONG, L. & CHUNG, Y.-S. 1996 Aerosol size distribution and elemental composition in urban areas of Northern China. *Atmospheric Environment* **30**, 2355–2362.

NORRA, S., FJER, N., LI, F., CHU, X., XIE, X. & STÜBEN, D. 2008 The influence of different land uses on mineralogical and chemical composition and horizonation of urban soil profiles in Qingdao, China. *Journal of Soils and Sediments* **8**, 4–16.

NORRA, S., HUNDT, B., STÜBEN, D., CEN, K., LIU, C., DIETZE, V. & SCHULTZ, E. 2007 Size, morphological, and chemical characterization of aerosols polluting the Beijing atmosphere in January/February 2005. *Highway and Urban Environment: Proceedings of the 8th Highway and Urban Environment Symposium* pp. 167–180.

NORRA, S., SCHLEICHER, N., STÜBEN, D., CHAI, F., CHEN, Y. & WANG, S. 2010 Assessment of aerosol concentration sampled at five sites in Beijing from 2005 till 2007. *Highway and Urban Environment, Alliance for Global Sustainability Bookseries* **17**, 133–140.

OBERDÖRSTER, G., CELEIN, R., FERIN, J. & WEISS, B. 1995 Association of particulate air pollution and acute mortality: involvement of ultrafine particles? *Inhalation Toxicology* **7**, 111–124.

OKUDA, T., KATO, J., MORI, J., TENMOKU, M., SUDA, Y., TANAKA, S., HE, K., MA, Y., YANG, F., YU, X., DUAN, F. & LEI, Y. 2004 Daily concentrations of trace metals in aerosols in Beijing, China, determined by using inductively coupled plasma mass spectrometry equipped with laser ablation analysis, and source identification of aerosols. *Science of the Total Environment* **330**, 145–158.

PEREIRA, P., LOPES, W., CARVALHO, L., DA ROCHA, G., DE CARVALHO BAHIA, N., LOYOLA, J., QUITERIO, S., ESCALEIRA, V., ARBILLA, G. & DE ANDRADE, J. 2007 Atmospheric concentrations and dry deposition fluxes of particulate trace metals in Salvador, Bahia, Brazil. *Atmospheric Environment* **41**, 7837–7850.

POPE, C. & DOCKERY, D. 2006 Health effects of fine particulate air pollution: lines that connect. *Journal of the Air & Waste Management Association* **56**, 709–742.

PÓSFAI, M. & BUSECK, P. 2010 Nature and climate effects of individual tropospheric aerosol particles. *Annual Review of Earth and Planetary Science* **38**, 17–43.

PÖYKIÖ, R., PERÄMÄKI, P. & RÖNKKÖMÄKI, H. 2003 The homogeneity of heavy metal deposition on glass fibre filters collected using a high-volume sampler in the vicinity of an opencast chrome mine complex at Kemi, Northern Finland. *Analytical and Bioanalytical Chemistry* **375**, 476–481.

PRAVEEN KUMAR, M., VENKATA MOHAN, S. & JAYARAMA REDDY, S. 2008 Chemical fractionation of heavy metals in airborne particulate matter (PM10) by sequential extraction procedure. *Toxicological & Environmental Chemistry* **90**, 31–41.

QIN, Y., KIM, E. & HOPKE, P. 2006 The concentrations and sources of PM2.5 in metropolitan New York City. *Atmospheric Environment* **40**, S312–S332.

QUITERIO, S., SOUSA DA SILVA, C., ARBILLA, G. & ESCALEIRA, V. 2004 Metals in airborne particulate matter in the industrial district of Santa Cruz, Rio de Janeiro, in an annual period. *Atmospheric Environment* **38**, 321–331.

RAGOSTA, M., CAGGIANO, R., D'EMILIO, M. & MACCHIATO, M. 2002 Source origin and parameters influencing levels of heavy metals in TSP, in an industrial background area of Southern Italy. *Atmospheric Environment* **36**, 3071–3087.

RAIZENNE, M., NEAS, L., DAMOKOSH, A., DOCKERY, D., SPENGLER, J., KOUTRAKIS, P., WARE, J. & SPEIZER, F. 1996 Health effects of acid aerosols on North American children: pulmonary function. *Environmental Health Perspectives* **104**, 506–514.

REED, S. 2006 *Electron Microprobe Analysis and Scanning Electron Microscopy in Geology*. Cambridge University Press, 206 p.

REIMANN, C. & CARITAT, P. 1998 *Chemical elements in the environment. Factsheets for the geochemist and environmental scientist.*. Springer, Berlin, Heidelberg, 398 p.

RICHTER, P., GRINO, P., AHUMADA, I. & GIORDANO, A. 2007 Total element concentration and chemical fractionation in airborne particulate matter from Santiago, Chile. *Atmospheric Environment* **41**, 6729–6738.

SCHLEICHER, N., NORRA, S., CHAI, F., CHEN, Y., WANG, S. & STÜBEN, D. 2010*a* Anthropogenic versus geogenic contribution to total suspended atmospheric particulate matter and its variations during a two-year sampling period in Beijing, China. *Journal of Environmental Monitoring* **12**, 434–441.

SCHLEICHER, N., NORRA, S., CHAI, F., CHEN, Y., WANG, S. & STÜBEN, D. 2010*b* Seasonal trend of water-soluble ions at one TSP and five PM2.5 sampling sites in Beijing, China. *Highway and Urban Environment, Alliance for Global Sustainability Bookseries* **17**, 87–95.

SCHLEICHER, N. & RECIO HERNÁNDEZ, C. 2010 Source identification of sulphate forming salts on sandstones from monuments in Salamanca, Spain – a stable isotope approach. *Environmental Science and Pollution Research* **17**, 770–778.

SCHWARTZ, J., DOCKERY, D. & NEAS, L. 1996 Is daily mortality associated specifically with fine particles? *Journal of the Air and Waste Management Association* **46**, 927–939.

SCHWARTZ, S. 2004 Uncertainty requirements in radiative forcing of climate change. *Journal of the Air and Waste Management Association* **54**, 1351–1359.

SEINFELD, J. & PANDIS, S. 2006 *Atmospheric chemistry and physics: from air pollution to climate change*. John Wiley & Sons, Inc., Hoboken, New Jersey.

SELLA, S., PEREIRA NETTO, A., DA SILVA FILHO, E. & ARAÚJO, M. 2004 Short-term and spatial variation of selected metals in the atmosphere of Niterói City, Brazil. *Microchemical Journal* **78**, 85–90.

SHEN, Z., CAO, J., ARIMOTO, R., HAN, Z., ZHANG, R., HAN, Y., LIU, S., OKUDA, T., NAKAO, S. & TANAKA, S. 2009 Ionic composition of TSP and PM2.5 during dust storms and air pollution episodes at Xi'an, China. *Atmospheric Environment* **43**, 2911–2918.

SHI, X., MAO, Y., KNAPTON, A., DING, M., ROJANASAKUL, Y., GANNETT, P., DALAL, N. & LIU, K. 1994 Reaction of Cr(VI) with ascorbate and hydrogen peroxide generates hydroxyl radical and causes DNA damage: role of a Cr(IV)-mediated Fenton-like reaction. *Carcinogenesis* **15**, 2475–2478.

SHU, J., DEARING, J., MORSE, A., YU, L. & YUAN, N. 2001 Determining the sources of atmospheric particles in Shanghai, China, from magnetic and geochemical properties. *Atmospheric Environment* **35**, 2615–2625.

SMICHOWSKI, P., POLLA, G. & GÓMEZ, D. 2005 Metal fractionation of atmospheric aerosols via sequential chemical extraction: a review. *Analytical and Bioanalytical Chemistry* **381**, 302–316.

SOLÉ, V., PAPILLON, E., COTTE, M., WALTER, P. & SUSINI, J. 2007 A multiplatform code for the analysis of energy-dispersive x-ray fluorescence spectra. *Spectrochimica Acta Part B* **62**, 63–68.

SPURGEON, D., JONES, O., DORNE, J.-L., SVENDSEN, C., SWAIN, S. & STÜRZENBAUM, S. 2010 Systems toxicology approaches for understanding the joint effects of environmental chemical mixtures. *Science of the Total Environment* **408**, 3725–3734.

STERNBECK, J., SJÖDIN, A. & ANDRÉASSON, K. 2002 Metal emissions from road traffic and the influence of resuspension – results from two tunnel studies. *Atmospheric Environment* **36**, 4735–4744.

STOLL, B. & JOCHUM, K. P. 1999 Mic-ssms analyses of eight new geological standard glasses. *Fresenius' Journal of Analytical Chemistry* **364**, 380–384.

STONE, R. 2008 Beijing's marathon run to clean foul air nears finish line. *Science* **321**, 636–637.

STREETS, D., GUPTA, S., WALDHOFF, S., WANG, M., BOND, T. & YIYUN, B. 2001 Black carbon emissions in China. *Atmospheric Environment* **35**, 4281–4296.

SUN, Y., ZHUANG, G., HUANG, K., LI, J., WANG, Q., WANG, Y., LIN, Y., FU, J., ZHANG, W., TANG, A. & ZHAO, X. 2010 Asian dust over northern China and its impact on the downstream aerosol chemistry in 2004. *Journal of Geophysical Research* **115**, D00K09.

SUN, Y., ZHUANG, G., TANG, A., WANG, Y. & AN, Z. 2006 Chemical characteristics of PM2.5 and PM10 in haze-fog episodes in Beijing. *Environmental Science and Technology* **40**, 3148–3155.

SUN, Y., ZHUANG, G., WANG, Y., HAN, L., GUO, J., DAN, M., ZHANG, W., WANG, Z. & HAO, Z. 2004 The air-borne particulate pollution in Beijing concentration, composition, distribution and sources. *Atmospheric Environment* **38**, 5991–6004.

TANNER, P. & WONG, A. 2000 Soluble trace metals and major ionic species in the bulk deposition and atmosphere of Hong Kong. *Water, Air, and Soil Pollution* **122**, 261–279.

TESSIER, A., CAMPBELL, P. & BISSON, M. 1979 Sequential extraction procedure for the speciation of particulate trace metals. *Analytical Chemistry* **51**, 844–851.

TIETZE, W. & DOMRÖS, M. 1987 The climate of China. *GeoJournal* **14.2**, 265–266.

URE, A., QUEVAUVILLER, P. & GRIEPINK, H. M. B. 1993 Speciation of heavy metals in soils and sediments. An account of the improvement and harmonization of extraction techniques undertaken under the auspices of the BCR of the Commission of the European Communities. *International Journal of Environmental Analytical Chemistry* **51**, 135–151.

VDI 1997 VDI-Guideline 2119, Part 4: Measurement of particulate precipitations – microscopic differentiation and size fractionated determination of particle deposition on adhesive collection plates – Sigma-2 sampler.

VDI 1999 VDI-Guideline 2465, Part 2: Measurement of soot (ambient air) – thermographical determination of elemental carbon after thermal desorption of organic carbon.

VECCHI, R., BERNARDONI, V., CRICCHIO, D., D'ALESSANDRO, A., FERMO, P., LUCARELLI, F., S. NAVA, A. P. & VALLI, G. 2008 The impact of fireworks on airborne particles. *Atmospheric Environment* **42**, 1121–1132.

VOUTSA, D., SAMARA, C., KOUIMTZIS, T. & OCHSENKÜHN, K. 2002 Elemental composition of airborne particulate matter in the multi-impacted urban area of Thessaloniki, Greece. *Atmospheric Environment* **36**, 4453–4462.

WÅHLIN, P., BERKOWICZ, R. & PALMGREN, F. 2006 Characterisation of traffic-generated particulate matter in Copenhagen. *Atmospheric Environment* **40**, 2151–2159.

WALDMANN, J., LIOY, P., ZELENKA, M., JING, L., LIN, Y., HE, Q., QIAN, Z., CHAPMAN, R. & WILSON, W. 1991 Wintertime measurements of aerosol acidity and trace elements in Wuhan, a city in central China. *Atmospheric Environment* **25**, 113–120.

WANG, G., HUANG, L., GAO, S., GAO, S. & WANG, L. 2002 Characterization of water-soluble species of PM10 and PM2.5 aerosols in urban area in Nanjing, China. *Atmospheric Environment* **36**, 1299–1307.

WANG, P., BI, S., ZHOU, Y., TAO, Q., GAN, W., XU, Y., HONG, Z. & CAI, W. 2007*a* Study of aluminium distribution and speciation in atmospheric particles of different diameters in Nanjing, China. *Atmospheric Environment* **41**, 5788–5796.

WANG, Q., LIU, Y. & PAN, X. 2008 Atmosphere pollutants and mortality rate of respiratory diseases in Beijing. *Science of the Total Environment* **391**, 143–148.

WANG, S., YUAN, W. & SHANG, K. 2006*a* The impacts of different kinds of dust events on PM_{10} pollution in northern China. *Atmospheric Environment* **40**, 7975–7982.

WANG, S., ZHAO, M., XING, J., WU, Y., ZHOU, Y., LEI, Y., HE, K., FU, L. & HAO, J. 2010*a* Quantifying the air pollutants emission reduction during the 2008 Olympic Games in Beijing. *Environmental Science and Technology* **44**, 2490–2496.

WANG, T., NIE, W., GAO, J., XUE, L., GAO, X., WANG, X., QIU, J., POON, C., MEINARDI, S., BLAKE, D., WANG, S., DING, A., CHAI, F., ZHANG, Q. & WANG, W. 2010*b* Air quality during the 2008 Beijing Olympics: secondary pollutants and regional impact. *Atmospheric Chemistry and Physics* **10**, 7603–7615.

WANG, Y., ZHUANG, G., SUN, Y. & AN, Z. 2006*b* The variation of characteristics and formation mechanisms of aerosols in dust, haze, and clear days in Beijing. *Atmospheric Environment* **40**, 6579–6591.

WANG, Y., ZHUANG, G., TANG, A., ZHANG, W., SUN, Y., WANG, Z. & AN, Z. 2007*b* The evolution of chemical components of aerosols at five monitoring sites of China during dust storms. *Atmospheric Environment* **41**, 1091–1106.

WANG, Y., ZHUANG, G., XU, C. & AN, Z. 2007*c* The air pollution caused by the burning of fireworks during the lantern festival in Beijing. *Atmospheric Environment* **41**, 417–431.

WATTS, J. 2005 China: the air pollution capital of the world. *The Lancet* pp. 1761–1762.

WECKWERTH, G. 2001 Verification of traffic emitted aerosol components in the ambient air of Cologne (Germany). *Atmospheric Environment* **35**, 5525–5536.

WHITEAKER, J., SUESS, D. & PRATHER, K. 2002 Effects of meteorological conditions on aerosol composition and mixing state in Bakersfield, CA. *Environmental Science and Technology* pp. 2345–2353.

WHO 2000 *Air quality guidelines for Europe*. 2nd edition.

WHO 2001 *Environment and people's health in China*.

WICHMANN, H.-E. & PETERS, A. 2000 Epidemiological evidence of the effects of ultrafine particle exposure. *Philosophical Transactions Of The Royal Society Of London – Series A* pp. 2751–2769.

WITHOLT, R., GWIADZA, R. & SMITH, D. 2000 The neurobehavioral effects of subchronic manganese exposure in the presence and absence of pre-parkinsonism. *Neurotoxicology and Teratology* **22**, 851–861.

WOO, K., CHEN, D., PUI, D. & MCMURRY, P. 2001 Measurement of Atlanta aerosol size distributions: Observations of ultrafine particle events. *Aerosol Science and Technology* **34**, 75–87.

WU, Y., LIU, C. & TU, C. 2008 Atmospheric deposition of metals in TSP of Guiyang, PR China. *Bulletin of Environmental Contamination and Toxicology* **80**, 465–468.

XIE, S., YU, T., ZHANG, Y., ZENG, L., QI, L. & TANG, X. 2005 Characteristics of PM_{10}, SO_2, NO_x and O_3 in ambient air during the dust storm period in Beijing. *Science of the Total Environment* **345**, 153–164.

XIN, J., WANG, Y., TANG, G., WANG, L., SUN, Y., WANG, Y., HU, B., SONG, T., JI, D., WANG, W., LI, L. & LIU, G. 2010 Variability and reduction of atmospheric pollutants in Beijing and its surrounding area during the Beijing 2008 Olympic Games. *Chinese Science Bulletin* **55**, 1937–1944.

XINHUA ONLINE 2010 .

XU, C. 2010 *Mineralogy and magnetic properties of airborne particulates in Beijing*. Master thesis, Karlsruhe Institute of Technology.

XU, M., YAN, R., ZHENG, C., QIAO, Y., HAN, J. & SHENG, C. 2003 Status of trace element emission in a coal combustion process: a review. *Fuel Processing Technology* **85**, 215–237.

XU, X., GAO, J., DOCKERY, D. & CHEN, Y. 1994 Air-pollution and daily mortality in residential areas of Beijing, China. *ARCHIVES OF ENVIRONMENTAL HEALTH* **49**, 216–222.

XU, X., WANG, L. & NIU, T. 1998 Air pollution and its health effects in Beijing. *Ecosystem Health* **4**, 199–209.

YANG, T., WANG, Z., ZHANG, B., WANG, X., WANG, W., GBAUIDI, A. & GONG, Y. 2010 Evaluation of the effect of air pollution control during the Beijing 2008 Olympic Games using Lidar data. *Chinese Science Bulletin* **55**, 1311–1316.

YAO, X., CHAN, C., FANG, M., CADLE, S., CHAN, T., MULAWA, P., HE, K. & YE, B. 2002 The water-soluble ionic composition of PM2.5 in Shanghai and Beijing, China. *Atmospheric Environment* **36**, 4223–4234.

YUE, W., LI, X., LIU, J., LI, Y., YU, X., DENG, B., WAN, T., ZHANG, G., HUANG, Y., HE, W., HUA, W., SHAO, L., LI, W. & YANG, S. 2006 Characterization of PM2.5 in the ambient air of Shanghai city by analyzing individual particles. *Science of the Total Environment* **368**, 916–925.

YUE, W., LI, Y., LI, X., YU, X., DENG, B., LIU, J., WAN, T., ZHANG, G., HUANG, Y., HE, W. & HUA, W. 2004 Source identification of PM10, collected at a heavy-trafifc roadside, by analyzing individual particles using synchrotron radiation. *Journal of Synchrotron Radiation* **11**, 428–431.

ZATKA, V., WARNER, J. & MASKERY, D. 1992 Chemical speciation of nickel in airborne dusts: analytical method and results of an interlaboratory test program. *Environmental Science and Technology* **26**, 138–144.

ZEREINI, F., SKERSTUPP, B., RANKENBURG, K., DIRKSEN, F., J.-M. BEYER, T. C. & URBAN, H. 2001 Anthropogenic emission of platinum-group elements into the environment. *Journal of Soils and Sediments* **1**, 44–49.

ZHANG, J., SONG, H., TONG, S., LI, L., LIU, B. & WANG, L. 2000 Ambient sulfate concentration and chronic disease mortality in Beijing. *The Science of the Total Environment* **262**, 63–71.

ZHANG, M., SONG, Y. & CAI, X. 2007 A health-based assessment of particulate air pollution in urban areas of Beijing in 2000-2004. *Science of the Total Environment* **376**, 100–108.

ZHANG, M., SONG, Y., CAI, X. & ZHOU, J. 2008 Economic assessment of the health effects related to particulate matter pollution in 111 chinese cities by using economic burden of disease analysis. *Journal of Environmental Management* **88**, 947–954.

ZHOU, Y., WU, Y., YANG, L., FU, L., HE, K., WANG, S., HAO, J., CHEN, J. & LI, C. 2010 The impact of transportation control measures on emission reductions during the 2008 Olympic Games in Beijing, China. *Atmospheric Environment* **44**, 285–293.

ZOLLER, W., GLADNEY, E. & DUCE, R. 1974 Atmospheric concentrations and sources of trace metals at the South Pole. *Science, New Series* **183**, 198–200.

In der Reihe "Karlsruher Geochemische Hefte" (ISSN 0943-8599) sind erschienen:

Band 1: U. Kramar (1993)
Methoden zur Interpretation von Daten der geochemischen Bachsedimentprospektion am Beispiel der Sierra de San Carlos/ Tamaulipas Mexiko

Band 2: Z. Berner (1993)
S-Isotopengeochemie in der KTB - Vorbohrung und Beziehungen zu den Spurenelementmustern der Pyrite

Band 3: J.-D. Eckhardt (1993)
Geochemische Untersuchungen an jungen Sedimenten von der Galapagos-Mikroplatte:
Hydrothermale und stratigraphisch signifikante Signale

Band 4: B. Bergfeldt (1994)
Lösungs- und Austauschprozesse in der ungesättigten Bodenwasserzone und Auswirkungen auf das Grundwasser

Band 5: M. Hodel (1994)
Untersuchungen zur Festlegung und Mobilisierung der Elemente As, Cd, Ni und Pb an ausgewählten Festphasen unter besonderer Berücksichtigung des Einflusses von Huminstoffen.

Band 6: T. Bergfeldt (1995)
Untersuchungen der Arsen- und Schwermetallmobilität in Bergbauhalden und kontaminierten Böden im Gebiet des Mittleren Schwarzwaldes.

Band 7: M. Manz (1995)
Umweltbelastungen durch Arsen und Schwermetalle in Böden, Halden, Pflanzen und Schlacken ehemaliger Bergbaugebiete des Mittleren und Südlichen Schwarzwaldes.

Band 8: J. Ritter (1995)
Genese der Mineralisation Herrmanngang im Albtalgranit (SE-Schwarzwald) und Wechselwirkungen mit dem Nebengestein.

Band 9: J. Castro (1995)
Umweltauswirkungen des Bergbaus im semiariden Gebiet von Santa Maria de la Paz, Mexiko.

Band 10: T. Rüde (1996)
Beiträge zur Geochemie des Arsens.

Band 11: J. Schäfer (1998)
Einträge und Kontaminationspfade Kfz-emittierter Platin-Gruppen-Elemente (PGE) in verschiedenen Umweltkompartimenten.

Band 12: M. A. Leosson (1999)
A Contribution to the Sulphur Isotope Geochemistry of the Upper Continental Crust: The KTB Main Hole - A Case Study

Band 13: B. A. Kappes (2000)
Mobilisierbarkeit von Schwermetallen und Arsen durch saure Grubenabwässer in Bergbaualtlasten der Ag-Pb-Zn-Lagerstätte in Wiesloch

Band 14: H. Philipp (2000)
The behaviour of platinum-group elements in petrogenetic processes:
A case study from the seaward-dipping reflector sequence (SDRS), Southeast Greenland volcanic rifted margin

Band 15: E. Walpersdorf (2000)
Nähr- und Spurenelementdynamik im Sediment/Wasser-Kontaktbereich nach einer Seekreideaufspülung - Pilotstudie Arendsee -

Band 16: R. H. Kölbl (2000)
Models of hydrothermal plumes by submarine diffuse venting in a coastal area: A case study for Milos, South Aegean Volcanic Arc, Greece

Band 17: U. Heiser (2000)
Calcium-rich Rhodochrosite layers in Sediments of the Gotland Deep, Baltic Sea, as Indicators for Seawater Inflow

In der Fortsetzungsreihe "Karlsruher Mineralogische und Geochemische Hefte" (ISSN 1618-2677) sind bisher erschienen:

Band 18: S. Norra (2001)
Umweltgeochemische Signale urbaner Systeme am Beispiel von Böden, Pflanzen, und Stäuben in Karlsruhe

Band 19: M. von Wagner, Alexander Salichow, Doris Stüben (2001)
Geochemische Reinigung kleiner Fließgewässer mit Mangankiesen, einem Abfallsprodukt aus Wasserwerken (GReiFMan)

Band 20: U. Berg (2002)
Die Kalzitapplikation als Restaurierungsmaßnahme für eutrophe Seen – ihre Optimierung und Bewertung

Band 21: Ch. Menzel (2002)
Bestimmung und Verteilung aquatischer PGE-Spezies in urbanen Systemen

Band 22: P. Graf (2002)
Meta-Kaolinit als latent-hydraulische Komponente in Kalkmörtel

Band 23: D. Buqezi-Ahmeti (2003)
Die Fluidgehalte der Mantel-Xenolithe des Alkaligesteins-Komplexes der Persani-Berge, Ostkarpaten, Rumänien: Untersuchungen an Fluid-Einschlüssen

Band 24: B. Scheibner (2003)
Das geochemische Verhalten der Platingruppenelemente bei der Entstehung und Differenzierung der Magmen der Kerguelen-Flutbasaltprovinz (Indischer Ozean)

Ab Band 25 erscheinen die Karlsruher Mineralogischen und Geochemischen Hefte bei KIT Scientific Publishing online unter der Internetadresse

http://www.ksp.kit.edu

Auf Wunsch sind bei KIT Scientific Publishing gedruckte Exemplare erhältlich („print on demand").

Band 25: A. Stögbauer (2005)
Schwefelisotopenfraktionierung in abwasserbelasteten Sedimenten - Biogeochemische Umsetzungen und deren Auswirkung auf den Schwermetallhaushalt

Band 26: X. Xie (2005)
Assessment of an ultramicroelectrode array (UMEA) sensor for the determination of trace concentrations of heavy metals in water

Band 27: F. Friedrich (2005)
Spectroscopic investigations of delaminated and intercalated phyllosilicates

Band 28: L. Niemann (2005)
Die Reaktionskinetik des Gipsabbindens: Makroskopische Reaktionsraten und Mechanismen in molekularem Maßstab

Band 29: F. Wagner (2005)
Prozessverständnis einer Naturkatastrophe: eine geo- und hydrochemische Untersuchung der regionalen Arsen-Anreicherung im Grundwasser West-Bengalens (Indien)

Band 30: F. Pujol (2005)
Chemostratigraphy of some key European Frasnian-Famennian boundary sections (Germany, Poland, France)

Band 31: Y. Dikikh (2006)
Adsorption und Mobilisierung wasserlöslicher Kfz-emittierter Platingruppenelemente (Pt, Pd, Rh) an verschiedenen bodentypischen Mineralen

Band 32: F. I. Müller (2007)
Influence of cellulose ethers on the kinetics of early Portland cement hydration

Band 33: H. Haile Tolera (2007)
Suitability of local materials to purify Akaki Sub-Basin water

Band 34: M. Memon (2008)
Role of Fe-oxides for predicting phosphorus sorption in calcareous soils

Band 35: S. Bleeck-Schmidt (2008)
Geochemisch-mineralogische Hochwassersignale in Auensedimenten und deren Relevanz für die Rekonstruktion von Hochwasserereignissen

Band 36: A. Steudel (2009)
Selection strategy and modification of layer silicates for technical applications

Band 37: E. Eiche (2009)
Arsenic mobilization processes in the red river delta, Vietnam: Towards a better understanding of the patchy distribution of dissolved arsenic in alluvial deposits